T0073469

INTRODUCTION TO THE
CALCULUS OF
VARIATIONS
3rd Edition

INTRODUCTION TO THE
CALCULUS OF
VARIATIONS

3rd Edition

BERNARD DACOROGNA

Ecole Polytechnique Fédérale Lausanne, Switzerland

Imperial College Press

Published by

Imperial College Press
57 Shelton Street
Covent Garden
London WC2H 9HE

Distributed by

World Scientific Publishing Co. Pte. Ltd.

5 Toh Tuck Link, Singapore 596224

USA office: 27 Warren Street, Suite 401-402, Hackensack, NJ 07601

UK office: 57 Shelton Street, Covent Garden, London WC2H 9HE

Library of Congress Cataloging-in-Publication Data
Dacorogna, Bernard, 1953–
   [Introduction au calcul des variations. English]
   Introduction to the calculus of variations / Bernard Dacorogna, Ecole Polytechnique Federale
Lausanne, Switzerland. -- 3rd edition.
    pages cm
   Also called: Third English edition.
   Includes bibliographical references and index.
   ISBN 978-1-78326-551-0 (hardcover : alk. paper) -- ISBN 978-1-78326-552-7 (pbk. : alk. paper)
   1. Calculus of variations. 2. Mathematical analysis. I. Title.
   QA315.D3413 2014
   515'.64--dc23
                              2014025436

British Library Cataloguing-in-Publication Data
A catalogue record for this book is available from the British Library.

Printed in Singapore by Mainland Press Pte Ltd.

Contents

Preface to the Third English Edition

The present book by now has a long history. It appeared first in French under the title "Introduction au calcul des variations" in 1992 with the Presses Polytechniques et Universitaires Romandes (PPUR). It was then translated into English and published by Imperial College Press (ICP) in 2004. A second edition was completed in 2009. Throughout the different editions, the size of the book has doubled, keeping the same structure but with more developments and exercises.

After several years of experience, I think that the present book can adequately serve as a concise and broad introduction to the calculus of variations. It can be used at undergraduate as well as at graduate level. Of course at a more advanced level it has to be complemented by more specialized materials and I have indicated, in every chapter, appropriate books for further reading. The numerous exercises (now 119), integrally solved in Chapter 7, will also be important to help understanding the subject better.

The calculus of variations is one of the classical subjects in mathematics. Several outstanding mathematicians have contributed, over centuries, to its development. It is still a very alive and evolving subject. Besides its mathematical importance and its links with other branches of mathematics, such as geometry or differential equations, it is widely used in physics, engineering, economics and biology. I have decided, in order to remain as unified and concise as possible, not to speak of any applications other than mathematical ones. Every interested reader, whether physicist, engineer or biologist, will easily see where, in their own subject, the results of the present monograph are used. This fact is clearly asserted by the numerous engineers and physicists that followed the course that resulted in the present book.

Let us now examine the content of the monograph. It should first be emphasized that it is not a reference book. Each individual chapter could, on its own, be the subject of a book. For example, I have written one that, essentially, covers the subject of Chapter 3. Furthermore, several aspects of the calculus of variations are not discussed here. One of the aims is to serve as a guide to the extensive existing literature. However, the main purpose is to help the non-specialist, whether mathematician, physicist, engineer, student or researcher, to discover the most important problems, results and techniques of the subject. Despite the aim of addressing the non-specialists, I have tried not to sacrifice the mathematical rigor. Most of the theorems are either fully proved or proved under stronger, but significant, assumptions than stated.

The different chapters may be read more or less independently. In Chapter 1, I have recalled some standard results on spaces of functions (Hölder, L^p or Sobolev spaces) and on convex analysis. The reader, familiar or not with these subjects, can, at first reading, omit this chapter and refer to it when needed in the next ones. They are much used in Chapters 3 and 4 but less in the others. All the chapters, besides numerous examples, contain exercises that are fully solved in Chapter 7.

In all the different editions, I benefited from many discussions with students and colleagues. I would like to thank particularly S. Bandyopadhyay, S. Basterrechea, O. Besson, S.D. Chatterji, G. Csato, G. Cupini, C. Hebeisen, O. Kneuss, M. M. Marques, F. Meylan, K.D. Semmler, J. Sesiano, S. Sil and F. Weissbaum. I also want to thank the staff of Imperial College Press and World Scientific for their very nice job.

Chapter 0

Introduction

0.1 Brief historical comments

The calculus of variations is one of the classical branches of mathematics. It was Euler who, looking at the work of Lagrange, gave the present name, not really self explanatory, to this field of mathematics.

In fact the subject is much older. It starts with one of the oldest problems in mathematics: the isoperimetric inequality. A variant of this inequality is known as the Dido problem (Dido was a semi historical Phoenician princess and later a Carthaginian queen). Several more or less rigorous proofs were known since the times of Zenodorus around 200 BC, who proved the inequality for polygons. There are also significant contributions by Archimedes and Pappus. Important attempts at proving the inequality are due to Euler, Galileo, Legendre, L'Huilier, Riccati, Simpson and Steiner. The first proof that agrees with modern standards is due to Weierstrass and it has been extended or proved with different tools by Blaschke, Bonnesen, Carathéodory, Edler, Frobenius, Hurwitz, Lebesgue, Liebmann, Minkowski, H.A. Schwarz, Sturm and Tonelli among others. We refer to Porter [97] for an interesting article on the history of the inequality.

Other important problems of the calculus of variations were considered in the seventeenth century in Europe, such as the work of Fermat on geometrical optics (1662), the problem of Newton (1685) for the study of bodies moving in fluids (see also Huygens in 1691 on the same problem) or the problem of the brachistochrone formulated by Galileo in 1638. This last problem had a very strong influence on the development of the calculus of variations. It was resolved by Johann Bernoulli in 1696 and almost immediately after also by his brother Jakob, Leibniz and Newton. A decisive step was achieved with the work of Euler and Lagrange, who found a systematic way of dealing with problems in this field

1

by introducing what is now known as the Euler-Lagrange equation. This work was then extended in many ways by Bliss, Bolza, Carathéodory, Clebsch, Hahn, Hamilton, Hilbert, Kneser, Jacobi, Legendre, Mayer and Weierstrass, just to quote a few. For an interesting historical book on the one dimensional problems of the calculus of variations, see Goldstine [61].

In the nineteenth century and in parallel to some of the works that were mentioned above, probably the most celebrated problem of the calculus of variations emerged, namely the study of the Dirichlet integral; a problem of multiple integrals. The importance of this problem was motivated by its relationship with the Laplace equation. Many important contributions were made by Dirichlet, Gauss, Thompson and Riemann, among others. It was Hilbert who, at the turn of the twentieth century, solved the problem and was immediately imitated by Lebesgue and then Tonelli. Their methods for solving the problem were, essentially, what are now known as the direct methods of the calculus of variations. We should also emphasize that the problem has been very important in the development of analysis in general and more notably functional analysis, measure theory, distribution theory, Sobolev spaces or partial differential equations. This influence is studied in the book by Monna [84].

The problem of minimal surfaces has also had, almost at the same time as the previous one, a strong influence on the calculus of variations. The problem was formulated by Lagrange in 1762. Many attempts to solve the problem were made by Ampère, Beltrami, Bernstein, Bonnet, Catalan, Darboux, Enneper, Haar, Korn, Legendre, Lie, Meusnier, Monge, Müntz, Riemann, H.A. Schwarz, Serret, Weierstrass, Weingarten and others. Douglas and Rado in 1930 gave, simultaneously and independently, the first complete proof. One of the first two Fields medals was awarded to Douglas in 1936 for having solved the problem. Immediately after the results of Douglas and Rado, many generalizations and improvements were made by Courant, Leray, MacShane, Morrey, Morse, Tonelli and many others since then. We refer for historical notes to Dierkes-Hildebrandt-Küster-Wohlrab [41] and Nitsche [89].

In 1900 at the International Congress of Mathematicians in Paris, Hilbert formulated 23 problems that he considered to be important for the development of mathematics in the twentieth century. Three of them (the 19th, 20th and 23rd) were devoted to the calculus of variations. These "predictions" of Hilbert have been amply justified throughout the twentieth century and the field is still as active as ever.

Finally we should mention that we will not speak of many important topics of the calculus of variations, such as, Morse or Liusternik-Schnirelman theories. The interested reader is referred to Ekeland [44], Mawhin-Willem [83], Struwe [104], Willem [110] or Zeidler [113].

0.2 Model problem and some examples

We now describe in more detail the problems that we consider. The model case takes the following form

$$(P) \quad \inf \left\{ I(u) = \int_\Omega f(x, u(x), \nabla u(x)) \, dx : u \in X \right\} = m.$$

This means that we want to minimize the integral, $I(u)$, among all functions $u \in X$ (and we call m the minimal value that can take such an integral), where

- $\Omega \subset \mathbb{R}^n$, $n \geq 1$, is a bounded open set, a point in Ω is denoted by $x = (x_1, \cdots, x_n)$;

- $u : \Omega \to \mathbb{R}^N$, $N \geq 1$, $u = (u^1, \cdots, u^N)$, and hence

$$\nabla u = \left(\frac{\partial u^j}{\partial x_i} \right)_{1 \leq i \leq n}^{1 \leq j \leq N} \in \mathbb{R}^{N \times n};$$

- $f : \overline{\Omega} \times \mathbb{R}^N \times \mathbb{R}^{N \times n} \longrightarrow \mathbb{R}$, $f = f(x, u, \xi)$, is continuous;

- X is the space of admissible functions (for example, $u \in C^1(\overline{\Omega})$ with $u = u_0$ on $\partial \Omega$).

We are concerned with finding a *minimizer* $\overline{u} \in X$ of (P), meaning that

$$I(\overline{u}) \leq I(u), \quad \forall u \in X.$$

Many problems coming from analysis, geometry or applied mathematics (in physics, economics or biology) can be formulated as above. Many other problems, even though not entering in this framework, can be solved by the very same techniques.

We now give several classical examples.

Example: Fermat principle. We want to find the trajectory that should follow a light ray in a medium with non-constant refractive index. We can formulate the problem in the above formalism. We have $n = N = 1$,

$$f(x, u, \xi) = g(x, u) \sqrt{1 + \xi^2}$$

and

$$(P) \quad \inf \left\{ I(u) = \int_a^b f(x, u(x), u'(x)) \, dx : u(a) = \alpha, \ u(b) = \beta \right\} = m.$$

Example: Newton problem. We seek a surface of revolution moving in a fluid with least resistance. The problem can be mathematically formulated as follows. Let $n = N = 1$,

$$f\left(x,u,\xi\right) = f\left(u,\xi\right) = 2\pi u\,\frac{\xi^3}{1+\xi^2}$$

and

$$(P)\quad \inf\left\{I\left(u\right) = \int_a^b f\left(u\left(x\right),u'\left(x\right)\right)dx : u\left(a\right) = \alpha,\ u\left(b\right) = \beta\right\} = m.$$

We will not treat this problem in the present book and we refer to Buttazzo-Kawohl [18] for a review article on this subject.

Example: brachistochrone. The aim is to find the shortest path between two points that follows a point mass moving under the influence of gravity. We place the initial point at the origin and the end one at $\left(b,-\beta\right)$, with $b,\beta > 0$. We let gravity act downwards along the $y-$axis and we represent any point along the path by $\left(x,-u\left(x\right)\right),\ 0\le x\le b$.

In terms of our notations, we have that $n = N = 1$ and the function under consideration is

$$f\left(x,u,\xi\right) = f\left(u,\xi\right) = \frac{\sqrt{1+\xi^2}}{\sqrt{2\,g\,u}}$$

and

$$(P)\quad \inf\left\{I\left(u\right) = \int_0^b f\left(u\left(x\right),u'\left(x\right)\right)dx : u\in X\right\} = m$$

where

$$X = \left\{u\in C^1\left(\left(0,b\right]\right) : u\left(0\right) = 0,\ u\left(b\right) = \beta \text{ and } u\left(x\right) > 0,\ \forall\,x\in\left(0,b\right]\right\}.$$

The shortest path turns out to be a *cycloid*.

Example: minimal surface of revolution. We have to determine among all surfaces of revolution of the form

$$v\left(x,y\right) = \left(x,u\left(x\right)\cos y,u\left(x\right)\sin y\right),$$

with fixed end points $u\left(a\right) = \alpha,\ u\left(b\right) = \beta$, one with minimal area. We still have $n = N = 1$,

$$f\left(x,u,\xi\right) = f\left(u,\xi\right) = 2\pi u\sqrt{1+\xi^2}$$

and

$$(P)\quad \inf\left\{I\left(u\right) = \int_a^b f\left(u\left(x\right),u'\left(x\right)\right)dx : u\left(a\right) = \alpha,\ u\left(b\right) = \beta,\ u > 0\right\} = m.$$

Solutions of this problem, when they exist, are *catenoids*. More precisely, the minimizer is given, $\lambda > 0$ and μ denoting some constants, by

$$u(x) = \lambda \cosh \frac{x + \mu}{\lambda}.$$

Example: mechanical system. Consider a mechanical system with M particles whose respective masses are m_i and positions at time t are

$$u^i(t) = \left(x^i(t), y^i(t), z^i(t)\right) \in \mathbb{R}^3, \quad 1 \le i \le M.$$

Let

$$T(u') = \frac{1}{2} \sum_{i=1}^{M} m_i \left|(u^i)'\right|^2 = \frac{1}{2} \sum_{i=1}^{M} m_i \left(\left((x^i)'\right)^2 + \left((y^i)'\right)^2 + \left((z^i)'\right)^2\right)$$

be the kinetic energy and denote the potential energy with $U = U(t, u)$. Finally let

$$f(t, u, \xi) = T(\xi) - U(t, u)$$

be the Lagrangian. In our formalism we have $n = 1$ and $N = 3M$.

Example: Dirichlet integral. This is the most celebrated problem of the calculus of variations. Here we have $n > 1$, $N = 1$ and

$$(P) \quad \inf\left\{I(u) = \frac{1}{2} \int_{\Omega} |\nabla u(x)|^2 \, dx : u = u_0 \text{ on } \partial\Omega\right\}.$$

As for every variational problem, we associate a differential equation that is nothing other than *Laplace equation*, namely $\Delta u = 0$.

Example: minimal surfaces. This problem is almost as famous as the preceding one. The question is to find among all surfaces $\Sigma \subset \mathbb{R}^3$ (or more generally in $\mathbb{R}^{n+1}$, $n \ge 2$) with prescribed boundary, $\partial\Sigma = \Gamma$, where Γ is a simple closed curve, one that is of minimal area. A variant of this problem is known as *Plateau problem*. One can experimentally realize such surfaces by dipping a wire loop into soapy water; the surface obtained when pulling the wire out from the water is a minimal surface.

The precise formulation of the problem depends on the kind of surfaces that we are considering. We have seen above how to write the problem for minimal surfaces of revolution. We now formulate the problem for more general surfaces.

Case 1: nonparametric surfaces. We consider (hyper) surfaces of the form

$$\Sigma = \left\{v(x) = (x, u(x)) \in \mathbb{R}^{n+1} : x \in \overline{\Omega}\right\}$$

with $u : \overline{\Omega} \to \mathbb{R}$ and where $\Omega \subset \mathbb{R}^n$ is a bounded connected open set. These surfaces are therefore graphs of functions. The fact that $\partial \Sigma$ is a preassigned curve, Γ, reads now as $u = u_0$ on $\partial \Omega$, where u_0 is a given function. The area of such a surface is given by

$$\text{Area}\,(\Sigma) = I\,(u) = \int_{\Omega} f\,(\nabla u\,(x))\,dx$$

where, for $\xi \in \mathbb{R}^n$, we have set

$$f\,(\xi) = \sqrt{1 + |\xi|^2}.$$

The problem is then written in the usual form

$$(P) \quad \inf \left\{ I\,(u) = \int_{\Omega} f\,(\nabla u\,(x))\,dx : u = u_0 \text{ on } \partial \Omega \right\}.$$

Associated with (P) we have the so-called *minimal surface equation*

$$(E) \quad Mu \equiv \left(1 + |\nabla u|^2\right) \Delta u - \sum_{i,j=1}^{n} u_{x_i} u_{x_j} u_{x_i x_j} = 0$$

which is the equation that any minimizer u of (P) should satisfy. In geometrical terms, this equation just expresses the fact that the corresponding surface Σ has everywhere vanishing *mean curvature*.

Case 2: parametric surfaces. Nonparametric surfaces are clearly too restrictive from the geometrical point of view and one is led to consider *parametric surfaces*. These are sets $\Sigma \subset \mathbb{R}^{n+1}$ so that there exist a connected open set $\Omega \subset \mathbb{R}^n$ and a map $v : \overline{\Omega} \to \mathbb{R}^{n+1}$ such that

$$\Sigma = v\,(\overline{\Omega}) = \{v\,(x) : x \in \overline{\Omega}\}.$$

For example, when $n = 2$ and $v = v\,(x_1, x_2) \in \mathbb{R}^3$, if we denote by $v_{x_1} \times v_{x_2}$ the normal to the surface (where $a \times b$ stands for the vectorial product of $a, b \in \mathbb{R}^3$ and $v_{x_1} = \partial v / \partial x_1$, $v_{x_2} = \partial v / \partial x_2$) we find that the area is given by

$$\text{Area}\,(\Sigma) = J\,(v) = \iint_{\Omega} |v_{x_1} \times v_{x_2}|\,dx_1 dx_2.$$

In terms of the notations introduced at the beginning of the present section we have $n = 2$ and $N = 3$.

Example: isoperimetric inequality. Let $A \subset \mathbb{R}^2$ be a bounded open set whose boundary, ∂A, is a sufficiently regular simple closed curve. Denote by

$L(\partial A)$ the length of the boundary and by $M(A)$ the measure (the area) of A. The isoperimetric inequality states that

$$[L(\partial A)]^2 - 4\pi M(A) \geq 0.$$

Furthermore, equality holds if and only if A is a disk (i.e. ∂A is a circle).

We can rewrite this into our formalism (here $n = 1$ and $N = 2$) by parametrizing the curve

$$\partial A = \left\{ u(x) = \left(u^1(x), u^2(x) \right) : x \in [a, b] \right\}$$

and setting

$$L(\partial A) = L(u) = \int_a^b \sqrt{\left((u^1)' \right)^2 + \left((u^2)' \right)^2},$$

$$M(A) = M(u) = \frac{1}{2} \int_a^b \left(u^1 \left(u^2 \right)' - u^2 \left(u^1 \right)' \right) = \int_a^b u^1 \left(u^2 \right)'.$$

The problem is then to show that

$$(P) \quad \inf \left\{ L(u) : M(u) = 1; \ u(a) = u(b) \right\} = 2\sqrt{\pi}.$$

The problem can then be generalized to open sets $A \subset \mathbb{R}^n$ with sufficiently regular boundary, ∂A, and it reads as

$$[L(\partial A)]^n - n^n \omega_n [M(A)]^{n-1} \geq 0$$

where ω_n is the measure of the unit ball of $\mathbb{R}^n$, $M(A)$ stands for the measure of A and $L(\partial A)$ for the $(n-1)$ measure of ∂A. Moreover, if A is sufficiently regular (for example, convex), there is equality if and only if A is a ball.

0.3 Presentation of the content of the monograph

To deal with problems of the type considered in the previous section, there are, roughly speaking, two ways of proceeding: the classical and the direct methods. Before describing a little more precisely these two methods, it might be enlightening to first discuss minimization problems in $\mathbb{R}^N$.

Let $X \subset \mathbb{R}^N$, $F : X \to \mathbb{R}$ and

$$(P) \quad \inf \left\{ F(x) : x \in X \right\}.$$

The first method consists, if F is continuously differentiable, of finding solutions $\overline{x} \in X$ of

$$F'(x) = 0, \quad x \in X.$$

Then, by analyzing the behavior of the higher derivatives of F, we determine whether $\overline{x}$ is a minimum (global or local), a maximum (global or local) or just a stationary point.

The second method consists of considering a minimizing sequence $\{x_\nu\} \subset X$ so that

$$F(x_\nu) \to \inf \{F(x) : x \in X\}.$$

We then, with appropriate hypotheses on F, prove that the sequence is compact in X, meaning that

$$x_\nu \to \overline{x} \in X, \quad \text{as } \nu \to \infty.$$

Finally, if F is lower semicontinuous, meaning that

$$\liminf_{\nu \to \infty} F(x_\nu) \geq F(\overline{x}),$$

we have indeed shown that $\overline{x}$ is a minimizer of (P).

We can proceed in a similar manner for problems of the calculus of variations. The first and second methods are then called, respectively, classical and direct methods. However, the problem is now considerably harder because we are working in infinite dimensional spaces.

Let us recall the problem under consideration

$$(P) \quad \inf \left\{ I(u) = \int_\Omega f(x, u(x), \nabla u(x)) \, dx : u \in X \right\} = m$$

where

 - $\Omega \subset \mathbb{R}^n$, $n \geq 1$, is a bounded open set, points in Ω are denoted by $x = (x_1, \cdots, x_n)$;

 - $u : \Omega \to \mathbb{R}^N$, $N \geq 1$, $u = (u^1, \cdots, u^N)$ and $\nabla u = \left(\frac{\partial u^j}{\partial x_i} \right)_{1 \leq i \leq n}^{1 \leq j \leq N} \in \mathbb{R}^{N \times n}$;

 - $f : \overline{\Omega} \times \mathbb{R}^N \times \mathbb{R}^{N \times n} \longrightarrow \mathbb{R}$, $f = f(x, u, \xi)$, is continuous;

 - X is a space of admissible functions that satisfy $u = u_0$ on $\partial \Omega$, where u_0 is a given function.

Here, contrary to the case of $\mathbb{R}^N$, we encounter a preliminary problem, namely: what is the best choice for the space X of admissible functions? A natural one seems to be $X = C^1(\overline{\Omega})$. There are several reasons, which will be clearer during the course of the book, which indicate that this is not the best choice. A better one is the *Sobolev space* $W^{1,p}(\Omega)$, $p \geq 1$. We say that $u \in W^{1,p}(\Omega)$, if u is (weakly) differentiable and if

$$\|u\|_{W^{1,p}} = \left[\int_\Omega (|u(x)|^p + |\nabla u(x)|^p) \, dx \right]^{1/p} < \infty.$$

The most important properties of these spaces are recalled in Chapter 1.

In Chapter 2, we briefly discuss the *classical methods* introduced by Euler, Hamilton, Hilbert, Jacobi, Lagrange, Legendre, Weierstrass and others. The most important tool is the *Euler-Lagrange equation*, the equivalent of $F'(x) = 0$ in the finite dimensional case, which should satisfy any $\overline{u} \in C^2(\overline{\Omega})$ minimizer of (P), namely (we write here the equation in the case $N = 1$)

$$(E) \quad \sum_{i=1}^{n} \frac{\partial}{\partial x_i} \left[f_{\xi_i}(x, \overline{u}, \nabla \overline{u}) \right] = f_u(x, \overline{u}, \nabla \overline{u}), \quad \forall x \in \overline{\Omega}$$

where $f_{\xi_i} = \partial f / \partial \xi_i$ and $f_u = \partial f / \partial u$.

In the case of the Dirichlet integral

$$(P) \quad \inf \left\{ I(u) = \frac{1}{2} \int_{\Omega} |\nabla u(x)|^2 \, dx : u = u_0 \text{ on } \partial \Omega \right\}$$

the Euler-Lagrange equation reduces to *Laplace equation*, namely $\Delta \overline{u} = 0$.

We immediately note that, in general, finding a C^2 solution of (E) is a difficult task, unless, perhaps, $n = 1$ or the equation (E) is linear. The next step is to know whether a solution $\overline{u}$ of (E), sometimes called a stationary point of I, is, in fact, a minimizer of (P). If $(u, \xi) \to f(x, u, \xi)$ is convex for every $x \in \Omega$ then $\overline{u}$ is indeed a minimizer of (P); in the above examples this happens for the Dirichlet integral or the problem of minimal surfaces in nonparametric form. If, however, $(u, \xi) \to f(x, u, \xi)$ is not convex, several criteria, especially in the case $n = 1$, can be used to determine the nature of the stationary point. Such criteria are, for example, Jacobi, Legendre, Weierstrass, Weierstrass-Erdmann conditions or the fields theories.

In Chapters 3 and 4 we present the *direct methods* introduced by Hilbert, Lebesgue and Tonelli. The idea is to split the discussion of the problem into two parts: the *existence* of minimizers in Sobolev spaces and then the *regularity* of the solution. We start by establishing, in Chapter 3, the existence of minimizers of (P) in Sobolev spaces $W^{1,p}(\Omega)$. In Chapter 4 we see that, sometimes, minimizers of (P) are more regular than in a Sobolev space, for example they are in C^1 or even in C^∞, if the data Ω, f and u_0 are sufficiently regular.

We now briefly describe the ideas behind the proof of the existence of minimizers in Sobolev spaces. As for the finite dimensional case we start by considering a minimizing sequence $\{u_\nu\} \subset W^{1,p}(\Omega)$, which means that

$$I(u_\nu) \to \inf \left\{ I(u) : u = u_0 \text{ on } \partial\Omega \text{ and } u \in W^{1,p}(\Omega) \right\} = m, \quad \text{as } \nu \to \infty.$$

The first step consists of showing that the sequence is compact, i.e. that the sequence converges to an element $\overline{u} \in W^{1,p}(\Omega)$. This, of course, depends on the topology that we have on $W^{1,p}$. The natural topology is the one induced by the norm, which we call *strong convergence* and denote by

$$u_\nu \to \bar{u} \quad \text{in } W^{1,p}.$$

However, it is, in general, not an easy matter to show that the sequence converges in such a strong topology. It is often better to weaken the notion of convergence and to consider the so-called *weak convergence*, denoted by $\rightharpoonup$. To obtain that

$$u_\nu \rightharpoonup \bar{u} \quad \text{in } W^{1,p}$$

is much easier and it is enough, for example if $p > 1$, to show (up to the extraction of a subsequence) that

$$\|u_\nu\|_{W^{1,p}} \leq \gamma$$

where γ is a constant independent of ν. Such an estimate follows, for instance, if we impose a *coercivity* assumption on the function f of the type

$$\lim_{|\xi| \to \infty} \frac{f(x, u, \xi)}{|\xi|} = +\infty, \quad \forall (x, u) \in \overline{\Omega} \times \mathbb{R}.$$

We observe that the Dirichlet integral, with

$$f(x, u, \xi) = \frac{1}{2} |\xi|^2,$$

satisfies this hypothesis but not the minimal surface in nonparametric form, where

$$f(x, u, \xi) = \sqrt{1 + |\xi|^2}.$$

The second step consists of showing that the functional I is lower semicontinuous with respect to weak convergence, namely

$$u_\nu \rightharpoonup \bar{u} \text{ in } W^{1,p} \quad \Rightarrow \quad \liminf_{\nu \to \infty} I(u_\nu) \geq I(\bar{u}).$$

We will see that this conclusion is true if

$$\xi \to f(x, u, \xi) \text{ is convex}, \quad \forall (x, u) \in \overline{\Omega} \times \mathbb{R}.$$

Since $\{u_\nu\}$ was a minimizing sequence, we deduce that $\bar{u}$ is indeed a minimizer of (P).

In Chapter 5 we consider the problem of minimal surfaces. The methods of Chapter 3 cannot be directly applied. In fact, the step of compactness of the minimizing sequences is much harder to obtain, for reasons that we explain in Chapter 5. There are, moreover, difficulties related to the geometrical nature of the problem; for instance, the type of surfaces that we consider, or the notion of area. We present a method due to Douglas and refined by Courant and Tonelli to deal with this problem. However, the techniques are, in essence, direct methods similar to those of Chapter 3.

In Chapter 6 we discuss the isoperimetric inequality in $\mathbb{R}^n$. Depending on the dimension, the way of solving the problem is very different. When $n = 2$, we present a proof that is essentially that of Hurwitz and is in the spirit of the techniques developed in Chapter 2. In higher dimensions the proof is more geometrical; it uses as a main tool the *Brunn-Minkowski theorem*.

Chapter 1

Preliminaries

1.1 Introduction

In this chapter we introduce several notions that are used throughout the book. Most of them are concerned with different spaces of functions. We recommend for first reading to omit this chapter and to refer to it only when needed in the following chapters.

In Section 1.2, we just fix the notations concerning spaces of $k-$times, $k \geq 0$ an integer, continuously differentiable functions, $C^k(\Omega)$. We next introduce the spaces of Hölder continuous functions, $C^{k,\alpha}(\Omega)$, where $k \geq 0$ is an integer and $0 < \alpha \leq 1$. We refer to Chapter 16 in Csató-Dacorogna-Kneuss [29] for a detailed study of Hölder spaces. The following books or paper also have extensive results on these spaces, see Edmunds-Evans [43], Gilbarg-Trudinger [57] or Hörmander [70].

In Section 1.3 we consider the Lebesgue spaces $L^p(\Omega)$, $1 \leq p \leq \infty$. We assume that the reader is familiar with Lebesgue integration and we do not recall theorems such as Fatou lemma, Lebesgue dominated convergence theorem or Fubini theorem. We do however state, mostly without proofs, some other important facts such as Hölder inequality, Riesz theorem and some density results. We also discuss the notion of weak convergence in L^p and the Riemann-Lebesgue theorem. We conclude with the fundamental lemma of the calculus of variations that is used throughout the book, in particular for deriving the Euler-Lagrange equations. There are many excellent books on this subject and we refer, for example, to Adams [1], Brézis [14], De Barra [38] and Willem [111].

In Section 1.4 we define the Sobolev spaces $W^{k,p}(\Omega)$, where $1 \leq p \leq \infty$ and $k \geq 1$ is an integer. We recall several important results concerning these spaces, notably the Sobolev imbedding theorem and Rellich-Kondrachov theorem. In some instances, we give some proofs for the one dimensional case in order to help the reader to become more familiar with these spaces. We recommend the books by Brézis [14], Evans [47] or Willem [111] for a clear introduction to the subject. The monographs of Edmunds-Evans [43] and Gilbarg-Trudinger [57] can also be of great help. The book by Adams [1] is surely one of the most complete in this field, but it is harder reading than the other four.

Finally, in Section 1.5 we gather some important properties of convex functions such as Jensen inequality, the Legendre transform and Carathéodory theorem. The book by Rockafellar [98] is classic in this field. One can also consult Dacorogna [32], Hörmander [71] or Webster [108].

1.2   Continuous and Hölder continuous functions

1.2.1   Space of continuous functions and notations

Definition 1.1 *Let $\Omega \subset \mathbb{R}^n$ be an open set.*

(i) $C^0(\Omega) = C(\Omega)$ *is the* set of continuous functions $u : \Omega \to \mathbb{R}$.

(ii) $C^0(\Omega; \mathbb{R}^N) = C(\Omega; \mathbb{R}^N)$ *is the* set of continuous maps $u : \Omega \to \mathbb{R}^N$, *meaning that if* $u = (u^1, \cdots, u^N)$, *then* $u^i \in C(\Omega)$, *for every* $i = 1, \cdots, N$.

(iii) $C^0(\overline{\Omega}) = C(\overline{\Omega})$ *is the set of bounded continuous functions* $u : \Omega \to \mathbb{R}$, *which are extended in a continuous and bounded way to* $\overline{\Omega}$.

(iv) When we are dealing with maps, $u : \Omega \to \mathbb{R}^N$, *we write, similarly as above,* $C^0(\overline{\Omega}; \mathbb{R}^N) = C(\overline{\Omega}; \mathbb{R}^N)$.

(v) The support *of a function* $u : \Omega \to \mathbb{R}$ *is defined as*

$$\operatorname{supp} u = \overline{\{x \in \Omega : u(x) \neq 0\}}.$$

(vi) $C_0(\Omega) = \{u \in C(\Omega) : \operatorname{supp} u \subset \Omega \text{ is compact}\}$.

(vii) We define the norm over $C(\overline{\Omega})$, *by*

$$\|u\|_{C^0} = \sup_{x \in \overline{\Omega}} |u(x)|.$$

Remark 1.2 **(i)** $C(\overline{\Omega})$ equipped with the norm $\|\cdot\|_{C^0}$ is a Banach space.

(ii) In the definition of $C(\overline{\Omega})$ we have required that the functions be bounded. If Ω is bounded, this is not a restriction. If Ω is unbounded, we could, as some authors do, require that the functions be only continuously extended to the boundary (and not necessarily bounded). In the framework of the present book the definition that we adopt is however more appropriate, since we want to define a norm on $C(\overline{\Omega})$ even if Ω is unbounded. Note that, according to our definition,

$C\left(\overline{\mathbb{R}^n}\right) \underset{\neq}{\subseteq} C\left(\mathbb{R}^n\right)$; indeed, the function $u\left(x\right) = x$ is such that $u \in C\left(\mathbb{R}\right)$, but $u \notin C\left(\overline{\mathbb{R}}\right)$.

Theorem 1.3 (Ascoli-Arzelà theorem) *Let $\Omega \subset \mathbb{R}^n$ be a bounded connected open set. Let $K \subset C\left(\overline{\Omega}\right)$ be bounded and such that the following property of equicontinuity holds: for every $\epsilon > 0$ there exists $\delta > 0$ so that*

$$|x - y| < \delta \quad \Rightarrow \quad |u\left(x\right) - u\left(y\right)| < \epsilon, \; \forall\, x, y \in \overline{\Omega} \; and \; \forall\, u \in K,$$

then $\overline{K}$ is compact.

We also use the following notations.

(i) If $u : \mathbb{R}^n \to \mathbb{R}$, $u = u\left(x_1, \cdots, x_n\right)$, we denote partial derivatives in either of the following ways

$$D_j u = u_{x_j} = \frac{\partial u}{\partial x_j},$$

$$\nabla u = \operatorname{grad} u = \left(\frac{\partial u}{\partial x_1}, \cdots, \frac{\partial u}{\partial x_n}\right) = \left(u_{x_1}, \cdots, u_{x_n}\right).$$

(ii) We now introduce the notations for the higher derivatives. Let $k \geq 1$ be an integer; an element of

$$\mathcal{A}_k = \left\{ a = \left(a_1, \cdots, a_n\right), \; a_j \geq 0 \text{ an integer and } \sum_{j=1}^{n} a_j = k \right\},$$

is called a multi-index of order k. On some occasions we also write for such elements

$$|a| = \sum_{j=1}^{n} a_j = k.$$

Let $a \in \mathcal{A}_k$, we write

$$D^a u = D_1^{a_1} \cdots D_n^{a_n} u = \frac{\partial^{|a|} u}{\partial x_1^{a_1} \cdots \partial x_n^{a_n}}.$$

We also let $\nabla^k u = \left(D^a u\right)_{a \in \mathcal{A}_k}$. In other words, $\nabla^k u$ contains all the partial derivatives of order k of the function u (for example $\nabla^0 u = u$, $\nabla^1 u = \nabla u$).

Definition 1.4 *Let $\Omega \subset \mathbb{R}^n$ be an open set and $k \geq 0$ be an integer.*

(i) The set of functions $u : \Omega \to \mathbb{R}$ that have all partial derivatives, $D^a u$, $a \in \mathcal{A}_m$, $0 \leq m \leq k$, continuous is denoted by $C^k\left(\Omega\right)$.

(ii) $C^k\left(\overline{\Omega}\right)$ *is the set of* $C^k\left(\Omega\right)$ *bounded functions whose derivatives up to order* k *can be extended continuously and in a bounded way to* $\overline{\Omega}$. *It is equipped with the following norm*

$$\|u\|_{C^k} = \max_{0\leq|a|\leq k} \sup_{x\in\overline{\Omega}} |D^a u\left(x\right)|.$$

(iii) $C_0^k\left(\Omega\right) = C^k\left(\Omega\right) \cap C_0\left(\Omega\right)$.

(iv) $C^\infty\left(\Omega\right) = \bigcap_{k=0}^{\infty} C^k\left(\Omega\right)$, $C^\infty\left(\overline{\Omega}\right) = \bigcap_{k=0}^{\infty} C^k\left(\overline{\Omega}\right)$.

(v) $C_0^\infty\left(\Omega\right) = \mathcal{D}\left(\Omega\right) = C^\infty\left(\Omega\right)\bigcap C_0\left(\Omega\right)$.

(vi) *When dealing with maps* $u : \Omega \to \mathbb{R}^N$, *we write, for example,* $C^k\left(\Omega;\mathbb{R}^N\right)$, *and similarly for the other cases.*

Remark 1.5 **(i)** $C^k\left(\overline{\Omega}\right)$ with its norm $\|\cdot\|_{C^k}$ is a Banach space.

(ii) When Ω is unbounded, see Remark 1.2.

We also need to define the set of piecewise continuous functions.

Definition 1.6 *Let* $\Omega \subset \mathbb{R}^n$ *be an open set.*

(i) Define $C_{piec}^0\left(\overline{\Omega}\right) = C_{piec}\left(\overline{\Omega}\right)$ *to be the* set of bounded piecewise continuous functions $u : \overline{\Omega} \to \mathbb{R}$. *This means that there exists a finite (or more generally a countable) partition of* Ω *into open sets* $\Omega_i \subset \Omega$, $i = 1, \cdots, I$, *so that*

$$\overline{\Omega} = \bigcup_{i=1}^{I} \overline{\Omega}_i, \quad \Omega_i \cap \Omega_j = \emptyset, \ if\ i \neq j,\ 1 \leq i, j \leq I$$

and $u|_{\overline{\Omega}_i}$ *is bounded and continuous.*

(ii) Similarly $C_{piec}^k\left(\overline{\Omega}\right)$, $k \geq 1$, *is the set of functions* $u \in C^{k-1}\left(\overline{\Omega}\right)$, *whose partial derivatives of order* k *are in* $C_{piec}^0\left(\overline{\Omega}\right)$.

1.2.2   Hölder continuous functions

We now turn to the notion of Hölder continuous functions (examples are given in exercises).

Definition 1.7 *Let* $D \subset \mathbb{R}^n$, $u : D \to \mathbb{R}$ *and* $0 < \alpha \leq 1$. *We let*

$$[u]_{C^{0,\alpha}(D)} = \sup_{\substack{x,y\in D \\ x\neq y}} \left\{ \frac{|u\left(x\right) - u\left(y\right)|}{|x - y|^\alpha} \right\}.$$

Let $\Omega \subset \mathbb{R}^n$ be an open set, $k \geq 0$ be an integer. We define the different spaces of Hölder continuous functions in the following way.

(i) $C^{0,\alpha}(\Omega)$ is the set of $u \in C(\Omega)$ so that

$$[u]_{C^{0,\alpha}(K)} = \sup_{\substack{x,y \in K \\ x \neq y}} \left\{ \frac{|u(x) - u(y)|}{|x - y|^\alpha} \right\} < \infty$$

for every compact set $K \subset \Omega$.

(ii) $C^{0,\alpha}(\overline{\Omega})$ is the set of functions $u \in C(\overline{\Omega})$ so that

$$[u]_{C^{0,\alpha}(\overline{\Omega})} < \infty.$$

It is equipped with the norm

$$\|u\|_{C^{0,\alpha}(\overline{\Omega})} = \|u\|_{C^0(\overline{\Omega})} + [u]_{C^{0,\alpha}(\overline{\Omega})}.$$

If there is no ambiguity, we drop the dependence on the set $\overline{\Omega}$ and write simply

$$\|u\|_{C^{0,\alpha}} = \|u\|_{C^0} + [u]_{C^{0,\alpha}}.$$

(iii) $C^{k,\alpha}(\Omega)$ is the set of $u \in C^k(\Omega)$ so that

$$[D^a u]_{C^{0,\alpha}(K)} < \infty$$

for every compact set $K \subset \Omega$ and every $a \in \mathcal{A}_k$.

(iv) $C^{k,\alpha}(\overline{\Omega})$ is the set of functions $u \in C^k(\overline{\Omega})$ so that

$$[D^a u]_{C^{0,\alpha}(\overline{\Omega})} < \infty$$

for every multi-index $a \in \mathcal{A}_k$. It is equipped with the following norm

$$\|u\|_{C^{k,\alpha}} = \|u\|_{C^k} + \max_{a \in \mathcal{A}_k} [D^a u]_{C^{0,\alpha}}.$$

(v) $C^{k,\alpha}(\overline{\Omega}; \mathbb{R}^N)$ stands for the set of $u : \Omega \to \mathbb{R}^N$, $u = (u^1, \cdots, u^N)$, so that $u^i \in C^{k,\alpha}(\overline{\Omega})$, for every $i = 1, \cdots, N$.

Remark 1.8 (i) $C^{k,\alpha}(\overline{\Omega})$ with its norm $\|\cdot\|_{C^{k,\alpha}}$ is a Banach space.

(ii) By abuse of notations we write $C^k(\Omega) = C^{k,0}(\Omega)$; or in other words, the set of continuous functions is identified with the set of Hölder continuous functions with exponent 0.

(iii) Similarly when $\alpha = 1$, we see that $C^{0,1}\left(\overline{\Omega}\right)$ is in fact the set of *Lipschitz continuous functions*, namely the set of functions u such that there exists a constant $\gamma > 0$ so that

$$|u\left(x\right) - u\left(y\right)| \leq \gamma\left|x - y\right|, \quad \forall\, x, y \in \overline{\Omega}.$$

The best such constant is $\gamma = [u]_{C^{0,1}}$.

(iv) In Exercise 1.2.1 (iii) we will see that Hölder continuity for $\alpha \in (0, 1)$ does not imply any differentiability of the function. In contrast when $\alpha = 1$, the classical Rademacher theorem (see for example Evans-Gariepy [48]) implies that the function is almost everywhere differentiable.

(v) When Ω is unbounded, see Remark 1.2.

Proposition 1.9 *Let $\Omega \subset \mathbb{R}^n$ be bounded and open, $0 \leq \alpha \leq \beta \leq 1$ and $k \geq 0$ be an integer. The following properties then hold.*

(i) *If $u, v \in C^{0,\alpha}\left(\overline{\Omega}\right)$ then $uv \in C^{0,\alpha}\left(\overline{\Omega}\right)$.*

(ii) *The following set of inclusions is valid*

$$C^k\left(\overline{\Omega}\right) \supset C^{k,\alpha}\left(\overline{\Omega}\right) \supset C^{k,\beta}\left(\overline{\Omega}\right) \supset C^{k,1}\left(\overline{\Omega}\right).$$

(iii) *If, in addition, Ω is convex, then*

$$C^{k,1}\left(\overline{\Omega}\right) \supset C^{k+1}\left(\overline{\Omega}\right).$$

Remark 1.10 (i) The proposition is proved in Exercise 1.2.2 below when $k = 0$.

(ii) The statement (i) extends to higher derivatives. More precisely, if $u, v \in C^{k,\alpha}\left(\overline{\Omega}\right)$ then $uv \in C^{k,\alpha}\left(\overline{\Omega}\right)$, provided the set Ω is not too wild, for example if it has Lipschitz boundary (see Definition 1.42 below for the precise meaning).

(iii) The inclusion in (iii) remains valid for very general sets Ω, for example those with Lipschitz boundary; however, it is false in general, see Exercise 1.2.3.

1.2.3   Exercises

Exercise 1.2.1 (i) Let $\Omega = (0, 1)$ and $u_\alpha\left(x\right) = x^\alpha$ with $\alpha \in (0, 1]$. Show that $u_\alpha \in C^{0,\alpha}\left([0, 1]\right)$.

(ii) Prove that

$$u\left(x\right) = \begin{cases} -1/\log x & \text{if } x \in (0, 1/2] \\ 0 & \text{if } x = 0 \end{cases}$$

is continuous but not in $C^{0,\alpha}\left([0, 1/2]\right)$ for any $\alpha \in (0, 1]$.

(iii) Let $\lambda \in (0, 1]$ and

$$u_\lambda (x) = \sum_{n=1}^\infty \frac{\cos (2^n x)}{2^{n\lambda}}.$$

Show that $u_\lambda \in C^{0,\alpha} ([0, \pi])$, for any $0 < \alpha < \lambda$.

Exercise 1.2.2 Show Proposition 1.9 when $k = 0$.

Exercise 1.2.3 Let $\frac{1}{2} < \beta < 1$,

$$\Omega = \left\{ (x_1, x_2) \in \mathbb{R}^2 : x_2 < \sqrt{|x_1|} \quad \text{and} \quad x_1^2 + x_2^2 < 1 \right\},$$

$$u (x_1, x_2) = \begin{cases} 0 & \text{if } (x_1, x_2) \in \Omega \text{ and } x_2 \leq 0 \\ x_1 x_2^{2\beta} / |x_1| & \text{if } (x_1, x_2) \in \Omega \text{ and } x_2 > 0. \end{cases}$$

Show that $u \in C^1 (\overline{\Omega})$ but $u \notin C^{0,\alpha} (\overline{\Omega})$ for every $\frac{1}{2} < \beta < \alpha \leq 1$.

Exercise 1.2.4 (i) Let $\Omega \subset \mathbb{R}^n$ be a bounded open set, $0 \leq \alpha, \beta \leq 1$. Show that

$$u \in C^{0,\alpha}(\overline{\Omega}; \mathbb{R}^n) \quad \text{and} \quad v \in C^{0,\beta}(u (\overline{\Omega})) \Rightarrow v \circ u \in C^{0,\alpha\beta}(\overline{\Omega})$$

and that, in general, even when $n = 1$, $u \in C^{0,\alpha}(\overline{\Omega}; \mathbb{R}^n)$ and $v \in C^{0,\alpha}(u (\overline{\Omega}))$ do not imply $v \circ u \in C^{0,\alpha}(\overline{\Omega})$.

(ii) Prove that if $\Omega \subset \mathbb{R}^n$ is a bounded convex open set, $0 \leq \alpha \leq 1$, $k \geq 1$ is an integer, then

$$u \in C^{k,\alpha}(\overline{\Omega}; \mathbb{R}^n) \quad \text{and} \quad v \in C^{k,\alpha}(u (\overline{\Omega})) \Rightarrow v \circ u \in C^{k,\alpha}(\overline{\Omega}).$$

Exercise 1.2.5 Let $\Omega \subset \mathbb{R}^n$ be a bounded open set, $0 < \alpha \leq 1$, $u \in C^{0,\alpha}(\overline{\Omega})$ and

$$\gamma = [u]_{C^{0,\alpha}(\overline{\Omega})}.$$

Show that the functions

$$u_+ (x) = \inf_{y \in \overline{\Omega}} \left\{ u (y) + \gamma |x - y|^\alpha \right\},$$

$$u_- (x) = \sup_{y \in \overline{\Omega}} \left\{ u (y) - \gamma |x - y|^\alpha \right\}$$

are extensions of u outside $\overline{\Omega}$ and

$$[u_+]_{C^{0,\alpha}(\mathbb{R}^n)} = [u_-]_{C^{0,\alpha}(\mathbb{R}^n)} = [u]_{C^{0,\alpha}(\overline{\Omega})} = \gamma.$$

Prove that any other extension v of u such that $[v]_{C^{0,\alpha}(\mathbb{R}^n)} = \gamma$ satisfies

$$u_- \leq v \leq u_+.$$

1.3 L^p spaces

1.3.1 Basic definitions and properties

Definition 1.11 *Let $\Omega \subset \mathbb{R}^n$ be an open set and $1 \leq p \leq \infty$. We say that a measurable function $u : \Omega \to \mathbb{R}$ belongs to $L^p(\Omega)$ if*

$$\|u\|_{L^p} = \begin{cases} \left(\int_\Omega |u(x)|^p \, dx \right)^{1/p} & \text{if } 1 \leq p < \infty \\ \inf \{\alpha : |u(x)| \leq \alpha \text{ a.e. in } \Omega\} & \text{if } p = \infty \end{cases}$$

is finite. As above, if $u : \Omega \to \mathbb{R}^N$, $u = \left(u^1, \cdots, u^N\right)$, is such that $u^i \in L^p(\Omega)$, for every $i = 1, \cdots, N$, we write $u \in L^p\left(\Omega; \mathbb{R}^N\right)$.

Remark 1.12 **(i)** The abbreviation "a.e." means that a property holds almost everywhere. For example, the function

$$\chi_{\mathbb{Q}}(x) = \begin{cases} 1 & \text{if } x \in \mathbb{Q} \\ 0 & \text{if } x \notin \mathbb{Q} \end{cases}$$

where $\mathbb{Q}$ is the set of rational numbers, is such that $\chi_{\mathbb{Q}} = 0$ a.e.

(ii) We adopt the convention that we identify two functions that coincide except on a set of measure zero. So, strictly speaking, the spaces L^p are equivalence classes.

(iii) In the following, we let p' be the *conjugate exponent* of p. It is defined by

$$\frac{1}{p} + \frac{1}{p'} = 1 \quad \Leftrightarrow \quad p' = \frac{p}{p - 1}$$

with the convention that if $p = 1$, respectively $p = \infty$, then $p' = \infty$, respectively $p' = 1$.

In the next theorem we summarize the most important properties of L^p spaces that we need. However, we do not recall Fatou lemma, the dominated convergence theorem and other basic theorems of Lebesgue integral.

Theorem 1.13 *Let $\Omega \subset \mathbb{R}^n$ be open and $1 \leq p \leq \infty$.*

(i) $\|\cdot\|_{L^p}$ *is a norm and $L^p(\Omega)$, equipped with this norm, is a Banach space. The space $L^2(\Omega)$ is a Hilbert space with inner product given by*

$$\langle u; v \rangle = \int_\Omega u(x) \, v(x) \, dx.$$

(ii) **Hölder inequality** *asserts that if* $u \in L^p(\Omega)$ *and* $v \in L^{p'}(\Omega)$ *where* $1/p + 1/p' = 1$, *then* $uv \in L^1(\Omega)$ *and moreover*

$$\|uv\|_{L^1} \leq \|u\|_{L^p} \|v\|_{L^{p'}} .$$

(iii) **Minkowski inequality** *asserts that*

$$\|u + v\|_{L^p} \leq \|u\|_{L^p} + \|v\|_{L^p} .$$

(iv) **Riesz theorem**: *the dual space of* L^p, *denoted by* $(L^p)'$, *can be identified with* $L^{p'}(\Omega)$ *where* $1/p + 1/p' = 1$ *provided* $1 \leq p < \infty$. *More precisely, if* $\varphi \in (L^p)'$ *with* $1 \leq p < \infty$, *then there exists a unique* $u \in L^{p'}$ *so that*

$$\langle \varphi; f \rangle = \varphi(f) = \int_\Omega u(x) f(x) \, dx, \quad \forall f \in L^p(\Omega)$$

and moreover

$$\|u\|_{L^{p'}} = \|\varphi\|_{(L^p)'} .$$

(v) L^p *is separable if* $1 \leq p < \infty$ *and reflexive if* $1 < p < \infty$.

(vi) *Let* $1 \leq p < \infty$. *The piecewise constant functions (also called step functions if* $\Omega \subset \mathbb{R}$) *or the* $C_0^\infty(\Omega)$ *functions are dense in* L^p. *More precisely, if* $u \in L^p(\Omega)$ *then there exist* $u_\nu \in C_0^\infty(\Omega)$ *(or* u_ν *piecewise constant) so that*

$$\lim_{\nu \to \infty} \|u_\nu - u\|_{L^p} = 0.$$

Remark 1.14 (i) In the case $p = 2$ and hence $p' = 2$, Hölder inequality is nothing but *Cauchy-Schwarz* inequality

$$\|uv\|_{L^1} \leq \|u\|_{L^2} \|v\|_{L^2} , \quad \text{i.e.} \quad \int_\Omega |uv| \leq \left(\int_\Omega u^2 \right)^{1/2} \left(\int_\Omega v^2 \right)^{1/2} .$$

(ii) In Riesz theorem the result is false if $p = \infty$ (and hence $p' = 1$).

(iii) In the following, we always make the identification $(L^p)' = L^{p'}$. Summarizing the results on duality we have

$$(L^p)' = L^{p'} \quad \text{if } 1 < p < \infty,$$

$$(L^2)' = L^2, \quad (L^1)' = L^\infty, \quad L^1 \subsetneq (L^\infty)'.$$

(iv) The meaning of L^p *reflexive* is that the bidual of L^p, $(L^p)''$, can be identified with L^p.

(v) The last statement in the theorem is false if $p = \infty$.

1.3.2 Weak convergence and Riemann-Lebesgue theorem

We now turn our attention to the notions of convergence in L^p spaces. The natural notion, called *strong convergence*, is the one induced by the $\|\cdot\|_{L^p}$ norm. We also often need a weaker notion of convergence known as *weak convergence*. We now define these notions.

Definition 1.15 *Let $\Omega \subset \mathbb{R}^n$ be an open set and $1 \leq p \leq \infty$.*

(i) A sequence u_ν is said to (strongly) converge to u in L^p if u_ν, $u \in L^p$ and if

$$\lim_{\nu \to \infty} \|u_\nu - u\|_{L^p} = 0.$$

We denote this convergence by $u_\nu \to u$ in L^p.

(ii) If $1 \leq p < \infty$, we say that the sequence u_ν weakly converges to u in L^p if u_ν, $u \in L^p$ and if

$$\lim_{\nu \to \infty} \int_\Omega [u_\nu(x) - u(x)] \varphi(x)\, dx = 0, \quad \forall \varphi \in L^{p'}(\Omega).$$

This convergence is denoted by $u_\nu \rightharpoonup u$ in L^p.

(iii) If $p = \infty$, the sequence u_ν is said to weak $*$ converge to u in L^∞ if u_ν, $u \in L^\infty$ and if

$$\lim_{\nu \to \infty} \int_\Omega [u_\nu(x) - u(x)] \varphi(x)\, dx = 0, \quad \forall \varphi \in L^1(\Omega)$$

and is denoted by: $u_\nu \overset{*}{\rightharpoonup} u$ in L^∞.

Remark 1.16 **(i)** We speak of weak $*$ convergence in L^∞ instead of weak convergence, because as seen above the dual of L^∞ is strictly larger than L^1. Formally, however, weak convergence in L^p and weak $*$ convergence in L^∞ take the same form.

(ii) The limit (weak or strong) is unique.

(iii) It is obvious that

$$u_\nu \to u \text{ in } L^p \quad \Rightarrow \quad \begin{cases} u_\nu \rightharpoonup u \text{ in } L^p & \text{if } 1 \leq p < \infty \\ u_\nu \overset{*}{\rightharpoonup} u \text{ in } L^\infty & \text{if } p = \infty. \end{cases}$$

Example 1.17 (see Exercise 1.3.2). Let $\Omega = (0,1)$, $\alpha \geq 0$ and

$$u_\nu(x) = \begin{cases} \nu^\alpha & \text{if } x \in (0, 1/\nu) \\ 0 & \text{if } x \in (1/\nu, 1). \end{cases}$$

If $1 < p < \infty$, we find

$$u_\nu \to 0 \text{ in } L^p \quad \Leftrightarrow \quad 0 \le \alpha < 1/p$$
$$u_\nu \rightharpoonup 0 \text{ in } L^p \quad \Leftrightarrow \quad 0 \le \alpha \le 1/p.$$

Example 1.18 Let $\Omega = (0, 2\pi)$ and $u_\nu(x) = \sin(\nu x)$, then

$$u_\nu \not\to 0 \text{ in } L^p, \quad \forall 1 \le p \le \infty$$
$$u_\nu \rightharpoonup 0 \text{ in } L^p, \quad \forall 1 \le p < \infty$$

and

$$u_\nu \overset{*}{\rightharpoonup} 0 \quad \text{in } L^\infty.$$

These facts are consequences of Riemann-Lebesgue theorem (see Theorem 1.22).

Example 1.19 Let $\Omega = (0, 1)$, $\alpha, \beta \in \mathbb{R}$

$$u(x) = \begin{cases} \alpha & \text{if } x \in (0, 1/2) \\ \beta & \text{if } x \in (1/2, 1). \end{cases}$$

Extend u by periodicity from $(0, 1)$ to $\mathbb{R}$ and define

$$u_\nu(x) = u(\nu x).$$

Note that u_ν takes only the values α and β and the sets where it takes such values are both sets of measure $1/2$. It is clear that $\{u_\nu\}$ cannot be compact in any L^p spaces; however, from Riemann-Lebesgue theorem (see Theorem 1.22), we find

$$u_\nu \rightharpoonup \frac{\alpha + \beta}{2} \text{ in } L^p, \ \forall 1 \le p < \infty \quad \text{and} \quad u_\nu \overset{*}{\rightharpoonup} \frac{\alpha + \beta}{2} \text{ in } L^\infty.$$

Theorem 1.20 *Let $\Omega \subset \mathbb{R}^n$ be a bounded open set. The following properties then hold.*

(i) If $u_\nu \overset{*}{\rightharpoonup} u$ in L^∞, then $u_\nu \rightharpoonup u$ in L^p, $\forall 1 \le p < \infty$.

(ii) If $1 \le p \le \infty$ and $u_\nu \to u$ in L^p, then

$$\|u\|_{L^p} = \lim_{\nu \to \infty} \|u_\nu\|_{L^p}.$$

(iii) If $1 \le p < \infty$ and if $u_\nu \rightharpoonup u$ in L^p, then there exists a constant $\gamma > 0$ so that

$$\|u_\nu\|_{L^p} \le \gamma \quad \text{and} \quad \|u\|_{L^p} \le \liminf_{\nu \to \infty} \|u_\nu\|_{L^p}.$$

The result remains valid if $p = \infty$ and if $u_\nu \overset{*}{\rightharpoonup} u$ in L^∞.

(iv) If $1 < p < \infty$ and if there exists a constant $\gamma > 0$ so that $\|u_\nu\|_{L^p} \leq \gamma$, then there exist a subsequence $\{u_{\nu_i}\}$ and $u \in L^p$ so that

$$u_{\nu_i} \rightharpoonup u \quad in \ L^p.$$

The result remains valid if $p = \infty$, the conclusion is then $u_{\nu_i} \overset{*}{\rightharpoonup} u$ in L^∞.

(v) Let $1 \leq p \leq \infty$ and $u_\nu \to u$ in L^p, then there exist a subsequence $\{u_{\nu_i}\}$ and $h \in L^p$ such that

$$u_{\nu_i} \to u \ a.e. \quad and \quad |u_{\nu_i}| \leq h \ a.e.$$

Remark 1.21 (i) Comparing (ii) and (iii) of the theorem, we see that weak convergence ensures the lower semicontinuity of the norm, while strong convergence guarantees its continuity.

(ii) The most interesting part of the theorem is (iv). We know that in $\mathbb{R}^n$, Bolzano-Weierstrass theorem ascertains that from any bounded sequence we can extract a convergent subsequence. This is false in L^p spaces (and more generally in infinite dimensional spaces); but it is true if we replace strong convergence by weak convergence.

(iii) The result (iv) is, however, false if $p = 1$; this is a consequence of the fact that L^1 is not a reflexive space. To deduce, up to the extraction of a subsequence, weak convergence, it is not sufficient to have $\|u_\nu\|_{L^1} \leq \gamma$, we need a condition known as "equiintegrability" (see, for example, [45]). This fact explains the difficulty of the minimal surface problem, which is discussed in Chapter 5.

We now turn to Riemann-Lebesgue theorem which allows one to easily construct weakly convergent sequences that do not converge strongly. This theorem is particularly useful when dealing with *Fourier series* (there $u(x) = \sin x$ or $\cos x$).

Theorem 1.22 (Riemann-Lebesgue theorem) *Let $1 \leq p \leq \infty$ and $u \in L^p(\Omega)$ where $\Omega = \prod_{i=1}^n (a_i, b_i)$. Let u be extended by periodicity from Ω to $\mathbb{R}^n$ and define*

$$u_\nu(x) = u(\nu x) \quad and \quad \overline{u} = \frac{1}{\text{meas}\,\Omega} \int_\Omega u(x)\,dx$$

then $u_\nu \rightharpoonup \overline{u}$ in L^p if $1 \leq p < \infty$ and, if $p = \infty$, $u_\nu \overset{*}{\rightharpoonup} \overline{u}$ in L^∞.

Proof To make the argument simpler we assume in the proof that $n = 1$, $\Omega = (0, 1)$ and $1 < p \leq \infty$. For the proof of the general case ($\Omega \subset \mathbb{R}^n$ or $p = 1$) see, for example, Theorem 2.1.5 in [32]. We also assume, without loss of generality, that

$$\overline{u} = \int_0^1 u(x)\,dx = 0.$$

Step 1. Observe that if $1 \leq p < \infty$, then

$$\|u_\nu\|_{L^p}^p = \int_0^1 |u_\nu(x)|^p \, dx = \int_0^1 |u(\nu x)|^p \, dx$$

$$= \frac{1}{\nu} \int_0^\nu |u(y)|^p \, dy = \int_0^1 |u(y)|^p \, dy.$$

The last identity being a consequence of the $1-$periodicity of u. We therefore find that

$$\|u_\nu\|_{L^p} = \|u\|_{L^p} . \tag{1.1}$$

The result is trivially true if $p = \infty$.

Step 2. (For a slightly different proof of this step see Exercise 1.3.6.) We therefore have that $u_\nu \in L^p$ and, since $\bar{u} = 0$, we have to show that

$$\lim_{\nu \to \infty} \int_0^1 u_\nu(x) \varphi(x) \, dx = 0, \quad \forall \varphi \in L^{p'}(0,1) . \tag{1.2}$$

Let $\epsilon > 0$ be arbitrary. Since $\varphi \in L^{p'}(0,1)$ and $1 < p \leq \infty$, which implies $1 \leq p' < \infty$ (i.e. $p' \neq \infty$), we have from Theorem 1.13 that there exists h a step function so that

$$\|\varphi - h\|_{L^{p'}} \leq \epsilon. \tag{1.3}$$

Since h is a step function, we can find $a_0 = 0 < a_1 < \cdots < a_I = 1$ and $\alpha_i \in \mathbb{R}$ such that

$$h(x) = \alpha_i \quad \text{if } x \in (a_{i-1}, a_i), \ 1 \leq i \leq I.$$

We now write

$$\int_0^1 u_\nu(x) \varphi(x) \, dx = \int_0^1 u_\nu(x) [\varphi(x) - h(x)] \, dx + \int_0^1 u_\nu(x) h(x) \, dx$$

and get that

$$\left| \int_0^1 u_\nu(x) \varphi(x) \, dx \right| \leq \int_0^1 |u_\nu(x)| \, |\varphi(x) - h(x)| \, dx + \left| \int_0^1 u_\nu(x) h(x) \, dx \right|.$$

Using Hölder inequality, (1.1) and (1.3) for the first term on the right-hand side of the inequality, we obtain

$$\left| \int_0^1 u_\nu(x) \varphi(x) \, dx \right| \leq \|u_\nu\|_{L^p} \|\varphi - h\|_{L^{p'}} + \sum_{i=1}^I |\alpha_i| \left| \int_{a_{i-1}}^{a_i} u_\nu(x) \, dx \right|$$

and thus

$$\left| \int_0^1 u_\nu(x) \varphi(x) \, dx \right| \leq \epsilon \|u\|_{L^p} + \sum_{i=1}^I |\alpha_i| \left| \int_{a_{i-1}}^{a_i} u_\nu(x) \, dx \right|. \tag{1.4}$$

To conclude we still have to evaluate

$$\int_{a_{i-1}}^{a_i} u_\nu(x)\, dx = \int_{a_{i-1}}^{a_i} u(\nu x)\, dx = \frac{1}{\nu} \int_{\nu a_{i-1}}^{\nu a_i} u(y)\, dy$$

$$= \frac{1}{\nu} \left\{ \int_{\nu a_{i-1}}^{[\nu a_{i-1}]+1} u\, dy + \int_{[\nu a_{i-1}]+1}^{[\nu a_i]} u\, dy + \int_{[\nu a_i]}^{\nu a_i} u\, dy \right\}$$

where $[a]$ stands for the integer part of $a \geq 0$. We now use the periodicity of u in the second term; this is legal since $[\nu a_i] - ([\nu a_{i-1}] + 1)$ is an integer. We therefore find that

$$\left| \int_{a_{i-1}}^{a_i} u_\nu \right| \leq \frac{2}{\nu} \int_0^1 |u| + \frac{[\nu a_i] - [\nu a_{i-1}] - 1}{\nu} \left| \int_0^1 u \right|.$$

Since $\bar{u} = \int_0^1 u = 0$, we have, using the above inequality and returning to (1.4),

$$\left| \int_0^1 u_\nu\, \varphi \right| \leq \epsilon \left\| u \right\|_{L^p} + \frac{2}{\nu} \left\| u \right\|_{L^1} \sum_{i=1}^I |\alpha_i|.$$

Letting $\nu \to \infty$, we hence obtain

$$0 \leq \limsup_{\nu \to \infty} \left| \int_0^1 u_\nu\, \varphi \right| \leq \epsilon \left\| u \right\|_{L^p}.$$

Since ϵ is arbitrary, we immediately have (1.2) and thus the result. ∎

1.3.3   The fundamental lemma of the calculus of variations

We conclude the present section with a result that will be used on several occasions when deriving the Euler-Lagrange equation associated with the problems of the calculus of variations. We start with a definition.

Definition 1.23 *Let $\Omega \subset \mathbb{R}^n$ be an open set and $1 \leq p \leq \infty$. We say that $u \in L_{loc}^p(\Omega)$ if $u \in L^p(\Omega')$ for every open set Ω' compactly contained in Ω (i.e. $\overline{\Omega'} \subset \Omega$ and $\overline{\Omega'}$ is compact).*

Theorem 1.24 (Fundamental lemma of the calculus of variations) *Let Ω be an open set of $\mathbb{R}^n$ and $u \in L_{loc}^1(\Omega)$ be such that*

$$\int_\Omega u(x)\, \psi(x)\, dx = 0, \quad \forall \psi \in C_0^\infty(\Omega) \tag{1.5}$$

then $u = 0$, a.e. in Ω.

Proof We prove the theorem under the stronger hypothesis that $u \in L^2(\Omega)$ and not only $u \in L^1_{\text{loc}}(\Omega)$ (recall that $L^2(\Omega) \subset L^1_{\text{loc}}(\Omega)$); for a proof in the general framework see, for example, Corollary 3.26 in Adams [1] or Lemma IV.2 in Brézis [14]. Let $\epsilon > 0$. Since $u \in L^2(\Omega)$, invoking Theorem 1.13, we can find $\psi \in C_0^\infty(\Omega)$ so that

$$\|u - \psi\|_{L^2} \leq \epsilon.$$

Using (1.5) we deduce that

$$\|u\|_{L^2}^2 = \int_\Omega u^2 = \int_\Omega u\,(u - \psi).$$

Combining the above identity and Hölder inequality, we find

$$\|u\|_{L^2}^2 \leq \|u\|_{L^2} \|u - \psi\|_{L^2} \leq \epsilon \|u\|_{L^2}.$$

Since $\epsilon > 0$ is arbitrary, we deduce that $\|u\|_{L^2} = 0$ and hence the claim. ∎

We next have as a consequence the following result (for a proof see Exercise 1.3.7 and for a generalization see Exercise 1.3.8).

Corollary 1.25 *Let $\Omega \subset \mathbb{R}^n$ be an open set and $u \in L^1_{\text{loc}}(\Omega)$ be such that*

$$\int_\Omega u\,(x)\,\psi\,(x)\,dx = 0, \quad \forall \psi \in C_0^\infty(\Omega) \text{ with } \int_\Omega \psi\,(x)\,dx = 0$$

then $u = $ constant, a.e. in Ω.

1.3.4 Exercises

Exercise 1.3.1 (i) Prove Hölder and Minkowski inequalities.

(ii) Show that if $p, q \geq 1$ with $pq/\,(p + q) \geq 1$, $u \in L^p$ and $v \in L^q$, then

$$uv \in L^{pq/p+q} \quad \text{and} \quad \|uv\|_{L^{pq/p+q}} \leq \|u\|_{L^p} \|v\|_{L^q}.$$

(iii) Deduce that if Ω is bounded, then

$$L^\infty(\Omega) \subset L^p(\Omega) \subset L^q(\Omega) \subset L^1(\Omega), \quad 1 \leq q \leq p \leq \infty.$$

Show that (iii) is, in general, false if Ω is unbounded.

Exercise 1.3.2 Establish the results in Example 1.17.

Exercise 1.3.3 (i) Prove that if $1 \leq p < \infty$, then

$$\left.\begin{array}{ll} u_\nu \rightharpoonup u & \text{in } L^p \\ v_\nu \to v & \text{in } L^{p'} \end{array}\right\} \quad \Rightarrow \quad u_\nu v_\nu \rightharpoonup uv \quad \text{in } L^1.$$

Find an example showing that the result is false if we replace $v_\nu \to v$ in $L^{p'}$ by $v_\nu \rightharpoonup v$ in $L^{p'}$.

(ii) Show that

$$\left.\begin{array}{ll} u_\nu \rightharpoonup u & \text{in } L^2 \\ u_\nu^2 \rightharpoonup u^2 & \text{in } L^1 \end{array}\right\} \quad \Rightarrow \quad u_\nu \to u \quad \text{in } L^2.$$

Exercise 1.3.4 Let $1 \leq p < \infty$ and $u_\nu \rightharpoonup u$ in $L^p(\Omega)$, Prove (using Example 1.53) that

$$\|u\|_{L^p} \leq \liminf_{\nu \to \infty} \|u_\nu\|_{L^p} .$$

Exercise 1.3.5 (Mollifier) Let $\varphi \in C_0^\infty(\mathbb{R}^n)$, $\varphi \geq 0$, $\varphi(x) = 0$ if $|x| > 1$ and $\int_{\mathbb{R}^n} \varphi = 1$, for example

$$\varphi(x) = \begin{cases} c \exp\left\{\dfrac{1}{|x|^2 - 1}\right\} & \text{if } |x| < 1 \\ 0 & \text{otherwise} \end{cases}$$

and c is chosen so that $\int_{\mathbb{R}^n} \varphi = 1$. Define

$$\varphi_\nu(x) = \nu^n \varphi(\nu x),$$

$$u_\nu(x) = (\varphi_\nu * u)(x) = \int_{\mathbb{R}^n} \varphi_\nu(x - y) u(y) \, dy.$$

(i) Show that if $1 \leq p \leq \infty$ then

$$\|u_\nu\|_{L^p} \leq \|u\|_{L^p} .$$

(ii) Prove that if $u \in L^p(\mathbb{R}^n)$, then $u_\nu \in C^\infty(\mathbb{R}^n)$.

(iii) Establish that if $u \in C(\mathbb{R}^n)$, then

$$u_\nu \to u \quad \text{uniformly on every compact set of } \mathbb{R}^n.$$

(iv) Show that if $u \in L^p(\mathbb{R}^n)$ and if $1 \leq p < \infty$, then

$$u_\nu \to u \quad \text{in } L^p(\mathbb{R}^n).$$

Exercise 1.3.6 In Step 2 of Theorem 1.22 use approximation by smooth functions instead of by step functions and anticipate (1.11) and (1.12) in Lemma 1.39.

Exercise 1.3.7 **(i)** Show Corollary 1.25.
   (ii) Prove that if $u \in L^1_{\text{loc}}(a,b)$ is such that

$$\int_a^b u(x)\,\varphi'(x)\,dx = 0, \quad \forall \varphi \in C_0^\infty(a,b)$$

then $u = \text{constant}$, a.e. in (a,b).

Exercise 1.3.8 Generalize Corollary 1.25 in the following manner. Let $\Omega \subset \mathbb{R}^n$ be an open set and $\alpha_1, \cdots, \alpha_N \in L^1_{\text{loc}}(\Omega)$. Let

$$X = \left\{ \psi \in C_0^\infty(\Omega) : \int_\Omega \alpha_i(x)\,\psi(x)\,dx = 0,\ i = 1, \cdots, N \right\}$$

and $u \in L^1_{\text{loc}}(\Omega)$ be such that

$$\int_\Omega u(x)\,\psi(x)\,dx = 0, \quad \forall \psi \in X.$$

Show that there exist constants $a_1, \cdots, a_N \in \mathbb{R}$ such that

$$u(x) = \sum_{i=1}^N a_i\,\alpha_i(x) \quad \text{a.e. } x \in \Omega.$$

Suggestion: use the elementary algebraic result (see Lemma 3.9 page 62 in [100]) which asserts that if X is a vector space and Λ, Λ_i, $i = 1, \cdots, N$, are linear functionals on X such that

$$\Lambda(x) = 0, \quad \text{for every } x \in X \text{ with } \Lambda_i(x) = 0,\ i = 1, \cdots, N;$$

then there exist constants $a_1, \cdots, a_N \in \mathbb{R}$ such that

$$\Lambda = \sum_{i=1}^N a_i\,\Lambda_i.$$

Exercise 1.3.9 Generalize Exercise 1.3.7 (ii) in the following way. Prove that if $n \geq 1$ is an integer and $u \in L^1_{\text{loc}}(a,b)$ is such that

$$\int_a^b u(x)\,\varphi^{(n)}(x)\,dx = 0, \quad \forall \varphi \in C_0^\infty(a,b)$$

then there exist constants $a_1, \cdots, a_n \in \mathbb{R}$ such that

$$u(x) = \sum_{k=1}^{n} a_k \, x^{k-1} \quad \text{a.e. } x \in (a, b).$$

Exercise 1.3.10 Let $\Omega \subset \mathbb{R}^n$ be an open set and $u \in L^1(\Omega)$. Show that for every $\epsilon > 0$, there exists $\delta > 0$ so that for any measurable set $E \subset \Omega$

$$\text{meas } E \le \delta \quad \Rightarrow \quad \int_E |u(x)| \, dx \le \epsilon.$$

Prove that if $u \in L^p(\Omega)$, for a certain $1 < p \le \infty$, then for any bounded measurable set $E \subset \Omega$

$$\int_E |u(x)| \, dx \le \|u\|_{L^p(\Omega)} \, (\text{meas } E)^{1/p'} .$$

1.4   Sobolev spaces

1.4.1   Definitions and first properties

Before giving the definition of Sobolev spaces, we need to weaken the notion of derivative. In doing so we want to keep the right to integrate by parts; this is one of the reasons for the following definition.

Definition 1.26 *Let $\Omega \subset \mathbb{R}^n$ be open and $u \in L^1_{loc}(\Omega)$. We say that $v \in L^1_{loc}(\Omega)$ is the* weak partial derivative *of u with respect to x_i if*

$$\int_\Omega v(x) \, \varphi(x) \, dx = - \int_\Omega u(x) \, \frac{\partial \varphi}{\partial x_i}(x) \, dx, \quad \forall \varphi \in C_0^\infty(\Omega).$$

By abuse of notations we write $v = \partial u / \partial x_i$ or u_{x_i}.

We say that u is weakly differentiable *if the weak partial derivatives $u_{x_1}, \cdots, u_{x_n}$ exist.*

Remark 1.27 **(i)** If such a weak derivative exists it is unique (a.e.), as a consequence of Theorem 1.24.

(ii) All the usual rules of differentiation are easily generalized to the present context of weak differentiability.

(iii) In a similar way we can introduce the higher derivatives.

(iv) If a function is C^1, then the usual notion of derivative and the weak one coincide.

(v) The advantage of this notion of weak differentiability will be obvious when defining Sobolev spaces. We can compute many more derivatives of functions than one can usually do. However, not all measurable functions can be differentiated in this way. In particular, a discontinuous function of $\mathbb{R}$ cannot be differentiated in the weak sense (see Example 1.29).

Example 1.28 Let $\Omega = \mathbb{R}$ and the function $u(x) = |x|$. Its weak derivative is then given by

$$u'(x) = \begin{cases} +1 & \text{if } x > 0 \\ -1 & \text{if } x < 0. \end{cases}$$

Example 1.29 (Dirac mass) Let

$$H(x) = \begin{cases} 1 & \text{if } x > 0 \\ 0 & \text{if } x < 0. \end{cases}$$

We now show that H has no weak derivative. Let $\Omega = (-1,1)$. Assume, for the sake of contradiction, that $H' = \delta \in L^1_{\text{loc}}(-1,1)$ and let us prove that this is absurd. Let $\varphi \in C_0^\infty(0,1)$ be arbitrary and extend it to $(-1,0)$ by $\varphi \equiv 0$. We therefore have by definition that

$$\int_{-1}^1 \delta(x)\varphi(x)\,dx = -\int_{-1}^1 H(x)\varphi'(x)\,dx = -\int_0^1 \varphi'(x)\,dx$$
$$= \varphi(0) - \varphi(1) = 0.$$

We hence find

$$\int_0^1 \delta(x)\varphi(x)\,dx = 0, \quad \forall \varphi \in C_0^\infty(0,1)$$

which, combined with Theorem 1.24, leads to $\delta = 0$ a.e. in $(0,1)$. With an analogous reasoning we would get that $\delta = 0$ a.e. in $(-1,0)$ and consequently $\delta = 0$ a.e. in $(-1,1)$. Let us show that we have reached the desired contradiction. Indeed, if this were the case we would have, for every $\varphi \in C_0^\infty(-1,1)$,

$$0 = \int_{-1}^1 \delta(x)\varphi(x)\,dx = -\int_{-1}^1 H(x)\varphi'(x)\,dx$$
$$= -\int_0^1 \varphi'(x)\,dx = \varphi(0) - \varphi(1) = \varphi(0).$$

This would imply that $\varphi(0) = 0$, for every $\varphi \in C_0^\infty(-1,1)$, which is clearly absurd. Thus H is not weakly differentiable.

Remark 1.30 By weakening even more the notion of derivative (for example, by no longer requiring that v is in L^1_{loc}), the theory of distributions can give a meaning at $H' = \delta$; it is then called the *Dirac mass*. We will however not need this theory subsequently, except, but only marginally, in the exercises of Section 3.5.

Definition 1.31 *Let $\Omega \subset \mathbb{R}^n$ be an open set and $1 \leq p \leq \infty$.*

(i) We let $W^{1,p}(\Omega)$ be the set of functions $u : \Omega \to \mathbb{R}$, $u \in L^p(\Omega)$, whose weak partial derivatives $u_{x_i} \in L^p(\Omega)$ for every $i = 1, \cdots, n$. We endow this space with the following norm

$$\|u\|_{W^{1,p}} = (\|u\|_{L^p}^p + \|\nabla u\|_{L^p}^p)^{1/p} \quad \text{if } 1 \leq p < \infty,$$

$$\|u\|_{W^{1,\infty}} = \max\{\|u\|_{L^\infty}, \|\nabla u\|_{L^\infty}\} \quad \text{if } p = \infty.$$

(ii) If $1 \leq p < \infty$, the set $W_0^{1,p}(\Omega)$ is defined as the closure of $C_0^\infty(\Omega)$ functions in $W^{1,p}(\Omega)$. By abuse of language, we often say, if Ω is bounded, that $u \in W_0^{1,p}(\Omega)$ is such that $u \in W^{1,p}(\Omega)$ and $u = 0$ on $\partial\Omega$.

(iii) We also write $u \in u_0 + W_0^{1,p}(\Omega)$, meaning that $u, u_0 \in W^{1,p}(\Omega)$ and $u - u_0 \in W_0^{1,p}(\Omega)$.

(iv) We let $W_0^{1,\infty}(\Omega) = W^{1,\infty}(\Omega) \cap W_0^{1,1}(\Omega)$.

(v) Analogously we define the Sobolev spaces with higher derivatives as follows. If $k > 0$ is an integer we let $W^{k,p}(\Omega)$ be the set of functions $u : \Omega \to \mathbb{R}$, whose weak partial derivatives $D^a u \in L^p(\Omega)$, for every multi-index $a \in \mathcal{A}_m$, $0 \leq m \leq k$. The norm is given by

$$\|u\|_{W^{k,p}} = \begin{cases} \left(\displaystyle\sum_{0 \leq |a| \leq k} \|D^a u\|_{L^p}^p \right)^{1/p} & \text{if } 1 \leq p < \infty \\ \displaystyle\max_{0 \leq |a| \leq k} (\|D^a u\|_{L^\infty}) & \text{if } p = \infty. \end{cases}$$

(vi) If $1 \leq p < \infty$, $W_0^{k,p}(\Omega)$ is the closure of $C_0^\infty(\Omega)$ in $W^{k,p}(\Omega)$ and $W_0^{k,\infty}(\Omega) = W^{k,\infty}(\Omega) \cap W_0^{k,1}(\Omega)$.

(vii) We define $W^{k,p}(\Omega; \mathbb{R}^N)$ to be the set of maps $u : \Omega \to \mathbb{R}^N$, $u = (u^1, \cdots, u^N)$, with $u^i \in W^{k,p}(\Omega)$ for every $i = 1, \cdots, N$ and similarly for $W_0^{k,p}(\Omega; \mathbb{R}^N)$.

Remark 1.32 **(i)** By abuse of notations we write $W^{0,p} = L^p$.

(ii) Roughly speaking, we can say that $W^{1,p}$ is an extension of C^1 similar to that of L^p as compared to C^0.

(iii) Note that if Ω is bounded, then

$$C^1\left(\overline{\Omega}\right) \underset{\neq}{\subset} W^{1,\infty}\left(\Omega\right) \underset{\neq}{\subset} W^{1,p}\left(\Omega\right) \underset{\neq}{\subset} L^p\left(\Omega\right)$$

for every $1 \leq p < \infty$.

(iv) If $p = 2$, the spaces $W^{k,2}\left(\Omega\right)$ and $W_0^{k,2}\left(\Omega\right)$ are sometimes respectively denoted by $H^k\left(\Omega\right)$ and $H_0^k\left(\Omega\right)$.

Example 1.33 The following cases are discussed in Exercise 1.4.1.

(i) Let $s > 0$,

$$\Omega = \{x \in \mathbb{R}^n : |x| < 1\} \quad \text{and} \quad \psi\left(x\right) = |x|^{-s}.$$

We then have

$$\psi \in L^p \Leftrightarrow sp < n \quad \text{and} \quad \psi \in W^{1,p} \Leftrightarrow (s+1)p < n.$$

(ii) Let $0 < s < 1/2$,

$$\Omega = \left\{x = (x_1, x_2) \in \mathbb{R}^2 : |x| < 1/2\right\} \quad \text{and} \quad \psi\left(x\right) = |\log|x||^s.$$

We have that $\psi \in W^{1,2}\left(\Omega\right)$, $\psi \in L^p\left(\Omega\right)$ for every $1 \leq p < \infty$, but $\psi \notin L^\infty\left(\Omega\right)$.

(iii) Let $n \geq 2$ and

$$\Omega = \{x \in \mathbb{R}^n : |x| < 1\} \quad \text{and} \quad u\left(x\right) = \frac{x}{|x|},$$

then $u \in W^{1,p}\left(\Omega; \mathbb{R}^n\right)$ for every $1 \leq p < n$.

Theorem 1.34 *Let $\Omega \subset \mathbb{R}^n$ be open, $1 \leq p \leq \infty$ and $k \geq 1$ an integer.*

(i) $W^{k,p}\left(\Omega\right)$ equipped with its norm $\|\cdot\|_{k,p}$ is a Banach space which is separable if $1 \leq p < \infty$ and reflexive if $1 < p < \infty$.

(ii) $W^{1,2}\left(\Omega\right)$ is a Hilbert space when endowed with the following inner product

$$\langle u; v \rangle_{W^{1,2}} = \int_\Omega u\left(x\right) v\left(x\right) dx + \int_\Omega \langle \nabla u\left(x\right); \nabla v\left(x\right) \rangle dx.$$

(iii) The $C^\infty\left(\Omega\right) \cap W^{k,p}\left(\Omega\right)$ functions are dense in $W^{k,p}\left(\Omega\right)$ provided $1 \leq p < \infty$. Moreover, if Ω is a bounded connected open set with Lipschitz boundary (see Definition 1.42), then $C^\infty\left(\overline{\Omega}\right)$ is also dense in $W^{k,p}\left(\Omega\right)$ provided $1 \leq p < \infty$.

(iv) $W_0^{k,p}\left(\mathbb{R}^n\right) = W^{k,p}\left(\mathbb{R}^n\right)$, whenever $1 \leq p < \infty$.

Remark 1.35 (i) Note that as for the case of L^p the space $W^{k,p}$ is reflexive only when $1 < p < \infty$ and hence $W^{1,1}$ is not reflexive; as already said, this is the main source of difficulties in the minimal surface problem.

(ii) The density result is due to Meyers and Serrin, see Theorem 3.16 in Adams [1], Section 5.3 in Evans [47] or Section 7.6 in Gilbarg-Trudinger [57].

(iii) In general, we have $W_0^{1,p}(\Omega) \subsetneq W^{1,p}(\Omega)$, but when $\Omega = \mathbb{R}^n$ both coincide (see Corollary 3.19 in Adams [1]).

1.4.2   Some further properties

We now give a simple characterization of $W^{1,p}$ which turns out to be particularly helpful when dealing with regularity problems (Chapter 4). It relates the weak derivative with the difference quotient that characterizes classical derivatives.

Notation 1.36 For $\tau \in \mathbb{R}^n$, $\tau \neq 0$, we let the *difference quotient* be defined by

$$(D_\tau u)(x) = \frac{u(x + \tau) - u(x)}{|\tau|}.$$

Theorem 1.37 *Let $\Omega \subset \mathbb{R}^n$ be open, $1 < p \leq \infty$ and $u \in L^p(\Omega)$. The following properties are then equivalent.*

(i) $u \in W^{1,p}(\Omega)$;

(ii) *there exists a constant $\gamma = \gamma(u, \Omega, p)$ so that*

$$\left| \int_\Omega u \, \varphi_{x_i} \right| \leq \gamma \|\varphi\|_{L^{p'}}, \quad \forall \varphi \in C_0^\infty(\Omega), \ \forall i = 1, 2, \cdots, n$$

(recalling that $1/p + 1/p' = 1$);

(iii) *there exists a constant $\gamma = \gamma(u, \Omega, p)$ so that for every open set $\omega \subset \overline{\omega} \subset \Omega$, with $\overline{\omega}$ compact, and for every $\tau \in \mathbb{R}^n$ with $0 \neq |\tau| < \text{dist}(\omega, \Omega^c)$ (where $\Omega^c = \mathbb{R}^n \setminus \overline{\Omega}$), then*

$$\|D_\tau u\|_{L^p(\omega)} \leq \gamma.$$

Furthermore, if (ii) or (iii) holds, then

$$\|\nabla u\|_{L^p(\Omega)} \leq \gamma.$$

If (i) holds, then γ in (ii) or (iii) can be chosen to be $\|\nabla u\|_{L^p}$ and in particular

$$\|D_\tau u\|_{L^p(\omega)} \leq \|\nabla u\|_{L^p(\Omega)}.$$

Remark 1.38 (i) As a consequence of the theorem, it can easily be proved that if Ω is bounded and open then

$$C^{0,1}\left(\overline{\Omega}\right) \subset W^{1,\infty}\left(\Omega\right)$$

where $C^{0,1}\left(\overline{\Omega}\right)$ has been defined in Section 1.2, and the inclusion is, in general, strict. If, however, the set Ω is also convex (or sufficiently regular, see Theorem 5.8.4 in Evans [47]), then these two sets coincide (as usual, up to the choice of a representative in $W^{1,\infty}\left(\Omega\right)$). In other words, we can say that the set of Lipschitz functions over $\overline{\Omega}$ can be identified, if Ω is convex or sufficiently regular, with the space $W^{1,\infty}\left(\Omega\right)$.

(ii) The theorem is false when $p = 1$. We then only have (i) $\Rightarrow$ (ii) $\Leftrightarrow$ (iii). The functions satisfying (ii) or (iii) are then called functions of bounded variations (see Exercise 1.4.17).

Proof We prove the theorem only when $n = 1$ and $\Omega = (a, b)$. For the more general case see, for example, Proposition IX.3 in Brézis [14] or Theorems 5.8.3 and 5.8.4 in Evans [47].

(i) $\Rightarrow$ **(ii)**. This follows from Hölder inequality and the fact that u has a weak derivative; indeed

$$\left| \int_a^b u\, \varphi' \right| = \left| \int_a^b u'\, \varphi \right| \leq \|u'\|_{L^p} \|\varphi\|_{L^{p'}} .$$

(ii) $\Rightarrow$ **(i)**. Let F be a linear functional defined by

$$F\left(\varphi\right) = \langle F; \varphi \rangle = \int_a^b u\, \varphi', \quad \forall \varphi \in C_0^\infty\left(a, b\right). \tag{1.6}$$

Note that, by (ii), it is continuous over $C_0^\infty\left(a, b\right)$. Since $C_0^\infty\left(a, b\right)$ is dense in $L^{p'}\left(a, b\right)$ (note that we use here the fact that $p \neq 1$ and hence $p' \neq \infty$), we can extend it, by continuity (or appealing to Hahn-Banach theorem), to the whole $L^{p'}\left(a, b\right)$; we have therefore defined a continuous linear operator F over $L^{p'}\left(a, b\right)$, with

$$|F\left(\varphi\right)| \leq \gamma \|\varphi\|_{L^{p'}} , \quad \forall \varphi \in L^{p'}\left(a, b\right). \tag{1.7}$$

From Riesz theorem (Theorem 1.13) we find that there exists $v \in L^p\left(a, b\right)$ so that

$$F\left(\varphi\right) = \langle F; \varphi \rangle = \int_a^b v\, \varphi, \quad \forall \varphi \in L^{p'}\left(a, b\right). \tag{1.8}$$

Combining (1.6) and (1.8) we get

$$\int_a^b \left(-v\right) \varphi = - \int_a^b u\, \varphi', \quad \forall \varphi \in C_0^\infty\left(a, b\right)$$

which exactly means that $u' = -v \in L^p(a,b)$ and hence $u \in W^{1,p}(a,b)$.

Note also that, since (1.7) and (1.8) hold, we infer

$$\|u'\|_{L^p(a,b)} = \|v\|_{L^p(a,b)} \le \gamma.$$

(iii) $\Rightarrow$ **(ii).** Let $\varphi \in C_0^\infty(a,b)$ and let $\omega \subset \bar{\omega} \subset (a,b)$ with $\bar{\omega}$ compact and such that $\operatorname{supp} \varphi \subset \omega$. Let $\tau \in \mathbb{R}$ so that $0 \ne |\tau| < \operatorname{dist}(\omega, (a,b)^c)$. We then have for $1 < p \le \infty$, appealing to Hölder inequality and to (iii),

$$\left| \int_a^b (D_\tau u)\, \varphi \right| \le \|D_\tau u\|_{L^p(\omega)} \|\varphi\|_{L^{p'}(a,b)} \le \gamma \|\varphi\|_{L^{p'}(a,b)} . \tag{1.9}$$

We know, by hypothesis, that $\varphi \equiv 0$ on $(a, a+\tau)$ and $(b-\tau, b)$ if $\tau > 0$ and we therefore find (letting $\varphi \equiv 0$ outside (a,b))

$$\int_a^b u(x+\tau)\, \varphi(x)\, dx = \int_{a+\tau}^{b+\tau} u(x+\tau)\, \varphi(x)\, dx = \int_a^b u(x)\, \varphi(x-\tau)\, dx. \tag{1.10}$$

Since a similar argument holds for $\tau < 0$, we deduce from (1.9) and (1.10) that, if $1 < p \le \infty$,

$$\left| \int_a^b u(x)\, [\varphi(x-\tau) - \varphi(x)]\, dx \right| \le \gamma |\tau| \|\varphi\|_{L^{p'}(a,b)} .$$

Letting $|\tau|$ tend to zero, we get

$$\left| \int_a^b u\varphi' \right| \le \gamma \|\varphi\|_{L^{p'}(a,b)} , \qquad \forall \varphi \in C_0^\infty(a,b)$$

which is exactly (ii).

Note, in passing, that the γ appearing in (iii) and in (ii) can be taken the same.

(i) $\Rightarrow$ **(iii).** From Lemma 1.39 below, we have for every $x \in \omega$

$$u(x+\tau) - u(x) = \int_x^{x+\tau} u'(t)\, dt = \tau \int_0^1 u'(x+s\tau)\, ds$$

and hence

$$|u(x+\tau) - u(x)| \le |\tau| \int_0^1 |u'(x+s\tau)|\, ds.$$

Let $1 < p < \infty$ (the conclusion is obvious if $p = \infty$), we have from Jensen inequality that

$$|u(x+\tau) - u(x)|^p \le |\tau|^p \int_0^1 |u'(x+s\tau)|^p\, ds$$

and hence after integration

$$\int_\omega |u(x+\tau) - u(x)|^p \, dx \leq |\tau|^p \int_\omega \int_0^1 |u'(x+s\tau)|^p \, ds \, dx$$

$$= |\tau|^p \int_0^1 \int_\omega |u'(x+s\tau)|^p \, dx \, ds.$$

Since $\omega + s\tau \subset (a,b)$, we find

$$\int_\omega |u'(x+s\tau)|^p \, dx = \int_{\omega + s\tau} |u'(y)|^p \, dy \leq \|u'\|_{L^p(a,b)}^p$$

and hence

$$\|D_\tau u\|_{L^p(\omega)} \leq \|u'\|_{L^p(a,b)}$$

which is the claim. ∎

In the proof of Theorem 1.37, we have used a result that, roughly speaking, says that functions in $W^{1,p}$ are continuous and are primitives of functions in L^p.

Lemma 1.39 *Let $u \in W^{1,p}(a,b)$, $1 \leq p \leq \infty$. Then there exists a function $\tilde{u} \in C([a,b])$ such that $u = \tilde{u}$ a.e. and*

$$\tilde{u}(x) - \tilde{u}(y) = \int_y^x u'(t) \, dt, \quad \forall \, x, y \in [a,b].$$

Remark 1.40 (i) As already repeated, we ignore the difference between u and $\tilde{u}$ and we say that if $u \in W^{1,p}(a,b)$ then $u \in C([a,b])$ and u is the primitive of u', i.e.

$$u(x) - u(y) = \int_y^x u'(t) \, dt.$$

(ii) Lemma 1.39 is a particular case of Sobolev imbedding theorem (see Theorem 1.44). It gives a non-trivial result, in the sense that it is not, a priori, obvious that a function $u \in W^{1,p}(a,b)$ is continuous. We can therefore say that

$$C^1([a,b]) \subset W^{1,p}(a,b) \subset C([a,b]), \quad 1 \leq p \leq \infty.$$

(iii) The inequality (1.13) in the proof of the lemma below shows that if $u \in W^{1,p}(a,b)$, $1 < p < \infty$, then $u \in C^{0,1/p'}([a,b])$ and hence u is Hölder continuous with exponent $1/p'$. We have already seen in Remark 1.38 that if $p = \infty$, then $C^{0,1}([a,b])$ and $W^{1,\infty}(a,b)$ can be identified.

Proof We divide the proof into two steps.

Step 1. Let $c \in (a, b)$ be fixed and define

$$v(x) = \int_c^x u'(t)\, dt, \quad x \in [a, b]. \tag{1.11}$$

Let us show that $v \in C([a, b])$ and

$$\int_a^b v(x)\, \varphi'(x)\, dx = -\int_a^b u'(x)\, \varphi(x)\, dx, \quad \forall \varphi \in C_0^\infty(a, b). \tag{1.12}$$

Indeed we have

$$\int_a^b v(x)\, \varphi'(x)\, dx = \int_a^b \left(\int_c^x u'(t)\, dt \right) \varphi'(x)\, dx$$

$$= \int_a^c dx \int_c^x u'(t)\, \varphi'(x)\, dt + \int_c^b dx \int_c^x u'(t)\, \varphi'(x)\, dt$$

which combined with Fubini theorem (which allows us to permute the integrals), leads to

$$\int_a^b v(x)\, \varphi'(x)\, dx = -\int_a^c u'(t)\, dt \int_a^t \varphi'(x)\, dx + \int_c^b u'(t)\, dt \int_t^b \varphi'(x)\, dx$$

$$= -\int_a^c u'(t)\, \varphi(t)\, dt + \int_c^b u'(t)\, (-\varphi(t))\, dt$$

$$= -\int_a^b u'(t)\, \varphi(t)\, dt$$

which is exactly (1.12). The fact that v is continuous follows from the observation that if $x \geq y$, then

$$|v(x) - v(y)| \leq \int_y^x |u'(t)|\, dt \leq \left(\int_y^x |u'(t)|^p\, dt \right)^{1/p} \left(\int_y^x 1^{p'}\, dt \right)^{1/p'}$$

and thus

$$|v(x) - v(y)| \leq \|u'\|_{L^p} |x - y|^{1/p'} \tag{1.13}$$

where we have used Hölder inequality. If $p = 1$, the inequality (1.13) does not imply that v is continuous; the continuity of v follows from classical results of Lebesgue integrals, see Exercise 1.3.10.

Step 2. We are now in a position to conclude. Since from (1.12) we have

$$\int_a^b v\, \varphi' = -\int_a^b u'\, \varphi, \quad \forall \varphi \in C_0^\infty(a, b)$$

and we know that $u \in W^{1,p}(a,b)$ (and hence $\int u\,\varphi' = -\int u'\varphi$), we deduce

$$\int_a^b (v-u)\,\varphi' = 0, \quad \forall\,\varphi \in C_0^\infty(a,b).$$

Applying Exercise 1.3.7, we find that $v - u = \gamma$ a.e., γ denoting a constant. Since v is continuous, we have that $\widetilde{u} = v - \gamma$ has all the desired properties. $\blacksquare$

When dealing with $W_0^{1,p}$ spaces, we have a very similar result to that of Theorem 1.37 (see Proposition IX.18 in Brézis [14] for a proof or Exercise 1.4.10 for the implication $(i) \Rightarrow (iii)$).

Theorem 1.41 *Let* $\Omega \subset \mathbb{R}^n$ *be a bounded open set with Lipschitz boundary (see Definition 1.42),* $1 < p < \infty$ *and* $u \in L^p(\Omega)$. *The following properties are then equivalent.*

(i) $u \in W_0^{1,p}(\Omega)$;

(ii) there exists a constant $\gamma = \gamma(u,\Omega,p)$ *so that*

$$\left| \int_\Omega u\,\varphi_{x_i} \right| \leq \gamma \|\varphi\|_{L^{p'}}, \quad \forall\,\varphi \in C_0^\infty(\mathbb{R}^n), \ \forall\,i = 1, 2, \cdots, n$$

(recalling that $1/p + 1/p' = 1$*);*

(iii) the function

$$\widetilde{u}(x) = \begin{cases} u(x) & \text{if } x \in \Omega \\ 0 & \text{if } x \notin \Omega \end{cases}$$

is in $W^{1,p}(\mathbb{R}^n)$ *and moreover, for every* $i = 1, 2, \cdots, n$,

$$\widetilde{u}_{x_i}(x) = \begin{cases} u_{x_i}(x) & \text{if } x \in \Omega \\ 0 & \text{if } x \notin \Omega. \end{cases}$$

1.4.3 Imbeddings and compact imbeddings

We are now in a position to state the main results concerning Sobolev spaces. They give some inclusions between these spaces, as well as some compact imbeddings. These results generalize to $\mathbb{R}^n$ what has already been seen in Lemma 1.39 for the one dimensional case. Before stating these results we need to define what kind of regularity is assumed on the boundary of the sets $\Omega \subset \mathbb{R}^n$ that we consider. When $\Omega = (a,b) \subset \mathbb{R}$, there was no restriction. We assume, for the sake of simplicity, that $\Omega \subset \mathbb{R}^n$ is bounded. The following definition expresses in precise terms the intuitive notion of regular boundary (C^∞, C^k or Lipschitz).

Definition 1.42 (i) *Let $\Omega \subset \mathbb{R}^n$ be open and bounded. We say that Ω is a bounded open set with C^k, $k \geq 1$, boundary if for every $x \in \partial\Omega$, there exist a neighborhood $U \subset \mathbb{R}^n$ of x and a one-to-one and onto map $H : Q \to U$, where*

$$Q = \{x \in \mathbb{R}^n : |x_j| < 1, \; j = 1, 2, \cdots, n\}$$

$$H \in C^k\left(\overline{Q}\right), \; H^{-1} \in C^k\left(\overline{U}\right), \; H\left(Q_+\right) = U \cap \Omega, \; H\left(Q_0\right) = U \cap \partial\Omega$$

with $Q_+ = \{x \in Q : x_n > 0\}$ and $Q_0 = \{x \in Q : x_n = 0\}$.

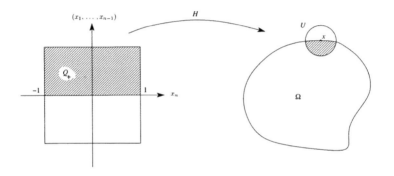

Figure 1.1: regular boundary

(ii) If H and H^{-1} are in $C^{k,\alpha}$, $0 < \alpha \leq 1$, we say that Ω is a bounded open set with $C^{k,\alpha}$ boundary.

(iii) If H and H^{-1} are only in $C^{0,1}$, we say that Ω is a bounded open set with Lipschitz boundary.

Remark 1.43 A polyhedron or a convex set has Lipschitz boundary, while the unit ball in $\mathbb{R}^n$ has a C^∞ boundary.

In the next two theorems (see for reference Theorems 5.4 and 6.2 in Adams [1]) we write some inclusions between spaces; they have to be understood up to a choice of a representative.

Theorem 1.44 (Sobolev imbedding theorem) *Let $\Omega \subset \mathbb{R}^n$ be a bounded open set with Lipschitz boundary.*

Case 1. If $1 \leq p < n$ then

$$W^{1,p}\left(\Omega\right) \subset L^q\left(\Omega\right)$$

for every $q \in [1, p^*]$ where

$$\frac{1}{p^*} = \frac{1}{p} - \frac{1}{n}, \quad i.e. \ p^* = \frac{np}{n-p}.$$

More precisely, for every $q \in [1, p^*]$ there exists $\gamma = \gamma(\Omega, p, q)$ so that

$$\|u\|_{L^q} \leq \gamma \|u\|_{W^{1,p}}.$$

Case 2. If $p = n$ then

$$W^{1,n}(\Omega) \subset L^q(\Omega), \quad for \ every \ q \in [1, \infty).$$

More precisely, for every $q \in [1, \infty)$ there exists $\gamma = \gamma(\Omega, q)$ so that

$$\|u\|_{L^q} \leq \gamma \|u\|_{W^{1,n}}.$$

Case 3. If $p > n$ then

$$W^{1,p}(\Omega) \subset C^{0,\alpha}(\overline{\Omega}), \quad for \ every \ \alpha \in [0, 1 - n/p].$$

In particular, there exists a constant $\gamma = \gamma(\Omega, p)$ so that

$$\|u\|_{L^\infty} \leq \gamma \|u\|_{W^{1,p}}.$$

The above theorem gives not only imbeddings but also compactness of these imbeddings under further restrictions. In the following we say that the imbedding of X into Y is *compact* if any bounded set of X is precompact (i.e. its closure is compact) in Y.

Theorem 1.45 (Rellich-Kondrachov theorem) *Let $\Omega \subset \mathbb{R}^n$ be a bounded open set with Lipschitz boundary.*

Case 1. If $1 \leq p < n$ then the imbedding of $W^{1,p}$ in L^q is compact, for every $q \in [1, p^*)$.

Case 2. If $p = n$ then the imbedding of $W^{1,n}$ in L^q is compact, for every $q \in [1, \infty)$.

Case 3. If $p > n$ then the imbedding of $W^{1,p}$ in $C^{0,\alpha}(\overline{\Omega})$ is compact, for every $0 \leq \alpha < 1 - n/p$.

In particular, in all cases (i.e. $1 \leq p \leq \infty$) the imbedding of $W^{1,p}(\Omega)$ in $L^p(\Omega)$ is compact.

Remark 1.46 (i) Let us examine the theorems when $\Omega = (a, b) \subset \mathbb{R}$. Only cases 2 and 3 apply and in fact we have an even better result (see Lemma 1.39), namely

$$C^1([a,b]) \subset W^{1,p}(a,b) \subset C^{0,1/p'}([a,b]) \subset C([a,b])$$

for every $p \geq 1$ (hence even when $p = 1$ we have that functions in $W^{1,1}$ are continuous). However, the imbedding of $W^{1,p}(a,b)$ into $C([a,b])$ is compact only when $p > 1$.

(ii) The result of case 1 in Theorem 1.45 is false if $q = p^*$.

(iii) In higher dimensions, $n \geq 2$, the case $p = n$ in Theorem 1.44 cannot be improved, in general. The functions in $W^{1,n}$ are in general not continuous and not even bounded (see Example 1.33).

(iv) If Ω is unbounded, for example $\Omega = \mathbb{R}^n$, we must be more careful; in particular, the compactness of the imbeddings is lost (see [1], [14], [43], [47], [57] and [111] for more details).

(v) If we consider $W_0^{1,p}$ instead of $W^{1,p}$ then the same imbeddings are valid, but no restriction on the regularity of $\partial\Omega$ is required.

(vi) Similar imbeddings can be obtained if we replace $W^{1,p}$ by $W^{k,p}$.

(vii) Recall that $W^{1,\infty}(\Omega)$, when Ω is bounded and convex, is identified with $C^{0,1}(\overline{\Omega})$.

(viii) We now try to summarize the results when $n = 1$. If we denote by $I = (a,b)$, we have, for $p \geq 1$,

$$\mathcal{D}(I) = C_0^\infty(I) \subset \cdots \subset W^{2,p}(I) \subset C^1(\overline{I}) \subset W^{1,p}(I)$$
$$\subset C(\overline{I}) \subset L^\infty(I) \subset \cdots \subset L^2(I) \subset L^1(I)$$

and furthermore C_0^∞ is dense in L^1, equipped with its norm.

Theorems 1.44 and 1.45 will not be proved; the first one has been discussed in the one dimensional case in Lemma 1.39. Concerning the compactness of the imbedding when $n = 1$, it is a consequence of Ascoli-Arzelà theorem (see Exercise 1.4.4 for more details).

Before proceeding further it is important to understand the significance of Theorem 1.45. We are going to formulate it for sequences, since it is in this framework that we use it. The corollary says that if a sequence converges weakly in $W^{1,p}$, it, in fact, converges strongly in L^p.

Corollary 1.47 *Let $\Omega \subset \mathbb{R}^n$ be a bounded open set with Lipschitz boundary and $1 \leq p < \infty$. If*

$$u_\nu \rightharpoonup u \quad \text{in } W^{1,p}(\Omega)$$

(this means that $u, u_\nu \in W^{1,p}(\Omega)$, $u_\nu \rightharpoonup u$ in L^p and $\nabla u_\nu \rightharpoonup \nabla u$ in L^p). Then

$$u_\nu \to u \quad \text{in } L^p(\Omega).$$

If $p = \infty$, $u_\nu \overset{*}{\rightharpoonup} u$ in $W^{1,\infty}$, then $u_\nu \to u$ in L^∞.

Example 1.48 Let $I = (0, 2\pi)$ and $u_\nu(x) = (1/\nu)\cos(\nu x)$. We have already seen that $u'_\nu \overset{*}{\rightharpoonup} 0$ in $L^\infty(0, 2\pi)$ and hence

$$u_\nu \overset{*}{\rightharpoonup} 0 \quad \text{in } W^{1,\infty}(0, 2\pi).$$

It is clear that we also have

$$u_\nu \to 0 \quad \text{in } L^\infty(0, 2\pi).$$

1.4.4 Poincaré inequality

The last theorem that we will often use is the following.

Theorem 1.49 (Poincaré inequality) *Let $\Omega \subset \mathbb{R}^n$ be a bounded open set and $1 \leq p \leq \infty$. Then there exists $\gamma = \gamma(\Omega, p) > 0$ so that*

$$\|u\|_{L^p} \leq \gamma \|\nabla u\|_{L^p}, \quad \forall u \in W_0^{1,p}(\Omega) \tag{1.14}$$

or equivalently

$$\|u\|_{W^{1,p}} \leq \gamma \|\nabla u\|_{L^p}, \quad \forall u \in W_0^{1,p}(\Omega).$$

Remark 1.50 (i) We need to impose a condition of the type $u = 0$ on $\partial\Omega$ (which comes from the hypothesis $u \in W_0^{1,p}$) to avoid constant functions u (which imply $\nabla u = 0$), otherwise the inequality would be trivially false.

(ii) Sometimes the Poincaré inequality appears under the following form (see Theorem 5.8.1 in Evans [47] and in the case $n = 1$ see the proof below). If $1 \leq p \leq \infty$, if $\Omega \subset \mathbb{R}^n$ is a bounded connected open set with Lipschitz boundary and if we denote by

$$u_\Omega = \frac{1}{\text{meas}\,\Omega} \int_\Omega u(x)\,dx$$

then there exists $\gamma = \gamma(\Omega, p) > 0$ so that

$$\|u - u_\Omega\|_{L^p} \leq \gamma \|\nabla u\|_{L^p}, \quad \forall u \in W^{1,p}(\Omega). \tag{1.15}$$

(iii) In the case $n = 1$, which is discussed in the proof, we do not really use that $u(a) = u(b) = 0$, but only that there exists $c \in [a, b]$ with $u(c) = 0$. The theorem thus remains valid under this weaker hypothesis.

(iv) We will often use Poincaré inequality under the following form. If $u_0 \in W^{1,p}(\Omega)$ and $u \in u_0 + W_0^{1,p}(\Omega)$, then there exist $\gamma_1, \gamma_2 > 0$ so that

$$\|\nabla u\|_{L^p} \geq \gamma_1 \|u\|_{W^{1,p}} - \gamma_2 \|u_0\|_{W^{1,p}}.$$

Proof We prove the inequalities (1.14) and (1.15) only when $n = 1$ and $\Omega = (a, b)$. We refer to Exercises 1.4.12 and 1.4.13 for a proof in the higher dimension of (1.14).

Step 1. Recall (see Lemma 1.39) that if $u \in W^{1,p}(a, b)$, then $u \in C([a, b])$. Let

$$X = \left\{ u \in W^{1,p}(a, b) : \text{there exists } c \in [a, b] \text{ with } u(c) = 0 \right\}.$$

We will prove that there exists $\gamma = \gamma(\Omega, p) > 0$ so that

$$\|u\|_{L^p} \leq \gamma \|\nabla u\|_{L^p}, \quad \forall u \in X.$$

This readily establishes (1.14), since, in this case, $u(a) = u(b) = 0$. But it also settles (1.15). Indeed, it is enough to prove (1.15) when $u_\Omega = 0$. In this last case, since u is continuous, we immediately infer that there exists $c \in [a, b]$ with $u(c) = 0$.

Step 2. We now prove that, for every $1 \leq p \leq \infty$ and every $u \in X$,

$$\|u\|_{L^p} \leq (b - a) \|u'\|_{L^p}. \tag{1.16}$$

Since $u \in X$, we have that there exists $c \in [a, b]$ with $u(c) = 0$. We therefore deduce that

$$|u(x)| = |u(x) - u(c)| = \left| \int_c^x u'(t)\, dt \right| \leq \left| \int_c^x |u'(t)|\, dt \right| \leq \int_a^b |u'(t)|\, dt = \|u'\|_{L^1}.$$

From this inequality we immediately get that (1.16) is true for $p = \infty$. When $p = 1$, we have after integration that

$$\|u\|_{L^1} = \int_a^b |u(x)|\, dx \leq (b - a) \|u'\|_{L^1}.$$

So it remains to prove (1.16) when $1 < p < \infty$. Applying Hölder inequality, we obtain

$$|u(x)| \leq \left(\int_a^b 1^{p'} \right)^{1/p'} \left(\int_a^b |u'|^p \right)^{1/p} = (b - a)^{1/p'} \|u'\|_{L^p}$$

and hence

$$\|u\|_{L^p} = \left(\int_a^b |u(x)|^p\, dx \right)^{1/p}$$

$$\leq \left((b - a)^{\frac{p}{p'}} \|u'\|_{L^p}^p \int_a^b dx \right)^{1/p} = (b - a) \|u'\|_{L^p}.$$

This concludes the proof of the theorem when $n = 1$. ∎

1.4.5 Exercises

Exercise 1.4.1 Let $1 \leq p < \infty$, $R > 0$ and $B_R = \{x \in \mathbb{R}^n : |x| < R\}$. Let for $f \in C^\infty (0, +\infty)$ and for $x \in B_R$

$$u(x) = f(|x|).$$

(i) Show that $u \in L^p(B_R)$ if and only if

$$\int_0^R r^{n-1} |f(r)|^p \, dr < \infty.$$

(ii) Assume that

$$\lim_{r \to 0} \left[r^{n-1} |f(r)| \right] = 0.$$

Prove that $u \in W^{1,p}(B_R)$ if and only if $u \in L^p(B_R)$ and

$$\int_0^R r^{n-1} |f'(r)|^p \, dr < \infty.$$

(iii) Discuss all the cases of Example 1.33.

Exercise 1.4.2 Let $AC([a,b])$ be the space of *absolutely continuous functions* on $[a,b]$. This means that a function $u \in AC([a,b])$, if for every $\epsilon > 0$ there exists $\delta > 0$ so that for every disjoint union of intervals $(a_k, b_k) \subset (a,b)$ the following implication is true

$$\sum_k |b_k - a_k| < \delta \quad \Rightarrow \quad \sum_k |u(b_k) - u(a_k)| < \epsilon.$$

    (i) Prove that $W^{1,1}(a,b) \subset AC([a,b]) \subset C([a,b])$, up to the usual selection of a representative.

    (ii) The converse $AC([a,b]) \subset W^{1,1}(a,b)$ is also true (see Section 2.2 in Buttazzo-Giaquinta-Hildebrandt [17] or Section 9.3 in De Barra [38]).

Exercise 1.4.3 Let $u \in W^{1,p}(a,b)$, $1 < p < \infty$ and $a < y < x < b$. Show that

$$u(x) - u(y) = o\left(|x - y|^{1/p'}\right)$$

where $o(t)$ stands for a function $f = f(t)$ so that $f(t)/t$ tends to 0 as t tends to 0.

Exercise 1.4.4 (Corollary 1.47 in dimension $n = 1$). Prove that if $1 < p < \infty$, then

$$u_\nu \rightharpoonup u \text{ in } W^{1,p}(a,b) \quad \Rightarrow \quad u_\nu \to u \text{ in } L^p(a,b)$$

and even $u_\nu \to u$ in $L^\infty(a,b)$.

Exercise 1.4.5 Show that if $\Omega \subset \mathbb{R}^n$ is a bounded open set with Lipschitz boundary, $1 < p < \infty$ and if there exists a constant $\gamma > 0$ so that

$$\|u_\nu\|_{W^{1,p}} \leq \gamma$$

then there exist a subsequence $\{u_{\nu_i}\}$ and $u \in W^{1,p}(\Omega)$ such that

$$u_{\nu_i} \rightharpoonup u \quad \text{in } W^{1,p}.$$

Exercise 1.4.6 Let $1 < p < \infty$, $u, u_\nu \in W^{1,p}(\Omega)$ and $\gamma > 0$ be such that

$$u_\nu \rightharpoonup u \text{ in } L^p \quad \text{and} \quad \|u\|_{W^{1,p}}, \|u_\nu\|_{W^{1,p}} \leq \gamma.$$

Prove that

$$u_\nu \rightharpoonup u \quad \text{in } W^{1,p}$$

and that the result is also valid if $p = \infty$ and weak convergence is replaced by weak $*$ convergence.

Exercise 1.4.7 Let $\Omega = (0,1) \times (0,1) \subset \mathbb{R}^2$ and

$$u_\nu(x_1, x_2) = \frac{1}{\sqrt{\nu}}(1 - x_2)^\nu \sin(\nu x_1).$$

Prove that $u_\nu \to 0$ in L^∞, $\|\nabla u_\nu\|_{L^2} \leq \gamma$ for some constant $\gamma > 0$ and

$$u_\nu \rightharpoonup 0 \quad \text{in } W^{1,2}.$$

Exercise 1.4.8 Let $u \in W^{1,p}(\Omega)$ and $\varphi \in W_0^{1,p'}(\Omega)$, where $1/p + 1/p' = 1$ and $p > 1$. Show that

$$\int_\Omega u_{x_i}\varphi = -\int_\Omega u\varphi_{x_i}, \quad i = 1, \cdots, n.$$

Exercise 1.4.9 Let $\varphi \in C_0^\infty(\mathbb{R}^n)$, $\varphi \geq 0$, $\varphi(x) = 0$ if $|x| > 1$ and $\int_{\mathbb{R}^n}\varphi = 1$ (for an example of such a function φ see Exercise 1.3.5). Let $1 \leq p \leq \infty$. Define

$$\varphi_\nu(x) = \nu^n\varphi(\nu x)$$

$$u_\nu(x) = (\varphi_\nu * u)(x) = \int_{\mathbb{R}^n} \varphi_\nu(x - y)u(y)\, dy.$$

(i) Prove that if $u \in W^{1,p}(\mathbb{R}^n)$, then

$$\nabla(\varphi_\nu * u) = \varphi_\nu * \nabla u.$$

(ii) Show that if $u \in W^{1,p}(\mathbb{R}^n)$, then $u_\nu \in C^\infty(\mathbb{R}^n)$ and

$$\|u_\nu\|_{W^{1,p}} \leq \|u\|_{W^{1,p}}.$$

(iii) Establish that if $u \in W^{1,p}(\mathbb{R}^n)$ and if $1 \leq p < \infty$, then

$$u_\nu \to u \quad \text{in } W^{1,p}(\mathbb{R}^n).$$

Exercise 1.4.10 Let $\Omega \subset \mathbb{R}^n$ be a bounded open set, $1 < p \leq \infty$ and $u \in W_0^{1,p}(\Omega)$. Show that

$$\widetilde{u}(x) = \begin{cases} u(x) & \text{if } x \in \Omega \\ 0 & \text{if } x \notin \Omega \end{cases}$$

is in $W^{1,p}(\mathbb{R}^n)$ and, moreover, for every $i = 1, 2, \cdots, n$,

$$\widetilde{u}_{x_i}(x) = \begin{cases} u_{x_i}(x) & \text{if } x \in \Omega \\ 0 & \text{if } x \notin \Omega. \end{cases}$$

Exercise 1.4.11 Let $\Omega \subset \mathbb{R}^n$ be a bounded open set with Lipschitz boundary, $1 < p < \infty$ and $u^\nu \in W_0^{1,p}(\Omega)$ be such that

$$u^\nu \rightharpoonup u \quad \text{in } W^{1,p}(\Omega).$$

Deduce that $u \in W_0^{1,p}(\Omega)$.

Exercise 1.4.12 Let $1 \leq p < \infty$. Prove Poincaré inequality in the following way.

(i) Show that it is sufficient to establish the inequality for functions in $C_0^\infty(\Omega)$.

(ii) Let $Q = (-R, R)^n$ and $u \in C^1(\overline{Q})$ be such that

$$u(-R, x_2, \cdots, x_n) = 0, \quad \text{for every } -R \leq x_2, \cdots, x_n \leq R.$$

Prove (using Jensen inequality, see Theorem 1.54) that

$$\|u\|_{L^p} \leq (2R) \|\nabla u\|_{L^p}.$$

(iii) Conclude.

Exercise 1.4.13 Prove Poincaré inequality when $p = \infty$.

Exercise 1.4.14 Let $\Omega \subset \mathbb{R}^n$ be a bounded open set, $1 \leq p \leq \infty$ and $u, v \in W^{1,p}(\Omega)$.

(i) Prove that if $u - v \in W_0^{1,p}(\Omega)$ and

$$\nabla u = \nabla v \quad \text{a.e. in } \Omega,$$

then $u = v$ a.e. in Ω.

(ii) Let Ω be a connected set with Lipschitz boundary and assume that

$$\nabla u = \nabla v \quad \text{a.e. in } \Omega.$$

Show that there exists a constant $c \in \mathbb{R}$ such that $u - v = c$ a.e. in Ω.

(iii) Let $A \subset \overline{A} \subset \Omega$ be a measurable set. Prove that if

$$u = v \quad \text{a.e. in } A,$$

then $\nabla u = \nabla v$ a.e. in A.

Exercise 1.4.15 Let $\Omega \subset \mathbb{R}^2$ be a bounded open set and $u \in W_0^{1,1}(\Omega)$, $u = u(x_1, x_2)$, with $u_{x_2} = 0$ a.e. in Ω. Show that $u = 0$.

Exercise 1.4.16 Let $n \geq 2$, $p > n/(n-1)$, $r > 0$ and $B_r = \{x \in \mathbb{R}^n : |x| < r\}$. Let $u \in L^p(B_1) \cap C^1\left(\overline{B_1} \setminus \{0\}\right)$ be such that, for a certain $i \in \{1, \cdots, n\}$,

$$u_{x_i} = 0 \quad \text{in } \overline{B_1} \setminus \{0\}.$$

Prove that

$$\int_{B_1} u\, \varphi_{x_i} = 0, \quad \text{for every } \varphi \in C_0^\infty(B_1).$$

Exercise 1.4.17 Let $\Omega \subset \mathbb{R}^n$ be an open set. Define for $u \in L^1(\Omega)$ the *variation of u in Ω* by

$$V(u, \Omega) = \sup\left\{\left|\int_\Omega u \operatorname{div} \varphi\right| : \varphi \in C_0^1(\Omega; \mathbb{R}^n) \text{ with } \|\varphi\|_{L^\infty} \leq 1\right\}.$$

We define the space of *functions of bounded variation* as

$$BV(\Omega) = \left\{u \in L^1(\Omega) : V(u, \Omega) < \infty\right\}.$$

(i) Prove that $W^{1,1}(\Omega) \subset BV(\Omega)$.

(ii) Let (see Example 1.29)

$$H(x) = \begin{cases} 1 & \text{if } x \in (0, 1) \\ 0 & \text{if } x \in (-1, 0). \end{cases}$$

Prove that $H \in BV(-1, 1)$, although $H \notin W^{1,1}(-1, 1)$.

(iii) Let

$$u(x) = \begin{cases} x \sin\left(\frac{\pi}{4x}\right) & \text{if } x \in (0, 1] \\ 0 & \text{if } x \in [-1, 0]. \end{cases}$$

Prove that $u \in C([-1, 1])$ and $u \notin BV(-1, 1)$.

1.5 Convex analysis

In this final section we recall the most important results concerning convex functions.

Definition 1.51 *(i) A set $\Omega \subset \mathbb{R}^n$ is said to be* convex *if for every $x, y \in \Omega$ and every $\lambda \in [0, 1]$, then $\lambda x + (1 - \lambda) y \in \Omega$.*

(ii) Let $\Omega \subset \mathbb{R}^n$ be convex. A function $f : \Omega \to \mathbb{R}$ is said to be convex *if for every $x, y \in \Omega$ and every $\lambda \in [0, 1]$, the following inequality holds*

$$f(\lambda x + (1 - \lambda) y) \leq \lambda f(x) + (1 - \lambda) f(y).$$

(iii) Let $\Omega \subset \mathbb{R}^n$ be convex. A function $f : \Omega \to \mathbb{R}$ is said to be strictly *convex if for every $x, y \in \Omega$, $x \neq y$, and every $\lambda \in (0, 1)$, the following strict inequality holds*

$$f(\lambda x + (1 - \lambda) y) < \lambda f(x) + (1 - \lambda) f(y).$$

We now give some criteria equivalent to the convexity.

Theorem 1.52 *Let $f : \mathbb{R}^n \to \mathbb{R}$, $f \in C^1(\mathbb{R}^n)$ and denote the scalar product in $\mathbb{R}^n$ by $\langle .;. \rangle$. The following assertions are then equivalent.*

(i) The function f is convex.

(ii) For every $x, y \in \mathbb{R}^n$, the following inequality holds

$$f(x) \geq f(y) + \langle \nabla f(y); x - y \rangle.$$

(iii) For every $x, y \in \mathbb{R}^n$, the following inequality is valid

$$\langle \nabla f(x) - \nabla f(y); x - y \rangle \geq 0.$$

If, moreover, $f \in C^2(\mathbb{R}^n)$, then the above statements are equivalent to

(iv) for every $x, v \in \mathbb{R}^n$, the following inequality holds

$$\langle \nabla^2 f(x) v; v \rangle \geq 0.$$

Example 1.53 Let $n = 1$, $1 \leq p < \infty$, $f(x) = |x|^p$ and

$$
x^*(y) = \begin{cases} p|y|^{p-2} y & \text{if } 1 < p < \infty \\ +1 & \text{if } p = 1 \text{ and } y \geq 0 \\ -1 & \text{if } p = 1 \text{ and } y < 0. \end{cases}
$$

It follows, trivially if $p = 1$ and from (ii) of the theorem otherwise, that, for every $x, y \in \mathbb{R}$,

$$
|x|^p \geq |y|^p + x^*(y)(x - y).
$$

Note that, when $p = 1$, we could have chosen $x^*(0)$ arbitrarily in $[-1, 1]$. Moreover, the quantity $x^*(y)$ is called, in convex analysis, the *subgradient* of f at y.

The following inequality is important (and is proved in a particular case in Exercise 1.5.3).

Theorem 1.54 (Jensen inequality) *Let $\Omega \subset \mathbb{R}^n$ be open and bounded, $u = (u^1, \cdots, u^N) \in L^1(\Omega; \mathbb{R}^N)$ and $f : \mathbb{R}^N \to \mathbb{R}$ be convex, then*

$$
f(u_\Omega) \leq \frac{1}{\operatorname{meas} \Omega} \int_\Omega f(u(x)) \, dx
$$

where

$$
u_\Omega = (u_\Omega^1, \cdots, u_\Omega^N) \quad \text{with} \quad u_\Omega^i = \frac{1}{\operatorname{meas} \Omega} \int_\Omega u^i(x) \, dx.
$$

We now need to introduce the notion of duality, also known as Legendre transform, for convex functions. It is convenient to accept, in the definition, functions that are allowed to take the value $+\infty$ (a function that takes only finite values is called finite).

Definition 1.55 (Legendre transform) *Let $f : \mathbb{R}^n \to \mathbb{R}$ (or $f : \mathbb{R}^n \to \mathbb{R} \cup \{+\infty\}$).*

(i) *The* Legendre transform, *or* dual, *of f is the function $f^* : \mathbb{R}^n \to \mathbb{R} \cup \{\pm\infty\}$ defined by*

$$
f^*(x^*) = \sup_{x \in \mathbb{R}^n} \{\langle x; x^* \rangle - f(x)\}
$$

where $\langle .; . \rangle$ denotes the scalar product in $\mathbb{R}^n$.

(ii) *The* bidual *of f is the function $f^{**} : \mathbb{R}^n \to \mathbb{R} \cup \{\pm\infty\}$ defined by*

$$
f^{**}(x) = \sup_{x^* \in \mathbb{R}^n} \{\langle x; x^* \rangle - f^*(x^*)\}.
$$

Remark 1.56 (i) In general, f^* takes the value $+\infty$, even if f takes only finite values.

(ii) If $f \not\equiv +\infty$, then $f^* > -\infty$.

Let us see some simple examples that are discussed in Exercise 1.5.5.

Example 1.57 (i) Let $n = 1$ and $f(x) = |x|^p /p$, where $1 < p < \infty$. We then find

$$f^*(x^*) = \frac{1}{p'} |x^*|^{p'}$$

where p' is, as usual, defined by $1/p + 1/p' = 1$.

(ii) Let $n = 1$ and $f(x) = \left(x^2 - 1\right)^2$. We then have

$$f^{**}(x) = \begin{cases} \left(x^2 - 1\right)^2 & \text{if } |x| \geq 1 \\ 0 & \text{if } |x| < 1. \end{cases}$$

(iii) Let $n = 1$ and

$$f(x) = \begin{cases} 0 & \text{if } x \in (0,1) \\ +\infty & \text{otherwise.} \end{cases}$$

We immediately find that

$$f^*(x^*) = \sup_{x \in (0,1)} \{xx^*\} = \begin{cases} x^* & \text{if } x^* \geq 0 \\ 0 & \text{if } x^* \leq 0. \end{cases}$$

f is often called the *indicator function* of $(0,1)$ and f^* *the* support function of $(0,1)$. We also have

$$f^{**}(x) = \begin{cases} 0 & \text{if } x \in [0,1] \\ +\infty & \text{otherwise} \end{cases}$$

and hence f^{**} is the indicator function of $[0,1]$.

(iv) Let $X \in \mathbb{R}^{2 \times 2}$, where $\mathbb{R}^{2 \times 2}$ is the set of 2×2 real matrices. We identify the set $\mathbb{R}^{2 \times 2}$ with $\mathbb{R}^4$ so that

$$X = \begin{pmatrix} X_1^1 & X_2^1 \\ X_1^2 & X_2^2 \end{pmatrix} \simeq \left(X_1^1, X_2^1, X_1^2, X_2^2\right) \quad \text{and} \quad \det X = X_1^1 X_2^2 - X_1^2 X_2^1.$$

Let $f(X) = \det X$, then

$$f^*(X^*) \equiv +\infty \quad \text{and} \quad f^{**}(X) \equiv -\infty.$$

We now gather some properties of the Legendre transform (for a proof see the exercises).

Theorem 1.58 *Let $f : \mathbb{R}^n \to \mathbb{R}$ (or $f : \mathbb{R}^n \to \mathbb{R} \cup \{+\infty\}$).*

(i) The function f^* is convex (even if f is not).

(ii) The function f^{**} is convex and $f^{**} \leq f$. If, furthermore, f is convex, bounded below and finite then $f^{**} = f$. More generally, if f is bounded below and finite but not necessarily convex, then f^{**} is its convex envelope (which means that it is the largest convex function that is smaller than f).

(iii) The following identity always holds: $f^{***} = f^*$.

(iv) If $f \in C^1(\mathbb{R}^n)$, convex and finite, then

$$f(x) + f^*(\nabla f(x)) = \langle \nabla f(x) ; x \rangle, \quad \forall x \in \mathbb{R}^n.$$

(v) If $f : \mathbb{R}^n \to \mathbb{R}$ is strictly convex and if

$$\lim_{|x| \to \infty} \frac{f(x)}{|x|} = +\infty$$

then $f^* \in C^1(\mathbb{R}^n)$. Moreover, if $f \in C^1(\mathbb{R}^n)$ and

$$f(x) + f^*(x^*) = \langle x^* ; x \rangle$$

then

$$x^* = \nabla f(x) \quad and \quad x = \nabla f^*(x^*).$$

We finally conclude with a theorem that allows us to compute the convex envelope without using duality (see Theorem 2.35 in [32, 2nd edition] or Corollary 17.1.5 in Rockafellar [98]).

Theorem 1.59 (Carathéodory theorem) *Let $f : \mathbb{R}^n \to \mathbb{R}$ be bounded below. Then*

$$f^{**}(x) = \inf \left\{ \sum_{i=1}^{n+1} \lambda_i f(x_i) : x = \sum_{i=1}^{n+1} \lambda_i x_i, \; \lambda_i \geq 0 \text{ and } \sum_{i=1}^{n+1} \lambda_i = 1 \right\}.$$

1.5.1   Exercises

Exercise 1.5.1 Prove Theorem 1.52.

Exercise 1.5.2 Let $f \in C^1(\mathbb{R}^n)$ and denote the scalar product in $\mathbb{R}^n$ by $\langle . ; . \rangle$. Show that f is strictly convex if and only if

$$f(x) > f(y) + \langle \nabla f(y) ; x - y \rangle \quad \forall x \neq y.$$

Exercise 1.5.3 Prove Jensen inequality, when $f \in C^1$. Discuss the case of equality, in particular what happens if f is strictly convex.

Exercise 1.5.4 Let $f(x) = \sqrt{1 + x^2}$. Compute f^*.

Exercise 1.5.5 Establish (i), (ii) and (iv) of Example 1.57.

Exercise 1.5.6 Let $X \in \mathbb{R}^{2 \times 2}$ be a 2×2 real matrix. Show that if $f(X) = (\det X)^2$, then $f^{**}(X) \equiv 0$.

Exercise 1.5.7 Prove (i), (iii) and (iv) of Theorem 1.58. For proofs of (ii) and (v) see the bibliography in the solutions of the present exercise and the exercise below.

Exercise 1.5.8 Show (v) of Theorem 1.58 under the further restrictions that $n = 1$, $f \in C^2(\mathbb{R})$ and
$$f''(x) > 0, \quad \forall x \in \mathbb{R}.$$
Prove in addition that $f^* \in C^2(\mathbb{R})$.

Exercise 1.5.9 Let $f \in C^1(\mathbb{R}^n)$ be convex, $p \geq 1$, $\alpha_1 > 0$ and
$$|f(x)| \leq \alpha_1 (1 + |x|^p), \quad \forall x \in \mathbb{R}^n. \tag{1.17}$$
Show that there exist $\alpha_2, \alpha_3 > 0$, so that
$$\left| \frac{\partial f}{\partial x_i}(x) \right| \leq \alpha_2 \left(1 + |x|^{p-1}\right), \quad \forall x \in \mathbb{R}^n \text{ and } \forall i = 1, \cdots, n \tag{1.18}$$
$$|f(x) - f(y)| \leq \alpha_3 \left(1 + |x|^{p-1} + |y|^{p-1}\right) |x - y|, \quad \forall x, y \in \mathbb{R}^n. \tag{1.19}$$
Note that (1.18) and (1.19) always imply (1.17) independently of the convexity of f.

Exercise 1.5.10 Show that a convex function $f : \mathbb{R}^n \to \mathbb{R}$ is continuous.

Chapter 2

Classical methods

2.1 Introduction

In this chapter we study the model problem

$$(P) \quad \inf \left\{ I(u) : u \in C^1([a,b]), \ u(a) = \alpha, \ u(b) = \beta \right\}$$

where $f \in C^2([a,b] \times \mathbb{R} \times \mathbb{R})$ and

$$I(u) = \int_a^b f(x, u(x), u'(x)) \, dx.$$

Before describing the results that we obtain, it might be useful to recall the analogy with minimizations in $\mathbb{R}^n$, namely

$$\inf \left\{ F(x) : x \in X \subset \mathbb{R}^n \right\}.$$

The methods that we call *classical* consist in finding $\overline{x} \in X$ satisfying $F'(\overline{x}) = 0$, and then analyzing the higher derivatives of F so as to determine the nature of the critical point $\overline{x}$: an absolute minimizer or maximizer, a local minimizer or maximizer or a saddle point.

In Section 2.2 we derive the *Euler-Lagrange equation* (analogous to $F'(\overline{x}) = 0$ in $\mathbb{R}^n$) that should satisfy any $C^2([a,b])$ minimizer, $\overline{u}$, of (P),

$$(E) \quad \frac{d}{dx}[f_\xi(x, \overline{u}(x), \overline{u}'(x))] = f_u(x, \overline{u}(x), \overline{u}'(x)), \quad x \in [a,b]$$

where for $f = f(x, u, \xi)$ we let $f_\xi = \partial f / \partial \xi$ and $f_u = \partial f / \partial u$.

In general (as in the case of $\mathbb{R}^n$), the solutions of (E) are not necessarily minimizers of (P); they are merely stationary points of I (see below for a more precise definition). However, if $(u, \xi) \to f(x, u, \xi)$ is convex for every $x \in [a, b]$, then every solution of (E) is automatically a minimizer of (P).

In Section 2.3 we show that any minimizer $\overline{u}$ of (P) satisfies a different form of the Euler-Lagrange equation. Namely, for every $x \in [a, b]$ the following differential equation holds

$$\frac{d}{dx} \left[f(x, \overline{u}(x), \overline{u}'(x)) - \overline{u}'(x) f_\xi(x, \overline{u}(x), \overline{u}'(x)) \right] = f_x(x, \overline{u}(x), \overline{u}'(x)).$$

This rewriting of the equation turns out to be particularly useful when f does not depend explicitly on the variable x. Indeed, we then have a first integral of (E) which is

$$f(\overline{u}(x), \overline{u}'(x)) - \overline{u}'(x) f_\xi(\overline{u}(x), \overline{u}'(x)) = \text{constant}, \quad \forall x \in [a, b].$$

In Section 2.4, we present the *Hamiltonian formulation* of the problem. Roughly speaking, the idea is that the solutions of (E) are also solutions (and conversely) of

$$(H) \quad \begin{cases} u'(x) = H_v(x, u(x), v(x)) \\ v'(x) = -H_u(x, u(x), v(x)) \end{cases}$$

where $v(x) = f_\xi(x, u(x), u'(x))$ and H is the Legendre transform of f, namely

$$H(x, u, v) = \sup_{\xi \in \mathbb{R}} \{v\xi - f(x, u, \xi)\}.$$

In classical mechanics f is called the *Lagrangian* and H the *Hamiltonian*.

In Section 2.5, we study the relationship between the solutions of (H) and those of a partial differential equation known as *Hamilton-Jacobi equation*

$$(HJ) \quad S_x(x, u) + H(x, u, S_u(x, u)) = 0, \quad \forall (x, u) \in [a, b] \times \mathbb{R}.$$

Finally, in Section 2.6, we present the *fields theories* introduced by Weierstrass and Hilbert which allow us, in certain cases, to decide whether a solution of (E) is a (local or global) minimizer of (P).

We conclude this introduction with some comments. The methods presented in this chapter can easily be generalized to vector valued functions of the form $u : [a, b] \longrightarrow \mathbb{R}^N$, with $N > 1$, to different boundary conditions, to integral constraints, or to higher derivatives. These extensions are considered in the exercises at the end of each section. However, except for Section 2.2, the remaining part of the chapter does not generalize easily and completely to the multidimensional case, $u : \Omega \subset \mathbb{R}^n \longrightarrow \mathbb{R}$, with $n > 1$, let alone the considerably harder case where $u : \Omega \subset \mathbb{R}^n \longrightarrow \mathbb{R}^N$, with $n, N > 1$.

Moreover, the classical methods suffer two main drawbacks. The first one is that they assume, implicitly, that the solutions of (P) are regular (C^1, C^2 or sometimes piecewise C^1); this is, in general, difficult (or even often false in the case of (HJ)) to prove. However, the main drawback is that they rely on the fact that we can solve either of the equations (E), (H) or (HJ), which is usually not the case. The main interest in the classical methods is, when they can be carried out completely, that we have an essentially explicit solution. The advantage of the direct methods presented in the next two chapters is that they do not assume any solvability of such equations.

We recall, once more, that our presentation is only brief and we have omitted several important classical conditions such as Legendre, Weierstrass, Weierstrass-Erdmann or Jacobi conditions. The fields theories as well as all the sufficient conditions for the existence of local minima have only been very briefly presented. We refer for more developments to the following books: Akhiezer [2], Bliss [12], Bolza [13], Buttazzo-Giaquinta-Hildebrandt [17], Carathéodory [19], Cesari [20], Courant [25], Courant-Hilbert [26], Gelfand-Fomin [54], Giaquinta-Hildebrandt [56], Hestenes [66], Pars [93], Rund [101], Troutman [107] or Weinstock [109].

2.2 Euler-Lagrange equation

2.2.1 The main theorem

The main result of this chapter is

Theorem 2.1 *Let $f \in C^2\left([a, b] \times \mathbb{R} \times \mathbb{R}\right)$, $f = f\left(x, u, \xi\right)$, and*

$$(P) \quad \inf_{u \in X} \left\{ I\left(u\right) = \int_a^b f\left(x, u\left(x\right), u'\left(x\right)\right) dx \right\} = m$$

where $X = \left\{ u \in C^1\left([a, b]\right) : u\left(a\right) = \alpha, \ u\left(b\right) = \beta \right\}$.

Part 1. *If (P) admits a minimizer $\overline{u} \in X \cap C^2\left([a, b]\right)$, then necessarily*

$$(E) \quad \frac{d}{dx}\left[f_\xi\left(x, \overline{u}\left(x\right), \overline{u}'\left(x\right)\right)\right] = f_u\left(x, \overline{u}\left(x\right), \overline{u}'\left(x\right)\right), \quad x \in (a, b)$$

or in other words

$$f_{\xi\xi}\left(x, \overline{u}\left(x\right), \overline{u}'\left(x\right)\right) \overline{u}''\left(x\right) + f_{u\xi}\left(x, \overline{u}\left(x\right), \overline{u}'\left(x\right)\right) \overline{u}'\left(x\right)$$
$$+ f_{x\xi}\left(x, \overline{u}\left(x\right), \overline{u}'\left(x\right)\right) = f_u\left(x, \overline{u}\left(x\right), \overline{u}'\left(x\right)\right)$$

where we denote by $f_\xi = \partial f / \partial \xi$, $f_u = \partial f / \partial u$, $f_{\xi\xi} = \partial^2 f / \partial \xi^2$, $f_{x\xi} = \partial^2 f / \partial x \partial \xi$ and $f_{u\xi} = \partial^2 f / \partial u \partial \xi$.

Part 2. *Conversely if $\overline{u}$ satisfies (E) and if $(u, \xi) \to f(x, u, \xi)$ is convex for every $x \in [a, b]$ then $\overline{u}$ is a minimizer of (P).*

Part 3. *If moreover the function $(u, \xi) \to f(x, u, \xi)$ is strictly convex for every $x \in [a, b]$ then the minimizer of (P), if it exists, is unique.*

Remark 2.2 (i) One should immediately draw attention to the fact that this theorem does not state any existence result.

(ii) As will be seen below it is not always reasonable to expect that the minimizers of (P) are $C^2([a, b])$ or even $C^1([a, b])$.

(iii) If $(u, \xi) \to f(x, u, \xi)$ is not convex (even if $\xi \to f(x, u, \xi)$ is convex for every $(x, u) \in [a, b] \times \mathbb{R}$), then a solution of (E) is not necessarily an absolute minimizer of (P). It can be a local minimizer, a local maximizer ... It is often said that such a solution of (E) is a *stationary point* of I.

(iv) The theorem easily generalizes, for example (see the exercises below), to the following cases:

- u is a vector, i.e. $u : [a, b] \to \mathbb{R}^N$, $N > 1$, the Euler-Lagrange equations are then a system of ordinary differential equations;

- $u : \Omega \subset \mathbb{R}^n \to \mathbb{R}$, $n > 1$, the Euler-Lagrange equation is then a single partial differential equation;

- $u : \Omega \subset \mathbb{R}^n \to \mathbb{R}^N$, $n, N > 1$, the Euler-Lagrange equations are then a system of partial differential equations;

- $f = f\left(x, u, u', u'', \cdots, u^{(n)}\right)$, the Euler-Lagrange equation is then an ordinary differential equation of $(2n)$th order;

- other types of boundary conditions such as $u'(a) = \alpha$, $u'(b) = \beta$;

- integral constraints of the form $\int_a^b g(x, u(x), u'(x)) \, dx = 0$.

Proof *Part 1.* Since $\overline{u}$ is a minimizer among all elements of X, we have

$$I(\overline{u}) \leq I(\overline{u} + hv)$$

for every $h \in \mathbb{R}$ and every $v \in C^1([a, b])$ with $v(a) = v(b) = 0$. In other words, setting $\Phi(h) = I(\overline{u} + hv)$, we have that $\Phi \in C^1(\mathbb{R})$ and that $\Phi(0) \leq \Phi(h)$ for every $h \in \mathbb{R}$. We therefore deduce that

$$\Phi'(0) = \left.\frac{d}{dh} I(\overline{u} + hv)\right|_{h=0} = 0$$

and hence

$$\int_a^b [f_\xi(x, \overline{u}(x), \overline{u}'(x)) v'(x) + f_u(x, \overline{u}(x), \overline{u}'(x)) v(x)] \, dx = 0. \tag{2.1}$$

Let us mention that the above integral form is called the *weak form of the Euler-Lagrange equation*. Integrating by parts (2.1) we obtain that the following identity holds for every $v \in C^1([a,b])$ with $v(a) = v(b) = 0$

$$\int_a^b \left[-\frac{d}{dx} \left[f_\xi \left(x, \overline{u}(x), \overline{u}'(x) \right) \right] + f_u \left(x, \overline{u}(x), \overline{u}'(x) \right) \right] v(x)\, dx = 0.$$

Applying the fundamental lemma of the calculus of variations (Theorem 1.24) we have indeed obtained the Euler-Lagrange equation (E).

Part 2. Let $\overline{u}$ be a solution of (E) with $\overline{u}(a) = \alpha$, $\overline{u}(b) = \beta$. Since $(u, \xi) \rightarrow f(x, u, \xi)$ is convex for every $x \in [a, b]$, we get from Theorem 1.52 that

$$f(x, u, u') \geq f(x, \overline{u}, \overline{u}') + f_u(x, \overline{u}, \overline{u}')(u - \overline{u}) + f_\xi(x, \overline{u}, \overline{u}')(u' - \overline{u}')$$

for every $u \in X$. Integrating the above inequality we get

$$I(u) \geq I(\overline{u}) + \int_a^b \left[f_u(x, \overline{u}, \overline{u}')(u - \overline{u}) + f_\xi(x, \overline{u}, \overline{u}')(u' - \overline{u}') \right] dx.$$

Integrating by parts the second term in the integral, bearing in mind that

$$u(a) - \overline{u}(a) = u(b) - \overline{u}(b) = 0,$$

we get

$$I(u) \geq I(\overline{u}) + \int_a^b \left[f_u(x, \overline{u}, \overline{u}') - \frac{d}{dx} \left[f_\xi(x, \overline{u}, \overline{u}') \right] \right] (u - \overline{u})\, dx.$$

Using (E) we indeed have $I(u) \geq I(\overline{u})$, which is the claimed result.

Part 3. Let $u, v \in X$ be two solutions of (P) (recall that m denotes the value of the minimum) and let us show that they are necessarily equal. Define

$$w = \frac{1}{2}u + \frac{1}{2}v$$

and observe that $w \in X$. Appealing to the convexity of $(u, \xi) \rightarrow f(x, u, \xi)$, we obtain

$$\frac{1}{2}f(x, u, u') + \frac{1}{2}f(x, v, v') \geq f\left(x, \frac{1}{2}u + \frac{1}{2}v, \frac{1}{2}u' + \frac{1}{2}v' \right) = f(x, w, w')$$

and hence

$$m = \frac{1}{2}I(u) + \frac{1}{2}I(v) \geq I(w) \geq m.$$

We therefore get

$$\int_a^b \left[\frac{1}{2}f(x, u, u') + \frac{1}{2}f(x, v, v') - f\left(x, \frac{1}{2}u + \frac{1}{2}v, \frac{1}{2}u' + \frac{1}{2}v' \right) \right] dx = 0.$$

Since the integrand is, by strict convexity of f, positive unless $u = v$ and $u' = v'$, we deduce that $u \equiv v$, as desired. ∎

2.2.2   Some important special cases

We now consider several particular cases and examples that are arranged in order of increasing difficulty.

Case 2.3 $f(x, u, \xi) = f(\xi)$.

This is the simplest case. The Euler-Lagrange equation is

$$\frac{d}{dx}\left[f'(u')\right] = 0, \quad \text{i.e. } f'(u') = \text{constant}.$$

Note that

$$\bar{u}(x) = \frac{\beta - \alpha}{b - a}(x - a) + \alpha \tag{2.2}$$

is a solution of the equation and furthermore satisfies the boundary conditions $\bar{u}(a) = \alpha$, $\bar{u}(b) = \beta$. It is therefore a stationary point of I. It is not, however, always a minimizer of (P) as will be seen in the second and third examples.

1. **f is convex.**

   If f is convex, the above $\bar{u}$ is indeed a minimizer. This follows from the theorem but it can be seen in a more elementary way (which is also valid even if $f \in C^0(\mathbb{R})$). From Jensen inequality (see Theorem 1.54) it follows that for any $u \in C^1([a,b])$ with $u(a) = \alpha$, $u(b) = \beta$

$$\frac{1}{b-a}\int_a^b f(u'(x))\, dx \geq f\left(\frac{1}{b-a}\int_a^b u'(x)\, dx\right) = f\left(\frac{u(b) - u(a)}{b-a}\right)$$

$$= f\left(\frac{\beta - \alpha}{b-a}\right) = f(\bar{u}'(x))$$

$$= \frac{1}{b-a}\int_a^b f(\bar{u}'(x))\, dx$$

   which is the claim. If f is not strictly convex, then, in general, there are other minimizers (see Exercise 2.2.9).

2. **f is non-convex.**

   If f is non-convex, then (P) has, in general, no solution and therefore the above $\bar{u}$ is not necessarily a minimizer (in the particular example below it is a maximizer of the integral). Consider $f(\xi) = e^{-\xi^2}$ and

$$(P) \quad \inf_{u \in X}\left\{I(u) = \int_0^1 f(u'(x))\, dx\right\} = m$$

where $X = \{u \in C^1([0,1]) : u(0) = u(1) = 0\}$. We have from (2.2) that $\overline{u} \equiv 0$ (and it is clearly a maximizer of I in the class of admissible functions X), however (P) has no minimizer, as we now show. Assume for a moment that $m = 0$, then, clearly, no function $u \in X$ can satisfy

$$\int_0^1 e^{-(u'(x))^2} \, dx = 0$$

and hence (P) has no solution. Let us now show that $m = 0$. Let $\nu \in \mathbb{N}$ and define

$$u_\nu(x) = \nu \left(x - \frac{1}{2} \right)^2 - \frac{\nu}{4}$$

then $u_\nu \in X$ and

$$I(u_\nu) = \int_0^1 e^{-4\nu^2 (x-1/2)^2} \, dx = \frac{1}{2\nu} \int_{-\nu}^{\nu} e^{-y^2} \, dy \to 0 \quad \text{as } \nu \to \infty.$$

Thus $m = 0$, as claimed.

3. **Solutions of (P) are not necessarily C^1.**

We now show that solutions of (P) are not necessarily C^1 even in the present simple case (another example with a similar property is given in Exercise 2.2.8). Let $f(\xi) = (\xi^2 - 1)^2$

$$(P) \quad \inf_{u \in X} \left\{ I(u) = \int_0^1 f(u'(x)) \, dx \right\} = m$$

where $X = \{u \in C^1([0,1]) : u(0) = u(1) = 0\}$. We associate with (P) the following problem

$$(P_{\text{piec}}) \quad \inf_{u \in X_{\text{piec}}} \left\{ I(u) = \int_0^1 f(u'(x)) \, dx \right\} = m_{\text{piec}}$$

$$X_{\text{piec}} = \{u \in C^1_{\text{piec}}([0,1]) : u(0) = u(1) = 0\}.$$

This last problem has clearly

$$v(x) = \begin{cases} x & \text{if } x \in [0, 1/2] \\ 1 - x & \text{if } x \in (1/2, 1] \end{cases}$$

as a solution since v is piecewise C^1 and satisfies $v(0) = v(1) = 0$ and $I(v) = 0$; thus $m_{\text{piec}} = 0$. Assume for a moment that we have already proved that not only $m_{\text{piec}} = 0$ but also $m = 0$. This readily implies that

(P), contrary to (P_{piec}), has no solution. Indeed, $I(u) = 0$ implies that $|u'| = 1$ almost everywhere and no function $u \in X$ can satisfy $|u'| = 1$ (since by continuity of the derivative we should have either $u' = 1$ everywhere or $u' = -1$ everywhere and this is incompatible with the boundary data).

We now prove that $m = 0$. We give a direct argument now and a more elaborate one in Exercise 2.2.6. Consider the following sequence

$$u_\nu(x) = \begin{cases} x & \text{if } x \in \left[0, \frac{1}{2} - \frac{1}{\nu}\right] \\ -2\nu^2\left(x - \frac{1}{2}\right)^3 - 4\nu\left(x - \frac{1}{2}\right)^2 - x + 1 & \text{if } x \in \left(\frac{1}{2} - \frac{1}{\nu}, \frac{1}{2}\right] \\ 1 - x & \text{if } x \in \left(\frac{1}{2}, 1\right]. \end{cases}$$

Observe that $u_\nu \in X$ and

$$I(u_\nu) = \int_0^1 f(u_\nu'(x))\, dx = \int_{\frac{1}{2} - \frac{1}{\nu}}^{\frac{1}{2}} f(u_\nu'(x))\, dx \leq \frac{4}{\nu} \to 0.$$

This implies that indeed $m = 0$.

We can also make the further observation that the Euler-Lagrange equation is

$$\frac{d}{dx}\left[u'\left((u')^2 - 1\right)\right] = 0.$$

It has $\bar{u} \equiv 0$ as a solution. However, since $m = 0$, it is not a minimizer $(I(0) = 1)$.

Case 2.4 $f(x, u, \xi) = f(x, \xi)$.

The Euler-Lagrange equation is

$$\frac{d}{dx}\left[f_\xi(x, u')\right] = 0, \quad \text{i.e.} \quad f_\xi(x, u') = \text{constant}.$$

The equation is already harder to solve than the preceding one and, in general, it does not have a solution as simple as the one in (2.2).

We now give an important example known as *Weierstrass example*. Let

$$f(x, \xi) = x\xi^2$$

(note that $\xi \to f(x, \xi)$ is convex for every $x \in [0, 1]$ and even strictly convex if $x \in (0, 1]$) and

$$(P) \quad \inf_{u \in X}\left\{I(u) = \int_0^1 f(x, u'(x))\, dx\right\} = m$$

where $X = \{u \in C^1([0,1]) : u(0) = 1,\ u(1) = 0\}$. We will show that (P) has no C^1 or piecewise C^1 solution (not even in any Sobolev space). The Euler-Lagrange equation is

$$(xu')' = 0 \quad \Rightarrow \quad u' = \frac{c}{x} \quad \Rightarrow \quad u(x) = c\log x + d,\ x \in (0,1)$$

where c and d are constants. Observe first that such a u cannot satisfy simultaneously $u(0) = 1$ and $u(1) = 0$.

We associate with (P) the following problem

$$(P_{\mathrm{piec}}) \quad \inf_{u \in X_{\mathrm{piec}}} \left\{ I(u) = \int_0^1 f(x, u'(x))\, dx \right\} = m_{\mathrm{piec}}$$

$$X_{\mathrm{piec}} = \{u \in C^1_{\mathrm{piec}}([0,1]) : u(0) = 1,\ u(1) = 0\}.$$

We next prove that neither (P) nor (P_{piec}) have a minimizer. For both cases it is sufficient to establish that $m_{\mathrm{piec}} = m = 0$. Let us postpone for a moment the proof of these facts and show the claim. If there exists a piecewise C^1 function v satisfying $I(v) = 0$, this would imply that $v' = 0$ a.e. in $(0,1)$. Since the function $v \in X_{\mathrm{piec}}$, it should be continuous and $v(1)$ should be equal to 0, we would then deduce that $v \equiv 0$, which does not verify the other boundary condition, namely $v(0) = 1$. Hence, neither (P) nor (P_{piec}) have a minimizer.

We first prove that $m_{\mathrm{piec}} = 0$. Let $\nu \in \mathbb{N}$ and consider the sequence

$$u_\nu(x) = \begin{cases} 1 & \text{if } x \in \left[0, \frac{1}{\nu}\right] \\ \frac{-\log x}{\log \nu} & \text{if } x \in \left(\frac{1}{\nu}, 1\right]. \end{cases}$$

Note that u_ν is piecewise C^1, $u_\nu(0) = 1$, $u_\nu(1) = 0$ and

$$I(u_\nu) = \frac{1}{\log \nu} \to 0 \quad \text{as } \nu \to \infty,$$

hence $m_{\mathrm{piec}} = 0$.

We finally prove that $m = 0$. This can be done in two different ways. A more sophisticated argument is given in Exercise 2.2.6 and it provides an interesting continuity argument. A possible approach is to consider the following sequence

$$u_\nu(x) = \begin{cases} \frac{-\nu^2}{\log \nu}x^2 + \frac{\nu}{\log \nu}x + 1 & \text{if } x \in \left[0, \frac{1}{\nu}\right] \\ \frac{-\log x}{\log \nu} & \text{if } x \in \left(\frac{1}{\nu}, 1\right]. \end{cases}$$

We easily have $u_\nu \in X$ and since

$$u_\nu'(x) = \begin{cases} \frac{\nu}{\log \nu}(1 - 2\nu x) & \text{if } x \in \left[0, \frac{1}{\nu}\right] \\ \frac{-1}{x\log \nu} & \text{if } x \in \left(\frac{1}{\nu}, 1\right] \end{cases}$$

we deduce that

$$0 \le I(u_\nu) = \frac{\nu^2}{\log^2 \nu} \int_0^{1/\nu} x(1 - 2\nu x)^2 \, dx + \frac{1}{\log^2 \nu} \int_{1/\nu}^1 \frac{dx}{x} \to 0, \quad \text{as } \nu \to \infty.$$

This indeed shows that $m = 0$.

Case 2.5 $f(x, u, \xi) = f(u, \xi)$.

Although this case is a lot harder to treat than the preceding ones, it has an important property that is not present in the most general case when $f = f(x, u, \xi)$. The Euler-Lagrange equation is

$$\frac{d}{dx}[f_\xi(u(x), u'(x))] = f_u(u(x), u'(x)), \quad x \in (a, b)$$

and, according to Theorem 2.8 below, it has a *first integral* that is given by

$$f(u(x), u'(x)) - u'(x) f_\xi(u(x), u'(x)) = \text{constant}, \quad x \in (a, b).$$

1. **Poincaré-Wirtinger inequality**

We will show, in several steps, that

$$\int_a^b (u')^2 \ge \left(\frac{\pi}{b-a}\right)^2 \int_a^b u^2$$

for every u satisfying $u(a) = u(b) = 0$. By a change of variable we immediately reduce the study to the case $a = 0$ and $b = 1$. We will also prove in Theorem 6.1 a slightly more general inequality known as *Wirtinger inequality*, which states that

$$\int_{-1}^1 (u')^2 \ge \pi^2 \int_{-1}^1 u^2$$

among all u satisfying $u(-1) = u(1)$ and $\int_{-1}^1 u = 0$.

We start by writing the problem under the above formalism and we let $\lambda \ge 0$, $f_\lambda(u, \xi) = (\xi^2 - \lambda^2 u^2)/2$ and

$$(P_\lambda) \quad \inf_{u \in X} \left\{ I_\lambda(u) = \int_0^1 f_\lambda(u(x), u'(x)) \, dx \right\} = m_\lambda$$

where $X = \{u \in C^1([0, 1]) : u(0) = u(1) = 0\}$. Observe that $\xi \to f_\lambda(u, \xi)$ is convex while $(u, \xi) \to f_\lambda(u, \xi)$ is not. The Euler-Lagrange equation and its first integral are

$$u'' + \lambda^2 u = 0 \quad \text{and} \quad (u')^2 + \lambda^2 u^2 = \text{constant}.$$

We have the following facts.

- If $\lambda \leq \pi$ (see Example 2.24, for a weaker result see Exercise 2.2.7 and for a different proof see Theorem 6.1), then $m_\lambda = 0$, which implies, in particular,

$$\int_0^1 (u')^2 \geq \pi^2 \int_0^1 u^2 .$$

Moreover, if $\lambda < \pi$, then $u_0 \equiv 0$ is the only minimizer of (P_λ), because it is the only solution of the Euler-Lagrange equation (see also Exercise 2.2.7) satisfying $u(0) = u(1) = 0$. If $\lambda = \pi$, then (P_λ) has infinitely many minimizers which are all of the form $u_\alpha(x) = \alpha \sin(\pi x)$ with $\alpha \in \mathbb{R}$.

- If $\lambda > \pi$ then $m_\lambda = -\infty$, which implies that (P_λ) has no solution. To see this fact it is enough to choose u_α as above and to observe that since $\lambda > \pi$, then

$$I_\lambda(u_\alpha) = \frac{\alpha^2}{2} \int_0^1 [\pi^2 \cos^2(\pi x) - \lambda^2 \sin^2(\pi x)]\, dx \to -\infty \quad \text{as } \alpha \to \infty.$$

2. **Brachistochrone**

The function under consideration is $f(u, \xi) = \sqrt{1 + \xi^2}/\sqrt{u}$, here (compared with Chapter 0) we take $g = 1/2$, and

$$(P) \quad \inf_{u \in X} \left\{ I(u) = \int_0^b f(u(x), u'(x))\, dx \right\} = m$$

where

$$X = \{ u \in C^1((0, b]) : u(0) = 0,\ u(b) = \beta \text{ and } u(x) > 0,\ \forall x \in (0, b] \}.$$

Here, because of the singularity at $u = 0$, we cannot use Theorem 2.1; we therefore proceed only formally. The Euler-Lagrange equation and its first integral are

$$\left[\frac{u'}{\sqrt{u}\sqrt{1 + (u')^2}} \right]' = -\frac{\sqrt{1 + (u')^2}}{2\sqrt{u^3}}$$

$$\frac{\sqrt{1 + (u')^2}}{\sqrt{u}} - u' \left[\frac{u'}{\sqrt{u}\sqrt{1 + (u')^2}} \right] = \text{constant}.$$

This leads (μ being a positive constant) to

$$u\left(1 + (u')^2\right) = 2\mu.$$

The solution is a *cycloid* and it is (formally) given in implicit form by

$$u(x) = \mu\left(1 - \cos\theta^{-1}(x)\right)$$

where

$$\theta(t) = \mu(t - \sin t).$$

Note that $u(0) = 0$. It therefore remains to choose μ so that $u(b) = \beta$.

3. **Minimal surfaces of revolution**

This example is treated in Chapter 5 (see Proposition 5.11 and Exercise 5.2.3). Let us briefly present it here. The function under consideration is $f(u, \xi) = 2\pi u\sqrt{1 + \xi^2}$ and the minimization problem (which corresponds to minimization of the area of a surface of revolution) is

$$(P) \quad \inf_{u \in X}\left\{ I(u) = \int_a^b f(u(x), u'(x))\, dx \right\} = m$$

where

$$X = \left\{ u \in C^1([a, b]) : u(a) = \alpha,\ u(b) = \beta,\ u > 0 \right\}$$

and $\alpha, \beta > 0$. The Euler-Lagrange equation and its first integral are

$$\left[\frac{u'u}{\sqrt{1 + (u')^2}}\right]' = \sqrt{1 + (u')^2} \quad \Leftrightarrow \quad u''u = 1 + (u')^2$$

$$u\sqrt{1 + (u')^2} - u'\frac{u'u}{\sqrt{1 + (u')^2}} = \lambda = \text{constant.}$$

This leads to

$$(u')^2 = \frac{u^2}{\lambda^2} - 1.$$

The solutions, if they exist (this depends on a, b, α and β, see Exercise 5.2.3), are of the form (μ being a constant)

$$u(x) = \lambda \cosh\left(\frac{x}{\lambda} + \mu\right).$$

We next consider a generalization of a classical example.

Example 2.6 (Fermat principle) The function is $f(x, u, \xi) = g(x, u)\sqrt{1 + \xi^2}$ and

$$(P) \quad \inf_{u \in X}\left\{ I(u) = \int_a^b f(x, u(x), u'(x))\, dx \right\} = m$$

where $X = \{u \in C^1([a,b]) : u(a) = \alpha,\ u(b) = \beta\}$. Therefore the Euler-Lagrange equation is

$$\frac{d}{dx}\left[g(x,u)\frac{u'}{\sqrt{1+(u')^2}}\right] = g_u(x,u)\sqrt{1+(u')^2}.$$

Observing that

$$\frac{d}{dx}\left[\frac{u'}{\sqrt{1+(u')^2}}\right] = \frac{u''}{\left(1+(u')^2\right)^{3/2}}$$

we get

$$g(x,u)u'' + [g_x(x,u)u' - g_u(x,u)]\left(1+(u')^2\right) = 0.$$

2.2.3 Lavrentiev phenomenon

We have seen and will see several examples where minimizers could exist in a certain space, for example the space C^1_{piec}, but not in a set of smoother functions, for example C^1. We now briefly present the so-called *Lavrentiev phenomenon*, where the situation is even worse (for more details on the following example see [32] and for more general considerations see [8], [17], [20]).

Example 2.7 (Mania example) Let $f(x,u,\xi) = (x - u^3)^2 \xi^6$,

$$X = \{u \in C^1([0,1]) : u(0) = 0,\ u(1) = 1\}$$

$$Y = \{u \in C^0([0,1]) \cap C^1((0,1]) : u(0) = 0,\ u(1) = 1\}$$

$$(P_X) \quad \inf_{u \in X}\left\{I(u) = \int_0^1 f(x,u(x),u'(x))\,dx\right\} = m_X$$

$$(P_Y) \quad \inf_{u \in Y}\left\{I(u) = \int_0^1 f(x,u(x),u'(x))\,dx\right\} = m_Y.$$

It can be proved (see [32]) that, if $\overline{u}(x) = x^{1/3}$, then

$$m_Y = I(\overline{u}) = 0 < m_X.$$

2.2.4   Exercises

Exercise 2.2.1 Generalize Theorem 2.1 to the case where $u : [a,b] \to \mathbb{R}^N$, $N \geq 1$.

Exercise 2.2.2 Generalize Theorem 2.1 to the case where $u : [a,b] \to \mathbb{R}$ and

$$(P) \quad \inf_{u \in X} \left\{ I(u) = \int_a^b f\left(x, u(x), \cdots, u^{(n)}(x)\right) dx \right\}$$

where $X = \left\{ u \in C^n([a,b]) : u^{(j)}(a) = \alpha_j, \ u^{(j)}(b) = \beta_j, \ 0 \leq j \leq n-1 \right\}$.

Exercise 2.2.3 (i) Find the appropriate formulation of Theorem 2.1 when $u : [a,b] \to \mathbb{R}$ and

$$(P) \quad \inf_{u \in X} \left\{ I(u) = \int_a^b f(x, u(x) \, u'(x)) \, dx \right\}$$

where $X = \left\{ u \in C^1([a,b]) : u(a) = \alpha \right\}$, i.e. we leave one of the end points free.

(ii) Similar question, when we leave both end points free; i.e. when we minimize I over $C^1([a,b])$.

(iii) Generalize the previous question to

$$(P) \quad \inf_{u \in C^1([a,b])} \left\{ I(u) = \int_a^b f(x, u(x) \, u'(x)) \, dx + g(u(a), u(b)) \right\}$$

where $g = g(y,z)$ is $C^1(\mathbb{R}^2)$.

Exercise 2.2.4 (Lagrange multiplier) Generalize Theorem 2.1 in the following case where $u : [a,b] \to \mathbb{R}$,

$$(P) \quad \inf_{u \in X} \left\{ I(u) = \int_a^b f(x, u(x), u'(x)) \, dx \right\},$$

$$X = \left\{ u \in C^1([a,b]) : u(a) = \alpha, \ u(b) = \beta, \ \int_a^b g(x, u(x), u'(x)) \, dx = 0 \right\}$$

where $g \in C^2([a,b] \times \mathbb{R} \times \mathbb{R})$.

Exercise 2.2.5 (Second variation of I) Let $f \in C^3 \left([a, b] \times \mathbb{R} \times \mathbb{R} \right)$ and

$$(P) \quad \inf_{u \in X} \left\{ I(u) = \int_a^b f(x, u(x), u'(x)) \, dx \right\}$$

where $X = \left\{ u \in C^1 \left([a, b] \right) : u(a) = \alpha, \ u(b) = \beta \right\}$. Let $\bar{u} \in X \cap C^2 \left([a, b] \right)$ be a minimizer for (P). Show that the following inequality

$$\int_a^b \left[f_{uu}(x, \bar{u}, \bar{u}') v^2 + 2 f_{u\xi}(x, \bar{u}, \bar{u}') vv' + f_{\xi\xi}(x, \bar{u}, \bar{u}') (v')^2 \right] dx \geq 0$$

holds for every $v \in C_0^1 (a, b)$ (i.e. $v \in C^1 \left([a, b] \right)$ and v has compact support in (a, b)). Setting

$$P(x) = f_{\xi\xi}(x, \bar{u}, \bar{u}'), \ Q(x) = f_{uu}(x, \bar{u}, \bar{u}') - \frac{d}{dx} [f_{u\xi}(x, \bar{u}, \bar{u}')]$$

rewrite the above inequality as

$$\int_a^b \left[P(x) (v')^2 + Q(x) v^2 \right] dx \geq 0.$$

Exercise 2.2.6 Prove the following two results (see Case 2.3 and Case 2.4).

(i) If $X = \left\{ u \in C^1 \left([0, 1] \right) : u(0) = u(1) = 0 \right\}$, then

$$(P) \quad \inf_{u \in X} \left\{ I(u) = \int_0^1 \left((u'(x))^2 - 1 \right)^2 dx \right\} = m = 0.$$

(ii) If $X = \left\{ u \in C^1 \left([0, 1] \right) : u(0) = 1, \ u(1) = 0 \right\}$, then

$$(P) \quad \inf_{u \in X} \left\{ I(u) = \int_0^1 x \left(u'(x) \right)^2 dx \right\} = m = 0.$$

Exercise 2.2.7 Show (see Poincaré-Wirtinger inequality), using Poincaré inequality (see Theorem 1.49), that for $\lambda \geq 0$ small enough then

$$(P_\lambda) \quad \inf_{u \in X} \left\{ I_\lambda(u) = \frac{1}{2} \int_0^1 \left[(u'(x))^2 - \lambda^2 (u(x))^2 \right] dx \right\} = m_\lambda = 0.$$

Deduce that $u \equiv 0$ is the unique solution of (P_λ) for $\lambda \geq 0$ small enough.

Exercise 2.2.8 Let $f(u, \xi) = u^2 (1 - \xi)^2$ and

$$(P) \quad \inf_{u \in X} \left\{ I(u) = \int_{-1}^{1} f(u(x), u'(x)) \, dx \right\} = m$$

where $X = \{u \in C^1([-1, 1]) : u(-1) = 0, \ u(1) = 1\}$. Show that (P) has no solution in X. Prove, however, that

$$\overline{u}(x) = \begin{cases} 0 & \text{if } x \in [-1, 0] \\ x & \text{if } x \in (0, 1] \end{cases}$$

is a solution of (P) among all piecewise C^1 functions.

Exercise 2.2.9 Let $X = \{u \in C^1([0, 1]) : u(0) = 0, \ u(1) = 1\}$ and

$$(P) \quad \inf_{u \in X} \left\{ I(u) = \int_{0}^{1} |u'(x)| \, dx \right\} = m.$$

Prove that (P) has infinitely many solutions.

Exercise 2.2.10 Let $p \geq 1$ and $a \in C^0(\mathbb{R})$, with $a(u) \geq a_0 > 0$. Let A be defined by

$$A'(u) = [a(u)]^{1/p}.$$

Show that a minimizer (which is unique if $p > 1$) of

$$(P) \quad \inf_{u \in X} \left\{ I(u) = \int_{a}^{b} a(u(x)) |u'(x)|^p \, dx \right\}$$

where $X = \{u \in C^1([a, b]) : u(a) = \alpha, \ u(b) = \beta\}$ is given by

$$u(x) = A^{-1} \left[\frac{A(\beta) - A(\alpha)}{b - a} (x - a) + A(\alpha) \right].$$

2.3 Second form of the Euler-Lagrange equation

The next theorem gives a different way of expressing the Euler-Lagrange equation; this new equation is sometimes called *DuBois-Reymond equation*. It turns out to be useful when f does not depend explicitly on x, as already seen in some of the above examples.

Theorem 2.8 *Let* $f \in C^2 ([a, b] \times \mathbb{R} \times \mathbb{R})$, $f = f(x, u, \xi)$, *and*

$$(P) \quad \inf_{u \in X} \left\{ I(u) = \int_a^b f(x, u(x), u'(x)) \, dx \right\} = m$$

where $X = \{u \in C^1 ([a, b]) : u(a) = \alpha, u(b) = \beta\}$. *Let* $u \in X \cap C^2 ([a, b])$ *be a minimizer of* (P), *then for every* $x \in [a, b]$ *the following equation holds*

$$\frac{d}{dx} \left[f(x, u(x), u'(x)) - u'(x) f_\xi (x, u(x), u'(x)) \right] = f_x (x, u(x), u'(x)). \quad (2.3)$$

Proof We give two different proofs of the theorem. The first one is very elementary and uses the Euler-Lagrange equation. The second one is more involved but has several advantages that we do not discuss now.

Proof 1. Observe first that for any $u \in C^2 ([a, b])$ we have, by straight differentiation,

$$\frac{d}{dx} [f(x, u, u') - u' f_\xi (x, u, u')]$$
$$= f_x (x, u, u') + u' \left[f_u (x, u, u') - \frac{d}{dx} [f_\xi (x, u, u')] \right].$$

By Theorem 2.1 we know that any minimizer u of (P) satisfies the Euler-Lagrange equation

$$\frac{d}{dx} [f_\xi (x, u(x), u'(x))] = f_u (x, u(x), u'(x))$$

hence, combining the two identities, we have the result.

Proof 2. We next use a technique known as *variations of the independent variables*, which we encounter again in Chapter 5; the classical derivation of Euler-Lagrange equation can be seen as a technique of *variations of the dependent variables*.

Let $\epsilon \in \mathbb{R}$, $\varphi \in C_0^\infty (a, b)$, $\lambda = (2 \|\varphi'\|_{L^\infty})^{-1}$ and

$$\xi(x, \epsilon) = x + \epsilon \lambda \varphi(x) = y.$$

Observe that for $|\epsilon| \leq 1$, then $\xi(., \epsilon) : [a, b] \to [a, b]$ is a diffeomorphism with $\xi(a, \epsilon) = a$, $\xi(b, \epsilon) = b$ and $\xi_x (x, \epsilon) > 0$. Let $\eta(., \epsilon) : [a, b] \to [a, b]$ be its inverse, i.e.

$$\xi(\eta(y, \epsilon), \epsilon) = y.$$

Since

$$\xi_x (\eta(y, \epsilon), \epsilon) \eta_y (y, \epsilon) = 1$$
$$\xi_x (\eta(y, \epsilon), \epsilon) \eta_\epsilon (y, \epsilon) + \xi_\epsilon (\eta(y, \epsilon), \epsilon) = 0$$

we find ($O\left(t\right)$ stands for a function f such that $\left|f\left(t\right)/t\right|$ is bounded in a neighborhood of $t=0$)

$$\eta_y\left(y,\epsilon\right)=1-\epsilon\lambda\varphi'\left(y\right)+O\left(\epsilon^2\right)$$
$$\eta_\epsilon\left(y,\epsilon\right)=-\lambda\varphi\left(y\right)+O\left(\epsilon\right).$$

Set for u a minimizer of (P)

$$u^\epsilon\left(x\right)=u\left(\xi\left(x,\epsilon\right)\right).$$

Note that, performing also a change of variables $y=\xi\left(x,\epsilon\right),$

$$
\begin{aligned}
I\left(u^\epsilon\right) &= \int_a^b f\left(x,u^\epsilon\left(x\right),\left(u^\epsilon\right)'\left(x\right)\right)\,dx \\
&= \int_a^b f\left(x,u\left(\xi\left(x,\epsilon\right)\right),u'\left(\xi\left(x,\epsilon\right)\right)\xi_x\left(x,\epsilon\right)\right)\,dx \\
&= \int_a^b f\left(\eta\left(y,\epsilon\right),u\left(y\right),u'\left(y\right)/\eta_y\left(y,\epsilon\right)\right)\eta_y\left(y,\epsilon\right)\,dy.
\end{aligned}
$$

Denoting by $g\left(\epsilon\right)$ the last integrand, we get

$$g'\left(\epsilon\right)=\eta_{y\epsilon}f+\left[f_x\eta_\epsilon-\frac{\eta_{y\epsilon}}{\eta_y^2}u'f_\xi\right]\eta_y$$

which leads to

$$g'\left(0\right)=\lambda\left[-f_x\varphi+\left(u'f_\xi-f\right)\varphi'\right].$$

Since by hypothesis u is a minimizer of (P) and $u^\epsilon\in X$ we have $I\left(u^\epsilon\right)\geq I\left(u\right)$ and hence

$$
\begin{aligned}
0 &= \frac{d}{d\epsilon}I\left(u^\epsilon\right)\bigg|_{\epsilon=0} \\
&= \lambda\int_a^b \{-f_x\left(x,u\left(x\right),u'\left(x\right)\right)\varphi\left(x\right) \\
&\qquad + \left[u'\left(x\right)f_\xi\left(x,u\left(x\right),u'\left(x\right)\right)-f\left(x,u\left(x\right),u'\left(x\right)\right)\right]\varphi'\left(x\right)\}\,dx \\
&= \lambda\int_a^b \{-f_x\left(x,u\left(x\right),u'\left(x\right)\right) \\
&\qquad + \frac{d}{dx}\left[-u'\left(x\right)f_\xi\left(x,u\left(x\right),u'\left(x\right)\right)+f\left(x,u\left(x\right),u'\left(x\right)\right)\right]\}\varphi\left(x\right)\,dx.
\end{aligned}
$$

Appealing, once more, to Theorem 1.24 we have indeed obtained the claim. ∎

2.3.1 Exercises

Exercise 2.3.1 Generalize Theorem 2.8 to the case where $u : [a, b] \to \mathbb{R}^N$, $N \geq 1$.

Exercise 2.3.2 Let

$$f(x, u, \xi) = f(u, \xi) = \frac{1}{2}\xi^2 - u.$$

Show that $u \equiv 1$ is a solution of (2.3), but not of the Euler-Lagrange equation (E).

2.4 Hamiltonian formulation

Recall that we are considering functions $f : [a, b] \times \mathbb{R} \times \mathbb{R} \to \mathbb{R}$, $f = f(x, u, \xi)$, and

$$I(u) = \int_a^b f(x, u(x), u'(x))\, dx.$$

The Euler-Lagrange equation is

$$(E) \quad \frac{d}{dx}\left[f_\xi(x, u, u')\right] = f_u(x, u, u'), \quad x \in [a, b].$$

We have seen in the preceding sections that a minimizer of I, if it is sufficiently regular, is also a solution of (E). The aim of this section is to show that, in certain cases, solving (E) is equivalent to finding stationary points of a different functional, namely

$$J(u, v) = \int_a^b \left[u'(x)v(x) - H(x, u(x), v(x))\right]\, dx$$

whose Euler-Lagrange equations are

$$(H) \quad \begin{cases} u'(x) = H_v(x, u(x), v(x)) \\ v'(x) = -H_u(x, u(x), v(x)). \end{cases}$$

The function H is called the *Hamiltonian* and it is the Legendre transform of f, which is defined as

$$H(x, u, v) = \sup_{\xi \in \mathbb{R}} \{v\xi - f(x, u, \xi)\}.$$

Sometimes the system (H) is called the *canonical form* of the Euler-Lagrange equation.

2.4.1 A technical lemma

We start our analysis with a lemma (for a special case see Exercise 1.5.8).

Lemma 2.9 *Let $f \in C^2 ([a, b] \times \mathbb{R} \times \mathbb{R})$, $f = f(x, u, \xi)$, such that*

$$f_{\xi\xi} (x, u, \xi) > 0, \quad \text{for every } (x, u, \xi) \in [a, b] \times \mathbb{R} \times \mathbb{R} \qquad (2.4)$$

$$f (x, u, \xi) \geq \omega (|\xi|) + g (x, u), \quad \text{for every } (x, u, \xi) \in [a, b] \times \mathbb{R} \times \mathbb{R} \qquad (2.5)$$

where $g : [a, b] \times \mathbb{R} \to \mathbb{R}$ is continuous and ω is a non-negative continuous and increasing function with $\lim_{t\to\infty} \omega (t) /t = \infty$. Let

$$H (x, u, v) = \sup_{\xi \in \mathbb{R}} \{v \xi - f (x, u, \xi)\}. \qquad (2.6)$$

Then $H \in C^2 ([a, b] \times \mathbb{R} \times \mathbb{R})$ and

$$H_x (x, u, v) = -f_x (x, u, H_v (x, u, v)) \qquad (2.7)$$

$$H_u (x, u, v) = -f_u (x, u, H_v (x, u, v)) \qquad (2.8)$$

$$H (x, u, v) = v H_v (x, u, v) - f (x, u, H_v (x, u, v)) \qquad (2.9)$$

$$v = f_\xi (x, u, \xi) \quad \Leftrightarrow \quad \xi = H_v (x, u, v). \qquad (2.10)$$

Remark 2.10 (i) The lemma remains partially true if we replace the hypothesis (2.4) by the weaker condition

$$\xi \to f (x, u, \xi) \quad \text{is strictly convex.}$$

In general, however, the function H is only C^1, as the following simple example shows

$$f (x, u, \xi) = \frac{1}{4} |\xi|^4 \quad \text{and} \quad H (x, u, v) = \frac{3}{4} |v|^{4/3}.$$

(See also Example 2.14.)

(ii) The lemma also remains valid if the hypothesis (2.5) does not hold but then, in general, H is no longer finite everywhere as the following simple example suggests. Consider the strictly convex function

$$f (x, u, \xi) = f (\xi) = \sqrt{1 + \xi^2}$$

and observe (see Exercise 1.5.4) that

$$H (v) = \begin{cases} -\sqrt{1 - v^2} & \text{if } |v| \leq 1 \\ +\infty & \text{if } |v| > 1. \end{cases}$$

(iii) The same proof leads to the fact that if $f \in C^k$, $k \geq 2$, then $H \in C^k$.

Proof We divide the proof into four steps.

Step 1. Fix $(x, u, v) \in [a, b] \times \mathbb{R} \times \mathbb{R}$. We first show that the supremum, in the definition of H, is attained. Assume, for the sake of contradiction, that there exists a sequence $\xi_\nu \in \mathbb{R}$ with $|\xi_\nu| \to \infty$ such that

$$-f(x, u, 0) \leq H(x, u, v) \leq \frac{1}{\nu} + v\xi_\nu - f(x, u, \xi_\nu).$$

Use (2.5) to get

$$-f(x, u, 0) \leq H(x, u, v) \leq \frac{1}{\nu} + |\xi_\nu| \left[v \frac{\xi_\nu}{|\xi_\nu|} - \frac{\omega(|\xi_\nu|)}{|\xi_\nu|} \right] - g(x, u).$$

Since the right-hand side of the last inequality tends to $-\infty$ as ν tends to ∞, we have obtained the desired contradiction. We therefore have that there exists $\xi = \xi(x, u, v)$ such that

$$\begin{cases} H(x, u, v) = v\xi - f(x, u, \xi) \\ \qquad v = f_\xi(x, u, \xi). \end{cases} \tag{2.11}$$

Let us now prove that given $R > 0$ we can find $R_1 > 0$ such that

$$|\xi(x, u, v)| \leq R_1, \quad \text{for every } x \in [a, b], \, |u|, \, |v| \leq R.$$

By hypotheses, we can find $R_2, R_3 > 0$ such that

$$\frac{\omega(|\xi|)}{|\xi|} \geq R + 1, \quad \text{for every } |\xi| \geq R_2$$

$$f(x, u, 0) - g(x, u) \leq R_3, \quad \text{for every } x \in [a, b], \, |u| \leq R.$$

We then choose

$$R_1 = \max\{R_2, R_3\}.$$

We thus deduce that for every $x \in [a, b], \, |u|, \, |v| \leq R$

$$\omega(|\xi|) - |v| \, |\xi| + g(x, u) \leq f(x, u, \xi) - v\xi = -H(x, u, v) \leq f(x, u, 0)$$

and hence

$$|\xi| \left[\frac{\omega(|\xi|)}{|\xi|} - R \right] \leq f(x, u, 0) - g(x, u) \leq R_3. \tag{2.12}$$

The claim, $|\xi(x, u, v)| \leq R_1$, then follows. Otherwise $|\xi| > R_1 \geq R_2$, which, combined with the definition of R_2 and (2.12), leads to $|\xi| \leq R_3 \leq R_1$ and thus a contradiction.

Step 2. The function H is easily seen to be continuous. Indeed, let (x, u, v), $(x', u', v') \in [a, b] \times \mathbb{R} \times \mathbb{R}$, using (2.11) we find $\xi = \xi(x, u, v)$ such that

$$H(x, u, v) = v\,\xi - f(x, u, \xi).$$

Appealing to the definition of H we also have

$$H(x', u', v') \geq v'\,\xi - f(x', u', \xi).$$

Combining the two facts we get

$$H(x, u, v) - H(x', u', v') \leq (v - v')\,\xi + f(x', u', \xi) - f(x, u, \xi).$$

Since the reverse inequality is obtained similarly, we deduce the continuity of H from that of f (in fact only in the variables (x, u)).

Step 3. The inverse function theorem, the fact that $f \in C^2$ and the inequality (2.4) imply that $\xi \in C^1$. But, as an exercise, we establish this fact again. First let us prove that ξ is continuous (in fact locally Lipschitz). Let $R > 0$ be fixed and $R_1 > 0$ so that (see Step 1)

$$|\xi(x, u, v)| \leq R_1, \quad \text{for every } x \in [a, b], \ |u|, |v| \leq R.$$

Since f_ξ is C^1, we can find $\gamma_1 > 0$ so that

$$|f_\xi(x, u, \xi) - f_\xi(x', u', \xi')| \leq \gamma_1 (|x - x'| + |u - u'| + |\xi - \xi'|) \qquad (2.13)$$

for every $x, x' \in [a, b]$, $|u|, |u'| \leq R$, $|\xi|, |\xi'| \leq R_1$.

From (2.4), we find that there exists $\gamma_2 > 0$ so that

$$f_{\xi\xi}(x, u, \xi) \geq \gamma_2, \quad \text{for every } x \in [a, b], \ |u| \leq R, \ |\xi| \leq R_1$$

and we thus have, for every $x \in [a, b]$, $|u| \leq R$, $|\xi|, |\xi'| \leq R_1$,

$$|f_\xi(x, u, \xi) - f_\xi(x, u, \xi')| \geq \gamma_2 |\xi - \xi'|. \qquad (2.14)$$

Let $x, x' \in [a, b]$, $|u|, |u'| \leq R$, $|v|, |v'| \leq R$. By definition of ξ we have

$$f_\xi(x, u, \xi(x, u, v)) = v$$

$$f_\xi(x', u', \xi(x', u', v')) = v',$$

which leads to

$$f_\xi(x, u, \xi(x', u', v')) - f_\xi(x, u, \xi(x, u, v))$$
$$= f_\xi(x, u, \xi(x', u', v')) - f_\xi(x', u', \xi(x', u', v')) + v' - v.$$

Combining this identity with (2.13) and (2.14) we get

$$\gamma_2 \left| \xi\left(x,u,v\right) - \xi\left(x',u',v'\right) \right| \leq \gamma_1 \left(\left|x - x'\right| + \left|u - u'\right| \right) + \left|v - v'\right|$$

which, indeed, establishes the continuity of ξ.

We now show that ξ is in fact C^1. From the equation $v = f_\xi\left(x,u,\xi\right)$ we deduce that

$$\begin{cases} f_{x\xi}\left(x,u,\xi\right) + f_{\xi\xi}\left(x,u,\xi\right)\xi_x = 0 \\ f_{u\xi}\left(x,u,\xi\right) + f_{\xi\xi}\left(x,u,\xi\right)\xi_u = 0 \\ f_{\xi\xi}\left(x,u,\xi\right)\xi_v = 1. \end{cases}$$

Since (2.4) holds and $f \in C^2$, we deduce that $\xi \in C^1\left(\left[a,b\right] \times \mathbb{R} \times \mathbb{R}\right)$.

Step 4. We therefore have that the functions

$$\left(x,u,v\right) \to \xi\left(x,u,v\right), \; f_x\left(x,u,\xi\left(x,u,v\right)\right), \; f_u\left(x,u,\xi\left(x,u,v\right)\right)$$

are C^1. We then immediately obtain (2.7), (2.8), and thus $H \in C^2$. Indeed we have, differentiating (2.11),

$$\begin{cases} H_x = v\xi_x - f_x - f_\xi\,\xi_x = \left(v - f_\xi\right)\xi_x - f_x = -f_x \\ H_u = v\xi_u - f_u - f_\xi\,\xi_u = \left(v - f_\xi\right)\xi_u - f_u = -f_u \\ H_v = \xi + v\xi_v - f_\xi\,\xi_v = \left(v - f_\xi\right)\xi_v + \xi = \xi \end{cases}$$

and in particular

$$\xi = H_v\left(x,u,v\right).$$

This achieves the proof of the lemma. ∎

2.4.2 The main theorem and some examples

We are now in a position to state the main theorem.

Theorem 2.11 *Let f and H be as in Lemma 2.9. Let $(u,v) \in C^2\left(\left[a,b\right]\right) \times C^2\left(\left[a,b\right]\right)$ satisfy, for every $x \in \left[a,b\right]$,*

$$(H) \quad \begin{cases} u'\left(x\right) = H_v\left(x,u\left(x\right),v\left(x\right)\right) \\ v'\left(x\right) = -H_u\left(x,u\left(x\right),v\left(x\right)\right). \end{cases}$$

Then u verifies

$$(E) \quad \frac{d}{dx}\left[f_\xi\left(x,u\left(x\right),u'\left(x\right)\right)\right] = f_u\left(x,u\left(x\right),u'\left(x\right)\right), \quad \forall x \in \left[a,b\right].$$

Conversely, if $u \in C^2\left(\left[a,b\right]\right)$ satisfies (E) then (u,v) are solutions of (H) where

$$v\left(x\right) = f_\xi\left(x,u\left(x\right),u'\left(x\right)\right), \quad \forall x \in \left[a,b\right].$$

Remark 2.12 The same remarks as in the lemma apply also to the theorem.

Proof *Part 1.* Let (u, v) satisfy (H). Using (2.10) and (2.8) we get

$$u' = H_v(x, u, v) \quad \Leftrightarrow \quad v = f_\xi(x, u, u')$$

$$v' = -H_u(x, u, v) = f_u(x, u, u')$$

and thus u satisfies (E).

Part 2. Conversely, by (2.10) and since $v = f_\xi(x, u, u')$ we get the first equation

$$u' = H_v(x, u, v).$$

Moreover, since $v = f_\xi(x, u, u')$ and u satisfies (E), we have

$$v' = \frac{dv}{dx} = \frac{d}{dx}[f_\xi(x, u, u')] = f_u(x, u, u').$$

The second equation follows then from the combination of the above identity and (2.8). ∎

Example 2.13 The present example is motivated by classical mechanics. Let $m > 0$, $g \in C^1([a, b])$ and

$$f(x, u, \xi) = \frac{m}{2}\xi^2 - g(x)u.$$

The integral under consideration is

$$I(u) = \int_a^b f(x, u(x), u'(x))\, dx$$

and the associated Euler-Lagrange equation is

$$mu''(x) = -g(x), \quad x \in (a, b).$$

The Hamiltonian is then

$$H(x, u, v) = \frac{v^2}{2m} + g(x)u$$

while the associated Hamiltonian system is

$$\begin{cases} u'(x) = v(x)/m \\ v'(x) = -g(x). \end{cases}$$

Example 2.14 We now generalize the preceding example (see Remark 2.10 (i)). Let $p > 1$ and $p' = p/(p-1)$,

$$f(x, u, \xi) = \frac{1}{p} |\xi|^p - g(x, u) \quad \text{and} \quad H(x, u, v) = \frac{1}{p'} |v|^{p'} + g(x, u).$$

The Euler-Lagrange equation and the associated Hamiltonian system are

$$\frac{d}{dx} \left[|u'|^{p-2} u' \right] = -g_u(x, u)$$

and

$$\begin{cases} u' = |v|^{p'-2} v \\ v' = -g_u(x, u). \end{cases}$$

Example 2.15 Consider the simplest case where $f(x, u, \xi) = f(\xi)$ with $f'' > 0$ (or more generally f is strictly convex) and $\lim_{|\xi| \to \infty} f(\xi)/|\xi| = +\infty$. The Euler-Lagrange equation and its integrated form are

$$\frac{d}{dx} [f'(u')] = 0 \quad \Rightarrow \quad f'(u') = \lambda = \text{constant}.$$

The Hamiltonian is given by

$$H(v) = f^*(v) = \sup_{\xi} \{v\xi - f(\xi)\}.$$

The associated Hamiltonian system is

$$\begin{cases} u' = (f^*)'(v) \\ v' = 0. \end{cases}$$

We find trivially that (λ and μ denoting some constants) $v' = \lambda$ and hence (compare with Case 2.3)

$$u(x) = (f^*)'(\lambda) x + \mu.$$

Example 2.16 We now look for the slightly more involved case $f(x, u, \xi) = f(x, \xi)$ with the appropriate hypotheses. The Euler-Lagrange equation and its integrated form are

$$\frac{d}{dx} [f_\xi(x, u')] = 0 \quad \Rightarrow \quad f_\xi(x, u') = \lambda = \text{constant}.$$

The Hamiltonian of f is given by

$$H(x, v) = \sup_{\xi} \{v\xi - f(x, \xi)\}.$$

The associated Hamiltonian system is

$$\begin{cases} u'(x) = H_v(x, v(x)) \\ \quad v' = 0. \end{cases}$$

The solution is then given by $v = \lambda = $ constant and $u'(x) = H_v(x, \lambda)$.

Example 2.17 We next consider the more difficult case $f(x, u, \xi) = f(u, \xi)$ with the hypotheses of the theorem. The Euler-Lagrange equation and its integrated form are

$$\frac{d}{dx}\left[f_\xi(u, u')\right] = f_u(u, u') \quad \Rightarrow \quad f(u, u') - u' f_\xi(x, u') = \lambda = \text{constant}.$$

The Hamiltonian of f is given by

$$H(u, v) = \sup_\xi \{v\xi - f(u, \xi)\} \quad \text{with } v = f_\xi(u, \xi).$$

The associated Hamiltonian system is

$$\begin{cases} u'(x) = H_v(u(x), v(x)) \\ v'(x) = -H_u(u(x), v(x)). \end{cases}$$

The Hamiltonian system also has a first integral given by

$$\frac{d}{dx}\left[H(u, v)\right] = H_u(u, v) u' + H_v(u, v) v' \equiv 0.$$

In physical terms we can say that if the Lagrangian f is independent of the variable x (which is here the time), the Hamiltonian H is constant along the trajectories.

2.4.3 Exercises

Exercise 2.4.1 Generalize Theorem 2.11 to the case where $u : [a, b] \to \mathbb{R}^N$, $N \geq 1$.

Exercise 2.4.2 Consider a mechanical system with N particles whose respective masses are m_i and positions at time t are $u_i(t) = (x_i(t), y_i(t), z_i(t)) \in \mathbb{R}^3$, $1 \leq i \leq N$. Let

$$T(u') = \frac{1}{2}\sum_{i=1}^N m_i |u_i'|^2 = \frac{1}{2}\sum_{i=1}^N m_i\left[(x_i')^2 + (y_i')^2 + (z_i')^2\right]$$

be the kinetic energy and denote by $U = U(t, u)$ the potential energy. Finally let

$$L(t, u, u') = T(u') - U(t, u)$$

be the Lagrangian. Also let H be the associated Hamiltonian. With the help of the preceding exercise show the following results.

(i) Write the Euler-Lagrange equations. Find the associated Hamiltonian system.

(ii) Show that, along the trajectories (i.e. when $v = L_\xi(t, u, u')$), the Hamiltonian can be written as (in mechanical terms it is the total energy of the system)

$$H(t, u, v) = T(u') + U(t, u).$$

Exercise 2.4.3 Let $f(x, u, \xi) = \sqrt{g(x, u)}\sqrt{1 + \xi^2}$. Write the associated Hamiltonian system and find a first integral of this system when g does not depend explicitly on x.

2.5 Hamilton-Jacobi equation

We now discuss the connection between finding stationary points of the functionals I and J considered in the preceding sections and solving a first order partial differential equation known as *Hamilton-Jacobi equation*. This equation also plays an important role in the fields theories developed in the next section (see Exercise 2.6.3).

Let us start with the main theorem.

Theorem 2.18 *Let $H \in C^1([a, b] \times \mathbb{R} \times \mathbb{R})$, $H = H(x, u, v)$. Assume that there exists $S \in C^2([a, b] \times \mathbb{R})$, $S = S(x, u)$, a solution of the* Hamilton-Jacobi *equation*

$$S_x + H(x, u, S_u) = 0, \quad \forall (x, u) \in [a, b] \times \mathbb{R} \qquad (2.15)$$

where $S_x = \partial S/\partial x$ and $S_u = \partial S/\partial u$. Assume also that there exists $u \in C^1([a, b])$ a solution of

$$u'(x) = H_v(x, u(x), S_u(x, u(x))), \quad \forall x \in [a, b]. \qquad (2.16)$$

Setting

$$v(x) = S_u(x, u(x)) \qquad (2.17)$$

then $(u, v) \in C^1([a, b]) \times C^1([a, b])$ is a solution of

$$\begin{cases} u'(x) = H_v(x, u(x), v(x)) \\ v'(x) = -H_u(x, u(x), v(x)). \end{cases} \qquad (2.18)$$

Moreover, if there is a one parameter family

$$S = S\left(x, u, \alpha\right), \quad S \in C^2\left([a, b] \times \mathbb{R} \times \mathbb{R}\right),$$

solving (2.15) for every $(x, u, \alpha) \in [a, b] \times \mathbb{R} \times \mathbb{R}$, *then any solution of (2.16) satisfies*

$$\frac{d}{dx}\left[S_\alpha\left(x, u\left(x\right), \alpha\right)\right] = 0, \quad \forall\left(x, \alpha\right) \in [a, b] \times \mathbb{R}$$

where $S_\alpha = \partial S/\partial \alpha$.

Remark 2.19 (i) If the Hamiltonian does not depend explicitly on x, we can consider the *reduced form* of the Hamilton-Jacobi equation

$$H\left(u, S_u\right) = 0, \quad \forall\, u \in \mathbb{R}. \tag{2.19}$$

Then every solution $S^*\left(u, \alpha\right)$ of

$$H\left(u, S_u^*\right) = \alpha, \quad \forall\left(u, \alpha\right) \in \mathbb{R} \times \mathbb{R}$$

leads immediately to a solution of (2.15), setting

$$S\left(x, u, \alpha\right) = S^*\left(u, \alpha\right) - \alpha x.$$

(ii) It is, in general, a difficult task to solve (2.15) and an extensive bibliography on the subject exists, see for example Evans [47], Lions [80].

Proof *Step 1.* We differentiate (2.17) to get

$$v'\left(x\right) = S_{xu}\left(x, u\left(x\right)\right) + u'\left(x\right) S_{uu}\left(x, u\left(x\right)\right), \quad \forall\, x \in [a, b].$$

Differentiating (2.15) with respect to u we find, for every $(x, u) \in [a, b] \times \mathbb{R}$,

$$S_{xu}\left(x, u\right) + H_u\left(x, u, S_u\left(x, u\right)\right) + H_v\left(x, u, S_u\left(x, u\right)\right) S_{uu}\left(x, u\right) = 0.$$

Combining the two identities (the second one evaluated at $u = u\left(x\right)$) and (2.16) with the definition of v, we have

$$v'\left(x\right) = -H_u\left(x, u\left(x\right), S_u\left(x, u\left(x\right)\right)\right) = -H_u\left(x, u\left(x\right), v\left(x\right)\right)$$

as wished.

Step 2. Since S is a solution of the Hamilton-Jacobi equation, we have, for every $(x, u, \alpha) \in [a, b] \times \mathbb{R} \times \mathbb{R}$,

$$\frac{d}{d\alpha}\left[S_x\left(x, u, \alpha\right) + H\left(x, u, S_u\left(x, u, \alpha\right)\right)\right]$$
$$= S_{x\alpha}\left(x, u, \alpha\right) + H_v\left(x, u, S_u\left(x, u, \alpha\right)\right) S_{u\alpha}\left(x, u, \alpha\right) = 0.$$

Since this identity is valid for every u, it is also valid for $u = u(x)$ satisfying (2.16) and thus

$$S_{x\alpha}(x, u(x), \alpha) + u'(x) S_{u\alpha}(x, u(x), \alpha) = 0.$$

This last identity can be rewritten as

$$\frac{d}{dx}[S_\alpha(x, u(x), \alpha)] = 0$$

which is the claim. ∎

The above theorem admits a converse.

Theorem 2.20 (Jacobi theorem) *Let $H \in C^1([a, b] \times \mathbb{R} \times \mathbb{R})$, $S = S(x, u, \alpha)$ be $C^2([a, b] \times \mathbb{R} \times \mathbb{R})$ with*

$$S_{u\alpha}(x, u, \alpha) \neq 0, \quad \forall (x, u, \alpha) \in [a, b] \times \mathbb{R} \times \mathbb{R}$$

and solving (2.15), namely

$$S_x + H(x, u, S_u) = 0, \quad \forall (x, u, \alpha) \in [a, b] \times \mathbb{R} \times \mathbb{R}.$$

If $u = u(x)$ satisfies

$$\frac{d}{dx}[S_\alpha(x, u(x), \alpha)] = 0, \quad \forall (x, \alpha) \in [a, b] \times \mathbb{R} \qquad (2.20)$$

then u necessarily verifies

$$u'(x) = H_v(x, u(x), S_u(x, u(x), \alpha)), \quad \forall (x, \alpha) \in [a, b] \times \mathbb{R}.$$

Thus, if $v(x) = S_u(x, u(x), \alpha)$, then $(u, v) \in C^1([a, b]) \times C^1([a, b])$ is a solution of (2.18), namely

$$\begin{cases} u'(x) = H_v(x, u(x), v(x)) \\ v'(x) = -H_u(x, u(x), v(x)). \end{cases}$$

Proof Since (2.20) holds we have, for every $(x, \alpha) \in [a, b] \times \mathbb{R}$,

$$0 = \frac{d}{dx}[S_\alpha(x, u(x), \alpha)] = S_{x\alpha}(x, u(x), \alpha) + S_{u\alpha}(x, u(x), \alpha) u'(x).$$

From (2.15) we obtain, for every $(x, u, \alpha) \in [a, b] \times \mathbb{R} \times \mathbb{R}$,

$$0 = \frac{d}{d\alpha}[S_x(x, u, \alpha) + H(x, u, S_u(x, u, \alpha))]$$
$$= S_{x\alpha}(x, u, \alpha) + H_v(x, u, S_u(x, u, \alpha)) S_{u\alpha}(x, u, \alpha).$$

Combining the two identities (the second one evaluated at $u = u(x)$), with the hypothesis $S_{u\alpha}(x, u, \alpha) \neq 0$, we get

$$u'(x) = H_v(x, u(x), S_u(x, u(x), \alpha)), \quad \forall (x, \alpha) \in [a, b] \times \mathbb{R}$$

as wished. We still need to prove that $v' = -H_u$. Differentiating v we have, for every $(x, \alpha) \in [a, b] \times \mathbb{R}$,

$$\begin{aligned} v'(x) &= S_{xu}(x, u(x), \alpha) + u'(x) S_{uu}(x, u(x), \alpha) \\ &= S_{xu}(x, u(x), \alpha) + H_v(x, u(x), S_u(x, u(x), \alpha)) S_{uu}(x, u(x), \alpha). \end{aligned}$$

Appealing, once more, to (2.15) we obtain, for every $(x, u, \alpha) \in [a, b] \times \mathbb{R} \times \mathbb{R}$,

$$\begin{aligned} 0 &= \frac{d}{du} [S_x(x, u, \alpha) + H(x, u, S_u(x, u, \alpha))] \\ &= S_{xu}(x, u, \alpha) + H_u(x, u, S_u(x, u, \alpha)) + H_v(x, u, S_u(x, u, \alpha)) S_{uu}(x, u, \alpha). \end{aligned}$$

Combining the two identities (the second one evaluated at $u = u(x)$) we infer the result, namely

$$v'(x) = -H_u(x, u(x), S_u(x, u(x), \alpha)) = -H_u(x, u(x), v(x)).$$

This achieves the proof of the theorem. ∎

Example 2.21 Let $g \in C^1(\mathbb{R})$ with $g(u) \geq g_0 > 0$. Let

$$H(u, v) = \frac{1}{2}v^2 - g(u)$$

be the Hamiltonian associated with

$$f(u, \xi) = \frac{1}{2}\xi^2 + g(u).$$

The Hamilton-Jacobi equation and its reduced form (see (2.19)) are given by

$$S_x + \frac{1}{2}(S_u)^2 - g(u) = 0 \quad \text{and} \quad \frac{1}{2}(S_u)^2 = g(u).$$

Therefore a solution of the equation is given by

$$S = S(x, u) = S(u) = \int_0^u \sqrt{2g(s)}\, ds.$$

We next solve

$$u'(x) = H_v(u(x), S_u(u(x))) = S_u(u(x)) = \sqrt{2g(u(x))}$$

which has a solution given implicitly by

$$\int_{u(0)}^{u(x)} \frac{ds}{\sqrt{2g(s)}} = x.$$

Setting $v(x) = S_u(u(x))$, we have indeed found a solution of the Hamiltonian system

$$\begin{cases} u'(x) = H_v(u(x), v(x)) = v(x) \\ v'(x) = -H_u(u(x), v(x)) = g'(u(x)). \end{cases}$$

Note also that such a function u solves

$$u''(x) = g'(u(x))$$

which is the Euler-Lagrange equation associated with the Lagrangian f.

2.5.1 Exercises

Exercise 2.5.1 Write the Hamilton-Jacobi equation when $u \in \mathbb{R}^N$, $N \geq 1$, and generalize Theorem 2.20 to this case.

Exercise 2.5.2 Let $f(x, u, \xi) = f(u, \xi) = \sqrt{g(u)}\sqrt{1 + \xi^2}$. Solve the Hamilton-Jacobi equation and find the stationary points of

$$I(u) = \int_a^b f(u(x), u'(x)) \, dx.$$

Exercise 2.5.3 Same exercise as the preceding one with

$$f(x, u, \xi) = f(u, \xi) = \frac{a(u)}{2}\xi^2$$

where $a(u) \geq a_0 > 0$. Compare the result with Exercise 2.2.10.

Exercise 2.5.4 Let $H \in C^2([a, b] \times \mathbb{R})$, $H = H(x, v)$.

(i) Find a one parameter family $S \in C^2([a, b] \times \mathbb{R} \times \mathbb{R})$, $S = S(x, u, \alpha)$, satisfying Hamilton-Jacobi equation

$$S_x + H(x, S_u) = 0, \quad \forall(x, u, \alpha) \in [a, b] \times \mathbb{R} \times \mathbb{R}$$

and such that

$$S_{u\alpha}(x, u, \alpha) \neq 0, \quad \forall(x, u, \alpha) \in [a, b] \times \mathbb{R} \times \mathbb{R}.$$

(ii) Find a solution of the associated Hamiltonian system (compare with Example 2.16).

Exercise 2.5.5 Let $\Omega \subset \mathbb{R}^n$ be the unit ball and let $H : \mathbb{R}^n \to \mathbb{R}$ be given by

$$H(v) = |v|^2 - 1.$$

Show that the function

$$S(u) = \text{dist}(u, \partial\Omega)$$

satisfies the multidimensional reduced Hamilton-Jacobi equation (called, in this context, the eikonal equation)

$$\begin{cases} H(\nabla S(u)) = H(S_{u_1}(u), \cdots, S_{u_n}(u)) = 0 & u \in \Omega \setminus \{0\} \\ \qquad\qquad S(u) = 0 & u \in \partial\Omega. \end{cases}$$

2.6   Fields theories

As already mentioned, we will give only a very brief account of the fields theories and we refer, for example, to [2], [54], [56] and [107] for more details. These theories are conceptually important but often difficult to manage for specific examples.

Let us recall the problem under consideration

$$(P) \quad \inf_{u \in X} \left\{ I(u) = \int_a^b f(x, u(x), u'(x)) \, dx \right\} = m$$

where $X = \{ u \in C^1([a,b]) : u(a) = \alpha, \ u(b) = \beta \}$. The Euler-Lagrange equation is

$$(E) \quad \frac{d}{dx}[f_\xi(x, u, u')] = f_u(x, u, u'), \quad x \in (a, b).$$

2.6.1   A simple case

We now try to explain the nature of the theory, starting with a particularly simple case. We have seen in Section 2.2 that a solution of (E) is not, in general, a minimizer for (P). However (see Theorem 2.1) if $(u, \xi) \to f(x, u, \xi)$ is convex for every $x \in [a, b]$ then any solution of (E) is necessarily a minimizer of (P). We first show that we can, sometimes, recover this result under the only assumption that $\xi \to f(x, u, \xi)$ is convex for every $(x, u) \in [a, b] \times \mathbb{R}$.

Theorem 2.22 *Let $f \in C^2([a,b] \times \mathbb{R} \times \mathbb{R})$. If there exists $\Phi \in C^3([a,b] \times \mathbb{R})$ with $\Phi(a, \alpha) = \Phi(b, \beta)$ such that*

$$(u, \xi) \to \widetilde{f}(x, u, \xi) \quad \text{is convex for every } x \in [a, b]$$

where

$$\widetilde{f}(x, u, \xi) = f(x, u, \xi) + \Phi_u(x, u)\xi + \Phi_x(x, u),$$

then any solution $\overline{u}$ of (E) is a minimizer of (P).

Remark 2.23 We should immediately point out that in order to have $(u, \xi) \to \widetilde{f}(x, u, \xi)$ convex for every $x \in [a, b]$ we should, at least, have that $\xi \to f(x, u, \xi)$ is convex for every $(x, u) \in [a, b] \times \mathbb{R}$. If $(u, \xi) \to f(x, u, \xi)$ is already convex, then choose $\Phi \equiv 0$ and apply Theorem 2.1.

Proof Define

$$\varphi(x, u, \xi) = \Phi_u(x, u)\xi + \Phi_x(x, u).$$

Observe that the following two identities (the first one uses that $\Phi(a, \alpha) = \Phi(b, \beta)$ and the second one is just straight differentiation)

$$\int_a^b \frac{d}{dx}[\Phi(x, u(x))]\, dx = \Phi(b, \beta) - \Phi(a, \alpha) = 0$$

$$\frac{d}{dx}[\varphi_\xi(x, u(x), u'(x))] = \varphi_u(x, u(x), u'(x)), \quad x \in [a, b]$$

hold for any $u \in X = \{u \in C^1([a, b]) : u(a) = \alpha,\ u(b) = \beta\}$. The first identity expresses that the integral is *invariant*, while the second one says that $\varphi(x, u, u')$ satisfies the Euler-Lagrange equation identically (it is then called a *null Lagrangian*).

With the help of the above observations we immediately obtain the result by applying Theorem 2.1 to $\widetilde{f}$. Indeed, we have that $(u, \xi) \to \widetilde{f}(x, u, \xi)$ is convex,

$$I(u) = \int_a^b \widetilde{f}(x, u(x), u'(x))\, dx = \int_a^b f(x, u(x), u'(x))\, dx$$

for every $u \in X$ and any solution $\overline{u}$ of (E) also satisfies

$$(\widetilde{E}) \qquad \frac{d}{dx}\left[\widetilde{f}_\xi(x, \overline{u}, \overline{u}')\right] = \widetilde{f}_u(x, \overline{u}, \overline{u}'), \quad x \in (a, b).$$

This concludes the proof. ∎

With the help of the above elementary theorem we can now fully handle the Poincaré-Wirtinger inequality.

Example 2.24 (Poincaré-Wirtinger inequality) Let $\lambda \geq 0$,

$$f_\lambda(u, \xi) = \frac{\xi^2 - \lambda^2 u^2}{2}$$

$$(P_\lambda) \quad \inf_{u \in X} \left\{ I_\lambda(u) = \int_0^1 f_\lambda(u(x), u'(x)) \, dx \right\} = m_\lambda$$

where $X = \{ u \in C^1([0,1]) : u(0) = u(1) = 0 \}$. Observe that $\xi \to f_\lambda(u, \xi)$ is convex while $(u, \xi) \to f_\lambda(u, \xi)$ is not. The Euler-Lagrange equation is

$$(E_\lambda) \quad u'' + \lambda^2 u = 0, \quad x \in (0, 1).$$

Note that $u_0 \equiv 0$ is a solution of (E_λ). Define, if $\lambda < \pi$,

$$\Phi(x, u) = \frac{\lambda}{2} \tan\left[\lambda \left(x - \frac{1}{2} \right) \right] u^2, \quad (x, u) \in [0, 1] \times \mathbb{R}$$

and observe that Φ satisfies all the properties of Theorem 2.22. The function $\widetilde{f}$ is then

$$\widetilde{f}(x, u, \xi) = \frac{1}{2} \xi^2 + \lambda \tan\left[\lambda \left(x - \frac{1}{2} \right) \right] u\xi + \frac{\lambda^2}{2} \tan^2\left[\lambda \left(x - \frac{1}{2} \right) \right] u^2.$$

It is easy to see that $(u, \xi) \to \widetilde{f}(x, u, \xi)$ is convex and therefore applying Theorem 2.22 we have that, for every $0 \le \lambda < \pi$,

$$I_\lambda(u) \ge I_\lambda(0) = 0, \quad \forall u \in X.$$

An elementary passage to the limit leads to Poincaré-Wirtinger inequality

$$\int_0^1 (u')^2 \ge \pi^2 \int_0^1 u^2, \quad \forall u \in X.$$

For a different proof of a slightly more general form of Poincaré-Wirtinger inequality see Theorem 6.1 and for another generalization see Exercise 2.6.4.

2.6.2 Exact fields and Hilbert theorem

The way of proceeding in Theorem 2.22 is, in general, too naive and can be done only locally; in fact one needs a similar but more subtle theory.

Definition 2.25 *Let $D \subset \mathbb{R}^2$ be a connected open set. We say that $\Phi \in C^0(D)$, $\Phi = \Phi(x, u)$, is an exact field for f covering D if there exists $S \in C^1(D)$ satisfying*

$$S_u(x, u) = f_\xi(x, u, \Phi(x, u)) = p(x, u)$$

$$S_x(x, u) = f(x, u, \Phi(x, u)) - p(x, u) \Phi(x, u) = h(x, u).$$

Remark 2.26 (i) If $f \in C^2$, then a necessary condition for Φ to be exact is that $p_x = h_u$. Conversely, if D is simply connected and if $p_x = h_u$ then such an S exists.

(ii) In the case where $u : [a, b] \to \mathbb{R}^N$, $N > 1$, we have to add to the preceding remark, not only that $p_x^i = h_{u^i}$, but also $p_{u^j}^i = p_{u^i}^j$, for every $1 \leq i, j \leq N$.

We start with an elementary result that is a first justification for defining such a notion.

Proposition 2.27 *Let* $f \in C^2 \left([a, b] \times \mathbb{R} \times \mathbb{R}\right)$, $f = f\left(x, u, \xi\right)$, *and*

$$I\left(u\right) = \int_a^b f\left(x, u\left(x\right), u'\left(x\right)\right) dx.$$

Let $\Phi \in C^1\left(D\right)$, $\Phi = \Phi\left(x, u\right)$ *be an exact field for* f *covering* D, $[a, b] \times \mathbb{R} \subset D$. *Then any solution* $u \in C^2\left([a, b]\right)$ *of*

$$u'\left(x\right) = \Phi\left(x, u\left(x\right)\right) \tag{2.21}$$

solves the Euler-Lagrange equation associated with the functional I, *namely*

$$(E) \quad \frac{d}{dx}\left[f_\xi\left(x, u\left(x\right), u'\left(x\right)\right)\right] = f_u\left(x, u\left(x\right), u'\left(x\right)\right), \quad x \in [a, b]. \tag{2.22}$$

Proof By definition of Φ and using the fact that $p = f_\xi$, we have, for any $(x, u) \in D$,

$$h_u = f_u\left(x, u, \Phi\right) + f_\xi\left(x, u, \Phi\right)\Phi_u - p_u\Phi - p\Phi_u = f_u\left(x, u, \Phi\right) - p_u\Phi$$

and hence

$$f_u\left(x, u, \Phi\right) = h_u\left(x, u\right) + p_u\left(x, u\right)\Phi\left(x, u\right).$$

We therefore get for every $x \in [a, b]$

$$\frac{d}{dx}\left[f_\xi\left(x, u, u'\right)\right] - f_u\left(x, u, u'\right) = \frac{d}{dx}\left[p\left(x, u\right)\right] - \left[h_u\left(x, u\right) + p_u\left(x, u\right)\Phi\left(x, u\right)\right]$$

$$= p_x + p_u u' - h_u - p_u\Phi = p_x - h_u = 0$$

since we have that $u' = \Phi$ and $p_x = h_u$, Φ being exact. Thus we have reached the claim. ∎

The next theorem is the main result of this section and was established by Weierstrass and Hilbert.

Theorem 2.28 (Hilbert theorem) *Let* $f \in C^2\left([a,b] \times \mathbb{R} \times \mathbb{R}\right)$ *with*

$$\xi \to f\left(x, u, \xi\right) \quad \text{convex for every } \left(x, u\right) \in [a, b] \times \mathbb{R}.$$

Let $D \subset \mathbb{R}^2$ *be a connected open set and* $\Phi \in C^0\left(D\right)$, $\Phi = \Phi\left(x, u\right)$, *be an exact field for* f *covering* D. *Assume that there exists* $u_0 \in C^1\left([a,b]\right)$ *satisfying*

$$\left(x, u_0\left(x\right)\right) \in D \quad \text{and} \quad u_0'\left(x\right) = \Phi\left(x, u_0\left(x\right)\right), \; \forall\, x \in [a, b],$$

then u_0 *is a minimizer for* I, *i.e.*

$$I\left(u\right) = \int_a^b f\left(x, u\left(x\right), u'\left(x\right)\right) dx \geq I\left(u_0\right), \quad \forall\, u \in X$$

where

$$X = \left\{ \begin{array}{c} u \in C^1\left([a,b]\right) : u\left(a\right) = u_0\left(a\right), \; u\left(b\right) = u_0\left(b\right) \\ \text{with } \left(x, u\left(x\right)\right) \in D, \; \forall\, x \in [a, b] \end{array} \right\}.$$

Remark 2.29 (i) Observe that according to the preceding proposition we have that such a u_0 is necessarily a solution of the Euler-Lagrange equation.

(ii) As already mentioned it might be very difficult to construct such exact fields. Moreover, in general, D does not contain the whole of $[a,b] \times \mathbb{R}$ and, consequently, the theorem provides only local minima. The construction of such fields is intimately linked with the so-called *Jacobi condition* concerning conjugate points (see, for example, [2], [54] and [56] for more details).

Proof Denote by E the *Weierstrass function* defined by

$$E\left(x, u, \eta, \xi\right) = f\left(x, u, \xi\right) - f\left(x, u, \eta\right) - f_\xi\left(x, u, \eta\right)\left(\xi - \eta\right)$$

or in other words

$$f\left(x, u, \xi\right) = E\left(x, u, \eta, \xi\right) + f\left(x, u, \eta\right) + f_\xi\left(x, u, \eta\right)\left(\xi - \eta\right).$$

Since $\xi \to f\left(x, u, \xi\right)$ is convex, the function E is always non-negative. Note also that since $u_0'\left(x\right) = \Phi\left(x, u_0\left(x\right)\right)$ then

$$E\left(x, u_0\left(x\right), \Phi\left(x, u_0\left(x\right)\right), u_0'\left(x\right)\right) = 0, \quad \forall\, x \in [a, b].$$

Now using the definition of exact field we get that, for every $u \in X$,

$$f\left(x, u\left(x\right), \Phi\left(x, u\left(x\right)\right)\right) + f_\xi\left(x, u\left(x\right), \Phi\left(x, u\left(x\right)\right)\right)\left(u'\left(x\right) - \Phi\left(x, u\left(x\right)\right)\right)$$

$$= f\left(x, u, \Phi\right) - p\,\Phi + p\,u' = S_x + S_u u' = \frac{d}{dx}\left[S\left(x, u\left(x\right)\right)\right].$$

Combining these facts we obtain

$$I(u) = \int_a^b f(x, u(x), u'(x)) \, dx$$

$$= \int_a^b \left\{ E(x, u(x), \Phi(x, u(x)), u'(x)) + \frac{d}{dx} [S(x, u(x))] \right\} dx$$

$$\geq \int_a^b \frac{d}{dx} [S(x, u(x))] \, dx = S(b, u(b)) - S(a, u(a)).$$

Since $E(x, u_0, \Phi(x, u_0), u_0') = 0$ we have that

$$I(u_0) = S(b, u_0(b)) - S(a, u_0(a)).$$

Moreover, since $u_0(a) = u(a)$, $u_0(b) = u(b)$ we deduce that $I(u) \geq I(u_0)$ for every $u \in X$. This achieves the proof of the theorem.

The quantity

$$\int_a^b \frac{d}{dx} [S(x, u(x))] \, dx$$

is called the *invariant Hilbert integral*. ∎

2.6.3   Exercises

Exercise 2.6.1 Generalize Theorem 2.22 to the case where $u : [a, b] \to \mathbb{R}^N$, $N \geq 1$.

Exercise 2.6.2 Generalize Hilbert theorem (Theorem 2.28) to the case where $u : [a, b] \to \mathbb{R}^N$, $N \geq 1$.

Exercise 2.6.3 (The present exercise establishes the connection between exact field and Hamilton-Jacobi equation.) Let $f = f(x, u, \xi)$ and $H = H(x, u, v)$ be as in Theorem 2.11 and Lemma 2.9.

(i) Show that if there exists an exact field Φ covering D, then

$$S_x + H(x, u, S_u) = 0, \quad \forall (x, u) \in D$$

where

$$S_u(x, u) = f_\xi(x, u, \Phi(x, u))$$
$$S_x(x, u) = f(x, u, \Phi(x, u)) - S_u(x, u) \Phi(x, u).$$

(ii) Conversely, if the Hamilton-Jacobi equation has a solution for every $(x, u) \in D$, prove that

$$\Phi(x, u) = H_v(x, u, S_u(x, u))$$

is an exact field for f covering D.

Exercise 2.6.4 Let $X = \left\{ u \in C^1\left([0,1]\right) : u\left(0\right) = u\left(1\right) = 0 \right\}$ and

$$f\left(u, \xi\right) = h\left(\xi\right) - g\left(u\right)$$

with $g, h \in C^2\left(\mathbb{R}\right)$ and

$$g'\left(0\right) = 0, \quad h_0 \geq 0 \quad \text{and} \quad \pi^2 h_0 - g_0 \geq 0$$

where

$$h_0 = \inf\left\{h''\left(u\right) : u \in \mathbb{R}\right\} \quad \text{and} \quad g_0 = \sup\left\{g''\left(u\right) : u \in \mathbb{R}\right\}.$$

Prove that

$$\int_0^1 \left[h\left(u'\left(x\right)\right) - h\left(0\right)\right] dx \geq \int_0^1 \left[g\left(u\left(x\right)\right) - g\left(0\right)\right] dx, \quad \forall\, u \in X.$$

Note that the present exercise is a generalization of Example 2.24 (i.e. Poincaré-Wirtinger inequality) if we set

$$h\left(\xi\right) = \frac{1}{2}\xi^2 \quad \text{and} \quad g\left(u\right) = \frac{\pi^2}{2}u^2.$$

Chapter 3

Direct methods: existence

3.1 Introduction

In this chapter we study the problem

$$(P) \quad \inf \left\{ I(u) = \int_{\Omega} f(x, u(x), \nabla u(x)) \, dx : u \in u_0 + W_0^{1,p}(\Omega) \right\} = m$$

where

- $\Omega \subset \mathbb{R}^n$ is a bounded open set;
- $f : \overline{\Omega} \times \mathbb{R} \times \mathbb{R}^n \to \mathbb{R},\ f = f(x, u, \xi)$;
- $u \in u_0 + W_0^{1,p}(\Omega)$ means that $u, u_0 \in W^{1,p}(\Omega)$ and $u - u_0 \in W_0^{1,p}(\Omega)$ (which roughly means that $u = u_0$ on $\partial\Omega$).

This is the fundamental problem of the calculus of variations. We show that the problem (P) has a solution $\overline{u} \in u_0 + W_0^{1,p}(\Omega)$ provided the two following main hypotheses are satisfied.

(H_1) *Convexity*: $\xi \to f(x, u, \xi)$ is convex for every $(x, u) \in \overline{\Omega} \times \mathbb{R}$;

(H_2) *Coercivity*: there exist $p > q \geq 1$ and $\alpha_1 > 0$, $\alpha_2, \alpha_3 \in \mathbb{R}$ such that

$$f(x, u, \xi) \geq \alpha_1 |\xi|^p + \alpha_2 |u|^q + \alpha_3, \quad \forall (x, u, \xi) \in \overline{\Omega} \times \mathbb{R} \times \mathbb{R}^n.$$

The *Dirichlet integral*, which has as integrand

$$f(x, u, \xi) = \frac{1}{2} |\xi|^2,$$

satisfies both hypotheses. However, the *minimal surface* problem whose integrand is given by

$$f(x, u, \xi) = \sqrt{1 + |\xi|^2}$$

satisfies (H_1) but verifies (H_2) only with $p = 1$. This problem requires special treatment (see Chapter 5).

It is interesting to compare the generality of the result with those of the preceding chapter. The main drawback of the present analysis is that we prove the existence of minimizers only in Sobolev spaces. In the next chapter we will see that, under some extra hypotheses, the solution is in fact more regular (for example it is C^1, C^2 or C^∞).

We now describe the content of the present chapter. In Section 3.2 we consider the model case, namely the Dirichlet integral. Although this is just an example of the more general theorem obtained in Section 3.3, we fully discuss the particular case because of its importance and to make the understanding of the method easier. Recall that the origin of the direct methods goes back to Hilbert, Lebesgue and Tonelli, while treating the Dirichlet integral. Let us briefly describe the two main steps in the proof.

Step 1 (Compactness). Let $u_\nu \in u_0 + W_0^{1,p}(\Omega)$ be a minimizing sequence of (P), this means that

$$I(u_\nu) \to \inf\{I(u)\} = m, \quad \text{as } \nu \to \infty.$$

It is easy, invoking (H_2) and Poincaré inequality (see Theorem 1.49), to obtain that there exist $\overline{u} \in u_0 + W_0^{1,p}(\Omega)$ and a subsequence (still denoted u_ν) so that u_ν converges weakly to $\overline{u}$ in $W^{1,p}$, i.e.

$$u_\nu \rightharpoonup \overline{u} \text{ in } W^{1,p}, \quad \text{as } \nu \to \infty.$$

Step 2 (Lower semicontinuity). We then show that (H_1) implies the (sequential) weak lower semicontinuity of I, namely

$$u_\nu \rightharpoonup \overline{u} \text{ in } W^{1,p} \quad \Rightarrow \quad \liminf_{\nu \to \infty} I(u_\nu) \geq I(\overline{u}).$$

Since $\{u_\nu\}$ is a minimizing sequence, we deduce that $\overline{u}$ is a minimizer of (P).

In Section 3.4 we derive the *Euler-Lagrange equation* associated with (P). Since the solution of (P) is only in a Sobolev space, we are able to write only a weak form of this equation.

In Section 3.5 we briefly describe the considerably harder case where the unknown function u is a vector, i.e. $u : \Omega \subset \mathbb{R}^n \to \mathbb{R}^N$, with $n, N > 1$.

In Section 3.6 we briefly explain what can be done, in some cases, when the hypothesis (H_1) of convexity fails to hold.

The interested reader is referred for further reading to the present author's book [32] or to Buttazzo [15], Buttazzo-Giaquinta-Hildebrandt [17], Cesari [20], Ekeland-Temam [45], Giaquinta [55], Giusti [59], Ioffe-Tihomirov [73], Morrey [86], Struwe [104] and Zeidler [113].

3.2   The model case: Dirichlet integral

The main result of the present section is

Theorem 3.1 *Let $\Omega \subset \mathbb{R}^n$ be a bounded open set with Lipschitz boundary and $u_0 \in W^{1,2}(\Omega)$. The problem*

$$(D) \quad \inf\left\{ I(u) = \frac{1}{2}\int_\Omega |\nabla u(x)|^2\, dx : u \in u_0 + W_0^{1,2}(\Omega) \right\} = m$$

has one and only one solution $\bar{u} \in u_0 + W_0^{1,2}(\Omega)$.

Furthermore, $\bar{u}$ satisfies the weak form of Laplace equation, namely

$$\int_\Omega \langle \nabla \bar{u}(x)\,; \nabla \varphi(x)\rangle\, dx = 0, \quad \forall \varphi \in W_0^{1,2}(\Omega) \tag{3.1}$$

where $\langle .;. \rangle$ denotes the scalar product in $\mathbb{R}^n$.

Conversely, if $\bar{u} \in u_0 + W_0^{1,2}(\Omega)$ satisfies (3.1), then it is a minimizer of (D).

Remark 3.2 (i) We should again emphasize the very weak hypotheses on u_0 and Ω and recall that $u \in u_0 + W_0^{1,2}(\Omega)$ means that $u = u_0$ on $\partial\Omega$ (in the sense of Sobolev spaces).

(ii) If the solution $\bar{u}$ turns out to be more regular, namely in $W^{2,2}(\Omega)$, then (3.1) can be integrated by parts and we get

$$\int_\Omega \Delta \bar{u}(x)\, \varphi(x)\, dx = 0, \quad \forall \varphi \in W_0^{1,2}(\Omega)$$

which combined with the fundamental lemma of the calculus of variations (Theorem 1.24) leads to $\Delta \bar{u} = 0$ a.e. in Ω. This extra regularity of $\bar{u}$ (which turns out to be even $C^\infty(\Omega)$) is proved in Sections 4.3 and 4.5.

(iii) As we already said, the above theorem was proved by Hilbert, Lebesgue and Tonelli, but it was expressed in a different way since Sobolev spaces did not exist then. Throughout the nineteenth century there were several attempts to establish a theorem of the above kind, notably by Dirichlet and Riemann.

Proof The proof is surprisingly simple.

Part 1 (Existence). We divide, as explained in the introduction, the proof into three steps.

Step 1 (Compactness). We start with the observation that since $u_0 \in u_0 + W_0^{1,2}(\Omega)$ we have

$$0 \le m \le I(u_0) < \infty.$$

Let $u_\nu \in u_0 + W_0^{1,2}(\Omega)$ be a minimizing sequence of (D), this means that

$$I(u_\nu) \to \inf\{I(u)\} = m, \quad \text{as } \nu \to \infty.$$

Observe that by Poincaré inequality (see Theorem 1.49) we can find constants $\gamma_1, \gamma_2 > 0$ so that

$$\sqrt{2I(u_\nu)} = \|\nabla u_\nu\|_{L^2} \geq \gamma_1 \|u_\nu\|_{W^{1,2}} - \gamma_2 \|u_0\|_{W^{1,2}}.$$

Since u_ν is a minimizing sequence and $m < \infty$ we deduce that there exists $\gamma_3 > 0$ so that

$$\|u_\nu\|_{W^{1,2}} \leq \gamma_3.$$

Applying Exercises 1.4.5 and 1.4.11, we deduce that there exists $\overline{u} \in u_0 + W_0^{1,2}(\Omega)$ and a subsequence (still denoted u_ν) so that

$$u_\nu \rightharpoonup \overline{u} \text{ in } W^{1,2}, \quad \text{as } \nu \to \infty.$$

Step 2 (Lower semicontinuity). We now show that I is (sequentially) weakly lower semicontinuous; this means that

$$u_\nu \rightharpoonup \overline{u} \text{ in } W^{1,2} \quad \Rightarrow \quad \liminf_{\nu \to \infty} I(u_\nu) \geq I(\overline{u}).$$

This step is independent of the fact that $\{u_\nu\}$ is a minimizing sequence. We trivially have that

$$|\nabla u_\nu|^2 = |\nabla \overline{u}|^2 + 2\langle \nabla \overline{u}; \nabla u_\nu - \nabla \overline{u}\rangle + |\nabla u_\nu - \nabla \overline{u}|^2$$
$$\geq |\nabla \overline{u}|^2 + 2\langle \nabla \overline{u}; \nabla u_\nu - \nabla \overline{u}\rangle.$$

Integrating this expression we have

$$I(u_\nu) \geq I(\overline{u}) + \int_\Omega \langle \nabla \overline{u}; \nabla u_\nu - \nabla \overline{u}\rangle \, dx.$$

To conclude we show that the second term on the right-hand side of the inequality tends to 0. Indeed, since $\nabla \overline{u} \in L^2$ and $\nabla u_\nu - \nabla \overline{u} \rightharpoonup 0$ in L^2 this implies, by definition of weak convergence in L^2, that

$$\lim_{\nu \to \infty} \int_\Omega \langle \nabla \overline{u}; \nabla u_\nu - \nabla \overline{u}\rangle \, dx = 0.$$

Therefore, returning to the above inequality we have indeed obtained that

$$\liminf_{\nu \to \infty} I(u_\nu) \geq I(\overline{u}).$$

Step 3. We now combine the two steps. Since $\{u_\nu\}$ is a minimizing sequence (i.e. $I(u_\nu) \to \inf\{I(u)\} = m$) and for such a sequence we have lower semicontinuity (i.e. $\liminf I(u_\nu) \geq I(\overline{u})$), we deduce that $I(\overline{u}) = m$, i.e. $\overline{u}$ is a minimizer of (D).

Part 2 (Uniqueness). Assume that there exist $\overline{u}, \overline{v} \in u_0 + W_0^{1,2}(\Omega)$ so that

$$I(\overline{u}) = I(\overline{v}) = m$$

and let us show that this implies $\overline{u} = \overline{v}$. Denote $\overline{w} = (\overline{u} + \overline{v})/2$ and observe that $\overline{w} \in u_0 + W_0^{1,2}(\Omega)$. The function $\xi \to |\xi|^2$ being convex, we can infer that $\overline{w}$ is also a minimizer since

$$m \leq I(\overline{w}) \leq \frac{1}{2}I(\overline{u}) + \frac{1}{2}I(\overline{v}) = m,$$

which readily implies that

$$\int_\Omega \left[\frac{1}{2}|\nabla\overline{u}|^2 + \frac{1}{2}|\nabla\overline{v}|^2 - \left|\frac{\nabla\overline{u} + \nabla\overline{v}}{2}\right|^2 \right] dx = 0.$$

Appealing once more to the convexity of $\xi \to |\xi|^2$, we deduce that the integrand is non-negative, while the integral is 0. This is possible only if

$$\frac{1}{2}|\nabla\overline{u}|^2 + \frac{1}{2}|\nabla\overline{v}|^2 - \left|\frac{\nabla\overline{u} + \nabla\overline{v}}{2}\right|^2 = 0 \quad \text{a.e. in } \Omega.$$

We now use the strict convexity of $\xi \to |\xi|^2$ to obtain that $\nabla\overline{u} = \nabla\overline{v}$ a.e. in Ω, which combined with the fact that the two functions agree on the boundary of Ω (since $\overline{u}, \overline{v} \in u_0 + W_0^{1,2}(\Omega)$) leads to the claimed uniqueness $\overline{u} = \overline{v}$ a.e. in Ω (see Exercise 1.4.14 (i)).

Part 3 (Euler-Lagrange equation). Let us now establish (3.1). Let $\epsilon \in \mathbb{R}$ and $\varphi \in W_0^{1,2}(\Omega)$ be arbitrary. Note that $\overline{u} + \epsilon\varphi \in u_0 + W_0^{1,2}(\Omega)$, which combined with the fact that $\overline{u}$ is the minimizer of (D) leads to

$$I(\overline{u}) \leq I(\overline{u} + \epsilon\varphi) = \int_\Omega \frac{1}{2}|\nabla\overline{u} + \epsilon\nabla\varphi|^2\, dx$$

$$= I(\overline{u}) + \epsilon \int_\Omega \langle \nabla\overline{u}; \nabla\varphi \rangle\, dx + \epsilon^2 I(\varphi).$$

The fact that ϵ is arbitrary leads immediately to (3.1), which expresses nothing other than

$$\frac{d}{d\epsilon}I(\overline{u} + \epsilon\varphi)\bigg|_{\epsilon=0} = 0.$$

Part 4 (Converse). We finally prove that if $\bar{u} \in u_0 + W_0^{1,2}(\Omega)$ satisfies (3.1) then it is necessarily a minimizer of (D). Let $u \in u_0 + W_0^{1,2}(\Omega)$ be any element and set $\varphi = u - \bar{u}$. Observe that $\varphi \in W_0^{1,2}(\Omega)$ and

$$I(u) = I(\bar{u} + \varphi) = \int_\Omega \frac{1}{2} |\nabla \bar{u} + \nabla \varphi|^2 \, dx$$

$$= I(\bar{u}) + \int_\Omega \langle \nabla \bar{u}; \nabla \varphi \rangle \, dx + I(\varphi) \geq I(\bar{u})$$

since the second term is 0 according to (3.1) and the last one is non-negative. This achieves the proof of the theorem. ∎

3.2.1   Exercises

Exercise 3.2.1 Let Ω be as in the theorem and $h \in L^2(\Omega)$. Show that

$$(P) \quad \inf \left\{ I(u) = \int_\Omega \left[\frac{1}{2} |\nabla u(x)|^2 - h(x) u(x) \right] dx : u \in W_0^{1,2}(\Omega) \right\} = m$$

has a unique solution $\bar{u} \in W_0^{1,2}(\Omega)$ which satisfies in addition

$$\int_\Omega \langle \nabla \bar{u}(x); \nabla \varphi(x) \rangle \, dx = \int_\Omega h(x) \varphi(x) \, dx, \quad \forall \varphi \in W_0^{1,2}(\Omega) .$$

Exercise 3.2.2 Let $\Omega \subset \mathbb{R}^n$ be a bounded connected open set with Lipschitz boundary, $h \in L^2(\Omega)$ and

$$X = \left\{ u \in W^{1,2}(\Omega) : \int_\Omega u = 0 \right\} .$$

(i) Prove that the Neumann problem

$$(N) \quad \inf \left\{ I(u) = \int_\Omega \left[\frac{1}{2} |\nabla u(x)|^2 - h(x) u(x) \right] dx : u \in X \right\} = m$$

has one and only one solution $\bar{u} \in X$.

(ii) Write the weak form of the Euler-Lagrange equation satisfied by $\bar{u}$.

(iii) If, in addition $\bar{u} \in C^2(\overline{\Omega})$ and $h \in C(\overline{\Omega})$, write (the strong form of) the Euler-Lagrange equation.

3.3   A general existence theorem

3.3.1   The main theorem and some examples

The main theorem of the present chapter is the following.

Theorem 3.3 Let $\Omega \subset \mathbb{R}^n$ be a bounded open set with Lipschitz boundary. Let $f \in C^0\left(\overline{\Omega} \times \mathbb{R} \times \mathbb{R}^n\right)$, $f = f(x, u, \xi)$, satisfy

(H_1) $\xi \to f(x, u, \xi)$ is convex for every $(x, u) \in \overline{\Omega} \times \mathbb{R}$;

(H_2) there exist $p > q \geq 1$ and $\alpha_1 > 0$, $\alpha_2, \alpha_3 \in \mathbb{R}$ such that

$$f(x, u, \xi) \geq \alpha_1 |\xi|^p + \alpha_2 |u|^q + \alpha_3, \quad \forall (x, u, \xi) \in \overline{\Omega} \times \mathbb{R} \times \mathbb{R}^n.$$

Let

$$(P) \quad \inf\left\{ I(u) = \int_\Omega f(x, u(x), \nabla u(x))\, dx : u \in u_0 + W_0^{1,p}(\Omega) \right\} = m$$

where $u_0 \in W^{1,p}(\Omega)$ with $I(u_0) < \infty$. Then there exists $\overline{u} \in u_0 + W_0^{1,p}(\Omega)$, a minimizer of (P).

Furthermore, if $(u, \xi) \to f(x, u, \xi)$ is strictly convex for every $x \in \overline{\Omega}$, then the minimizer is unique.

Remark 3.4 (i) The hypotheses of the theorem are nearly optimal, in the sense that the weakening of any of them leads to a counterexample to the existence of minimizers (see below). The only hypothesis that can be slightly weakened is the continuity of f (see the above-mentioned literature). Indeed, it is enough to require that f is a *Carathéodory function*, which means that

$$x \to f(x, u, \xi) \quad \text{is measurable for every } (u, \xi)$$

$$(u, \xi) \to f(x, u, \xi) \quad \text{is continuous for almost every } x.$$

This weakening is particularly interesting when one considers, for example, functions of the form (see Exercise 3.2.1)

$$f(x, u, \xi) = \frac{1}{2} |\xi|^2 - h(x)\, u$$

with $h \in L^2(\Omega)$.

(ii) The theorem remains valid in the vectorial case, where $u : \Omega \subset \mathbb{R}^n \to \mathbb{R}^N$, with $n, N > 1$. However, the hypothesis (H_1) is then far from being optimal (see Section 3.5).

(iii) This theorem has a long history and we refer to [32] for details. Tonelli was the first to notice the importance of the convexity of f.

Before proceeding with the proof of the theorem, we discuss several examples, emphasizing the optimality of the hypotheses.

Example 3.5 (i) The Dirichlet integral considered in the preceding section enters, of course, in the framework of the present theorem; indeed, we have that

$$f(x, u, \xi) = f(\xi) = \frac{1}{2} |\xi|^2$$

satisfies all the hypotheses of the theorem with $p = 2$.

    (ii) The natural generalization of the preceding example is

$$f(x, u, \xi) = \frac{1}{p} |\xi|^p + g(x, u)$$

where g is continuous and non-negative and $p > 1$.

Example 3.6 The minimal surface problem has an integrand given by

$$f(x, u, \xi) = f(\xi) = \sqrt{1 + |\xi|^2}$$

that satisfies all the hypotheses of the theorem but (H_2), this hypothesis is only verified with $p = 1$. We will see in Chapter 5 that this failure may lead to non-existence of a minimizer for the corresponding (P). The reason why $p = 1$ is not allowed is that the corresponding Sobolev space $W^{1,1}$ is not reflexive (see Chapter 1).

Example 3.7 This example is of the minimal surface type but easier; it also shows that all the hypotheses of the theorem are satisfied, except (H_2) which is true with $p = 1$. This weakening of (H_2) leads to the following counterexample. Let $n = 1$,

$$f(x, u, \xi) = f(u, \xi) = \sqrt{u^2 + \xi^2}$$

$$(P) \quad \inf \left\{ I(u) = \int_0^1 f(u(x), u'(x))\, dx : u \in X \right\} = m$$

where $X = \{ u \in W^{1,1}(0, 1) : u(0) = 0,\ u(1) = 1 \}$. Let us prove that (P) has no solution. We first show that $m = 1$ and start by observing that $m \geq 1$ since

$$I(u) \geq \int_0^1 |u'(x)|\, dx \geq \int_0^1 u'(x)\, dx = u(1) - u(0) = 1.$$

To establish that $m = 1$, we construct a minimizing sequence $u_\nu \in X$ (ν being an integer) as follows

$$u_\nu(x) = \begin{cases} 0 & \text{if } x \in \left[0, 1 - \frac{1}{\nu} \right] \\ 1 + \nu(x - 1) & \text{if } x \in \left(1 - \frac{1}{\nu}, 1 \right]. \end{cases}$$

We therefore have $m = 1$ since

$$1 \leq I(u_\nu) = \int_{1-\frac{1}{\nu}}^{1} \sqrt{(1 + \nu(x-1))^2 + \nu^2} \, dx$$

$$\leq \frac{1}{\nu} \sqrt{1 + \nu^2} \to 1, \quad \text{as } \nu \to \infty.$$

Assume now, for the sake of contradiction, that there exists $\overline{u} \in X$ a minimizer of (P). We should then have, as above,

$$1 = I(\overline{u}) = \int_0^1 \sqrt{\overline{u}^2 + (\overline{u}')^2} \, dx \geq \int_0^1 |\overline{u}'| \, dx$$

$$\geq \int_0^1 \overline{u}' \, dx = \overline{u}(1) - \overline{u}(0) = 1.$$

This implies that $\overline{u} = 0$ a.e. in $(0, 1)$. Since elements of X are continuous we have that $\overline{u} \equiv 0$ and this is incompatible with the boundary data. Thus (P) has no solution.

Example 3.8 (Weierstrass example) We have seen this example in Section 2.2. Recall that $n = 1$,

$$f(x, u, \xi) = f(x, \xi) = x\xi^2$$

$$(P) \quad \inf\left\{I(u) = \int_0^1 f(x, u'(x)) \, dx : u \in X\right\} = m_X$$

where $X = \{u \in W^{1,2}(0, 1) : u(0) = 1, \, u(1) = 0\}$. All the hypotheses of the theorem are verified with the exception of (H_2) which is satisfied only with $\alpha_1 = 0$. This is enough to show that (P) has no minimizer in X. Indeed, we have seen in Exercise 2.2.6 that (P) has no solution in $Y = X \cap C^1([0, 1])$ and that the corresponding value of the infimum, let us denote it by m_Y, is 0. Since trivially $0 \leq m_X \leq m_Y$, we deduce that $m_X = 0$. Now assume, by absurd hypothesis, that (P) has a solution $\overline{u} \in X$; we should then have $I(\overline{u}) = 0$, but since the integrand is non-negative, we deduce that $\overline{u}' = 0$ a.e. in $(0, 1)$. Since elements of X are continuous we have that $\overline{u}$ is constant, and this is incompatible with the boundary data. Hence (P) has no solution.

Example 3.9 The present example (see Poincaré-Wirtinger inequality) shows that we cannot allow, in general, that $q = p$ in (H_2). Let $n = 1$, $\lambda > \pi$ and

$$f(x, u, \xi) = f(u, \xi) = \frac{1}{2}\left(\xi^2 - \lambda^2 u^2\right).$$

We have seen in Section 2.2 that if

$$(P) \quad \inf \left\{ I(u) = \int_0^1 f(u(x), u'(x)) \, dx : u \in W_0^{1,2}(0,1) \right\} = m$$

then $m = -\infty$, which means that (P) has no solution.

Example 3.10 (Bolza example) We now show that, as a general rule, one cannot weaken (H_1) either. One such example has already been seen in Section 2.2 where we had $f(x, u, \xi) = f(\xi) = e^{-\xi^2}$ (which satisfies neither (H_1) nor (H_2)). Let $n = 1$,

$$f(x, u, \xi) = f(u, \xi) = \left(\xi^2 - 1\right)^2 + u^4$$

$$(P) \quad \inf \left\{ I(u) = \int_0^1 f(u(x), u'(x)) \, dx : u \in W_0^{1,4}(0,1) \right\} = m.$$

Assume for a moment that we have already proved that $m = 0$ and let us show that (P) has no solution, using an argument by contradiction. Let $\overline{u} \in W_0^{1,4}(0,1)$ be a minimizer of (P), i.e. $I(\overline{u}) = 0$. This implies that $\overline{u} = 0$ and $|\overline{u}'| = 1$ a.e. in $(0,1)$. Since the elements of $W^{1,4}$ are continuous, we have that $\overline{u} \equiv 0$ and hence $\overline{u}' \equiv 0$, which is clearly absurd.

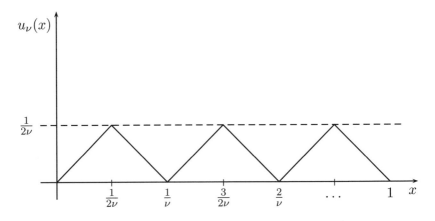

Figure 3.1: minimizing sequence

So let us show that $m = 0$ by constructing an appropriate minimizing sequence (as shown in Figure 3.1). Let $u_\nu \in W_0^{1,4}$ ($\nu \geq 2$ being an integer) defined on each interval $[k/\nu, (k+1)/\nu]$, $0 \leq k \leq \nu - 1$, as follows

$$u_\nu(x) = \begin{cases} x - \dfrac{k}{\nu} & \text{if } x \in \left[\dfrac{2k}{2\nu}, \dfrac{2k+1}{2\nu}\right] \\ -x + \dfrac{k+1}{\nu} & \text{if } x \in \left(\dfrac{2k+1}{2\nu}, \dfrac{2k+2}{2\nu}\right]. \end{cases}$$

Observe that $|u'_\nu| = 1$ a.e. and $|u_\nu| \leq 1/(2\nu)$ leading therefore to the desired convergence, namely

$$0 \leq I(u_\nu) \leq \frac{1}{(2\nu)^4} \to 0, \quad \text{as } \nu \to \infty.$$

3.3.2 Proof of the main theorem

We now turn to the proof of the theorem.

Proof We do not prove the theorem in its full generality. We refer to the literature and in particular to Corollary 3.24 in [32, 2nd edition] for a general proof; see also the exercises below. We prove it under the following stronger hypotheses. We assume that $f \in C^1(\overline{\Omega} \times \mathbb{R} \times \mathbb{R}^n)$, instead of C^0, and

(H_{1+}) $(u, \xi) \to f(x, u, \xi)$ is convex for every $x \in \overline{\Omega}$;

(H_{2+}) there exist $p > 1$ and $\alpha_1 > 0$, $\alpha_3 \in \mathbb{R}$ such that

$$f(x, u, \xi) \geq \alpha_1 |\xi|^p + \alpha_3, \quad \forall (x, u, \xi) \in \overline{\Omega} \times \mathbb{R} \times \mathbb{R}^n.$$

(H_3) there exists a constant $\beta \geq 0$ so that for every $(x, u, \xi) \in \overline{\Omega} \times \mathbb{R} \times \mathbb{R}^n$

$$|f_u(x, u, \xi)|, \; |f_\xi(x, u, \xi)| \leq \beta \left(1 + |u|^{p-1} + |\xi|^{p-1}\right)$$

where $f_\xi = (f_{\xi_1}, \cdots, f_{\xi_n})$, $f_{\xi_i} = \partial f / \partial \xi_i$ and $f_u = \partial f / \partial u$.

Once these hypotheses are made, the proof is very similar to that of Theorem 3.1. Note also that the function $f(x, u, \xi) = f(\xi) = |\xi|^2 / 2$ satisfies the above stronger hypotheses.

Part 1 (Existence). The proof is divided into three steps.

Step 1 (Compactness). Recall that by assumption on u_0 and by (H_{2+}) we have

$$-\infty < m \leq I(u_0) < \infty.$$

Let $u_\nu \in u_0 + W_0^{1,p}(\Omega)$ be a minimizing sequence of (P), i.e.

$$I(u_\nu) \to \inf\{I(u)\} = m, \quad \text{as } \nu \to \infty.$$

We therefore have from (H_{2+}) that for ν large enough

$$m + 1 \geq I(u_\nu) \geq \alpha_1 \|\nabla u_\nu\|_{L^p}^p - |\alpha_3| \, \text{meas} \, \Omega$$

and hence there exists $\alpha_4 > 0$ so that

$$\|\nabla u_\nu\|_{L^p} \leq \alpha_4.$$

Appealing to Poincaré inequality (see Theorem 1.49), we can find constants $\alpha_5, \alpha_6 > 0$ so that

$$\alpha_4 \geq \|\nabla u_\nu\|_{L^p} \geq \alpha_5 \|u_\nu\|_{W^{1,p}} - \alpha_6 \|u_0\|_{W^{1,p}}$$

and hence we can find $\alpha_7 > 0$ so that

$$\|u_\nu\|_{W^{1,p}} \leq \alpha_7 .$$

Applying Exercises 1.4.5 and 1.4.11 (it is only here that we use the fact that $p > 1$), we deduce that there exists $\overline{u} \in u_0 + W_0^{1,p}(\Omega)$ and a subsequence (still denoted u_ν) so that

$$u_\nu \rightharpoonup \overline{u} \text{ in } W^{1,p}, \quad \text{as } \nu \to \infty.$$

Step 2 (Lower semicontinuity). We now show that I is (sequentially) weakly lower semicontinuous; this means that

$$u_\nu \rightharpoonup \overline{u} \text{ in } W^{1,p} \quad \Rightarrow \quad \liminf_{\nu \to \infty} I(u_\nu) \geq I(\overline{u}). \tag{3.2}$$

This step is independent of the fact that $\{u_\nu\}$ is a minimizing sequence. Using the convexity of f and the fact that it is C^1 we get

$$
\begin{aligned}
f(x, u_\nu, \nabla u_\nu) \geq & \\
f(x, \overline{u}, \nabla \overline{u}) + f_u(x, \overline{u}, \nabla \overline{u})(u_\nu - \overline{u}) & + \langle f_\xi(x, \overline{u}, \nabla \overline{u}); \nabla u_\nu - \nabla \overline{u} \rangle .
\end{aligned}
\tag{3.3}
$$

Before proceeding further we need to show that the combination of (H_3) and $\overline{u} \in W^{1,p}(\Omega)$ leads to

$$f_u(x, \overline{u}, \nabla \overline{u}) \in L^{p'}(\Omega) \quad \text{and} \quad f_\xi(x, \overline{u}, \nabla \overline{u}) \in L^{p'}(\Omega; \mathbb{R}^n) \tag{3.4}$$

where $1/p + 1/p' = 1$ (i.e. $p' = p/(p-1)$). Indeed, let us prove the first statement, the other one being shown similarly. We have (β_1 being a constant)

$$
\int_\Omega |f_u(x, \overline{u}, \nabla \overline{u})|^{p'} \, dx \leq \beta^{p'} \int_\Omega \left(1 + |\overline{u}|^{p-1} + |\nabla \overline{u}|^{p-1}\right)^{\frac{p}{p-1}} \, dx
$$
$$
\leq \beta_1 \left(1 + \|\overline{u}\|_{W^{1,p}}^p\right) < \infty.
$$

Using Hölder inequality and (3.4) we find that for $u_\nu \in W^{1,p}(\Omega)$

$$f_u(x, \overline{u}, \nabla \overline{u})(u_\nu - \overline{u}), \ \langle f_\xi(x, \overline{u}, \nabla \overline{u}); \nabla u_\nu - \nabla \overline{u} \rangle \in L^1(\Omega).$$

We next integrate (3.3) to get

$$I(u_\nu) \geq I(\overline{u}) + \int_\Omega [f_u(x, \overline{u}, \nabla \overline{u})(u_\nu - \overline{u}) + \langle f_\xi(x, \overline{u}, \nabla \overline{u}); \nabla u_\nu - \nabla \overline{u} \rangle] \, dx.$$

$$\tag{3.5}$$

Since $u_\nu - \overline{u} \rightharpoonup 0$ in $W^{1,p}$ (i.e. $u_\nu - \overline{u} \rightharpoonup 0$ in L^p and $\nabla u_\nu - \nabla \overline{u} \rightharpoonup 0$ in L^p) and (3.4) holds, we deduce, from the definition of weak convergence in L^p, that

$$\lim_{\nu \to \infty} \int_\Omega f_u(x, \overline{u}, \nabla \overline{u})(u_\nu - \overline{u})\, dx = \lim_{\nu \to \infty} \int_\Omega \langle f_\xi(x, \overline{u}, \nabla \overline{u}) ; \nabla u_\nu - \nabla \overline{u} \rangle\, dx = 0.$$

Therefore, returning to (3.5) we have indeed obtained that

$$\liminf_{\nu \to \infty} I(u_\nu) \geq I(\overline{u}).$$

Step 3. We now combine the two steps. Since $\{u_\nu\}$ is a minimizing sequence (i.e. $I(u_\nu) \to \inf\{I(u)\} = m$) and for such a sequence we have lower semicontinuity (i.e. $\liminf I(u_\nu) \geq I(\overline{u})$), we deduce that $I(\overline{u}) = m$, i.e. $\overline{u}$ is a minimizer of (P).

Part 2 (Uniqueness). The proof is almost identical to those of Theorem 3.1 and Theorem 2.1. Assume that there exist $\overline{u}, \overline{v} \in u_0 + W_0^{1,p}(\Omega)$ so that

$$I(\overline{u}) = I(\overline{v}) = m$$

and we prove that this implies $\overline{u} = \overline{v}$. Denote

$$\overline{w} = \frac{\overline{u} + \overline{v}}{2}$$

and observe that $\overline{w} \in u_0 + W_0^{1,p}(\Omega)$. The function $(u, \xi) \to f(x, u, \xi)$ being convex, we can infer that $\overline{w}$ is also a minimizer since

$$m \leq I(\overline{w}) \leq \frac{1}{2} I(\overline{u}) + \frac{1}{2} I(\overline{v}) = m,$$

which readily implies that

$$\int_\Omega \left[\frac{f(x, \overline{u}, \nabla \overline{u})}{2} + \frac{f(x, \overline{v}, \nabla \overline{v})}{2} - f\left(x, \frac{\overline{u} + \overline{v}}{2}, \frac{\nabla \overline{u} + \nabla \overline{v}}{2}\right) \right] dx = 0.$$

The convexity of $(u, \xi) \to f(x, u, \xi)$ implies that the integrand is non-negative, while the integral is 0. This is possible only if

$$\frac{1}{2} f(x, \overline{u}, \nabla \overline{u}) + \frac{1}{2} f(x, \overline{v}, \nabla \overline{v}) - f\left(x, \frac{\overline{u} + \overline{v}}{2}, \frac{\nabla \overline{u} + \nabla \overline{v}}{2}\right) = 0 \quad \text{a.e. in } \Omega.$$

We now use the strict convexity of $(u, \xi) \to f(x, u, \xi)$ to obtain that $\overline{u} = \overline{v}$ and $\nabla \overline{u} = \nabla \overline{v}$ a.e. in Ω, which implies the desired uniqueness, namely $\overline{u} = \overline{v}$ a.e. in Ω. ∎

3.3.3   Exercises

Exercise 3.3.1 Prove Theorem 3.3 under the hypotheses (H_{1+}), (H_2) and (H_3).

Exercise 3.3.2 Prove Theorem 3.3 if

$$f(x, u, \xi) = g(x, u) + h(x, \xi)$$

where $g \in C^0 \left(\overline{\Omega} \times \mathbb{R} \right)$, $g \geq 0$ and $h \in C^1 \left(\overline{\Omega} \times \mathbb{R}^n \right)$ with

$$\xi \to h(x, \xi) \quad \text{is convex } \forall x \in \overline{\Omega},$$

and there exist $p > 1$ and $\alpha_1 > 0$, $\beta, \alpha_3 \in \mathbb{R}$ such that

$$h(x, \xi) \geq \alpha_1 |\xi|^p + \alpha_3, \quad \forall (x, \xi) \in \overline{\Omega} \times \mathbb{R}^n$$

$$|h_\xi(x, \xi)| \leq \beta \left(1 + |\xi|^{p-1} \right), \quad \forall (x, \xi) \in \overline{\Omega} \times \mathbb{R}^n.$$

Exercise 3.3.3 Prove Theorem 3.3 in the following framework. Let $\alpha, \beta \in \mathbb{R}^N$, $N \geq 1$ and

$$(P) \quad \inf_{u \in X} \left\{ I(u) = \int_a^b f(x, u(x), u'(x)) \, dx \right\} = m$$

where $X = \left\{ u \in W^{1,p} \left((a, b) ; \mathbb{R}^N \right) : u(a) = \alpha, \ u(b) = \beta \right\}$ and

(i) $f \in C^1 \left([a, b] \times \mathbb{R}^N \times \mathbb{R}^N \right)$, $(u, \xi) \to f(x, u, \xi)$ is convex for every $x \in [a, b]$;

(ii) there exist $p > q \geq 1$ and $\alpha_1 > 0$, $\alpha_2, \alpha_3 \in \mathbb{R}$ such that

$$f(x, u, \xi) \geq \alpha_1 |\xi|^p + \alpha_2 |u|^q + \alpha_3, \quad \forall (x, u, \xi) \in [a, b] \times \mathbb{R}^N \times \mathbb{R}^N;$$

(iii) for every $R > 0$, there exists $\beta = \beta(R)$ such that

$$|f_u(x, u, \xi)| \leq \beta (1 + |\xi|^p) \quad \text{and} \quad |f_\xi(x, u, \xi)| \leq \beta \left(1 + |\xi|^{p-1} \right)$$

for every $x \in [a, b]$ and every $u, \xi \in \mathbb{R}^N$ with $|u| \leq R$.

Exercise 3.3.4 Let

$$f(x, u, \xi) = g(x, u) + h(x, \xi)$$

with $u \to g(x, u)$, $\xi \to h(x, \xi)$ convex and at least one of them strictly convex. Prove that the uniqueness in Theorem 3.3 is conserved.

3.4 Euler-Lagrange equation

3.4.1 The main theorem and its proof

We now derive the Euler-Lagrange equation associated with (P). The way of proceeding is identical to that of Section 2.2, but we have to be more careful. Indeed, we assumed there that the minimizer $\bar{u}$ was C^2, while here we only know that it is in the Sobolev space $W^{1,p}$.

Theorem 3.11 *Let $\Omega \subset \mathbb{R}^n$ be a bounded open set with Lipschitz boundary. Let $p \geq 1$ and $f \in C^1\left(\overline{\Omega} \times \mathbb{R} \times \mathbb{R}^n\right)$, $f = f(x, u, \xi)$, satisfy*

(H_3) *there exists $\beta \geq 0$ so that for every $(x, u, \xi) \in \overline{\Omega} \times \mathbb{R} \times \mathbb{R}^n$*

$$|f_u(x, u, \xi)|, \ |f_\xi(x, u, \xi)| \leq \beta\left(1 + |u|^{p-1} + |\xi|^{p-1}\right)$$

where $f_\xi = (f_{\xi_1}, \cdots, f_{\xi_n})$, $f_{\xi_i} = \partial f / \partial \xi_i$ and $f_u = \partial f / \partial u$.
Let $\bar{u} \in u_0 + W_0^{1,p}(\Omega)$ be a minimizer of

$$(P) \quad \inf\left\{I(u) = \int_\Omega f(x, u(x), \nabla u(x)) \, dx : u \in u_0 + W_0^{1,p}(\Omega)\right\} = m$$

where $u_0 \in W^{1,p}(\Omega)$, then $\bar{u}$ satisfies the weak form of the Euler-Lagrange equation

$$(E_w) \quad \int_\Omega \left[f_u(x, \bar{u}, \nabla\bar{u})\varphi + \langle f_\xi(x, \bar{u}, \nabla\bar{u}) ; \nabla\varphi\rangle\right] dx = 0, \quad \forall \varphi \in W_0^{1,p}(\Omega).$$

Moreover, if $f \in C^2\left(\overline{\Omega} \times \mathbb{R} \times \mathbb{R}^n\right)$ and $\bar{u} \in C^2\left(\overline{\Omega}\right)$, then $\bar{u}$ satisfies the Euler-Lagrange equation

$$(E) \quad \sum_{i=1}^n \frac{\partial}{\partial x_i}\left[f_{\xi_i}(x, \bar{u}, \nabla\bar{u})\right] = f_u(x, \bar{u}, \nabla\bar{u}), \quad \forall x \in \overline{\Omega}.$$

Conversely, if $(u, \xi) \to f(x, u, \xi)$ is convex for every $x \in \overline{\Omega}$ and if $\bar{u}$ is a solution of either (E_w) or (E), then it is a minimizer of (P).

Remark 3.12 (i) A more condensed way of writing (E) is

$$(E) \quad \operatorname{div}\left[f_\xi(x, \bar{u}, \nabla\bar{u})\right] = f_u(x, \bar{u}, \nabla\bar{u}), \quad \forall x \in \overline{\Omega}.$$

(ii) The hypothesis (H_3) is necessary for giving a meaning to (E_w); more precisely for ensuring that

$$f_u\varphi, \ \langle f_\xi; \nabla\varphi\rangle \in L^1(\Omega).$$

It can be weakened, but only slightly by the use of Sobolev imbedding theorem (see Exercise 3.4.1), or by requiring only measurability in x. In this last case one can assume that f is a Carathéodory function (see Remark 3.4 (i)) that satisfies for almost every $x \in \Omega$ and every $(u, \xi) \in \mathbb{R} \times \mathbb{R}^n$

$$|f_u(x, u, \xi)|, \; |f_\xi(x, u, \xi)| \leq \beta_1(x) + \beta_2\left(|u|^{p-1} + |\xi|^{p-1}\right)$$

where $\beta_1 \in L^{p'}(\Omega)$ and $\beta_2 \geq 0$.

(iii) Of course any solution of (E) is a solution of (E_w). The converse is true only if $\overline{u}$ is sufficiently regular.

(iv) In the statement of the theorem we do not need hypothesis (H_1) or (H_2) of Theorem 3.3. Therefore we do not use the convexity of f (naturally for the converse we need the convexity of f). However, we require that a minimizer of (P) does exist.

(v) The theorem remains valid in the vectorial case, where $u : \Omega \subset \mathbb{R}^n \to \mathbb{R}^N$, with $n, N > 1$. The Euler-Lagrange equation now becomes a system of partial differential equations and reads as follows

$$(E) \quad \sum_{i=1}^{n} \frac{\partial}{\partial x_i}\left[f_{\xi_i^j}(x, \overline{u}, \nabla \overline{u})\right] = f_{u^j}(x, \overline{u}, \nabla \overline{u}), \quad \forall x \in \overline{\Omega}, \; j = 1, \cdots, N$$

where $f : \overline{\Omega} \times \mathbb{R}^N \times \mathbb{R}^{N \times n} \to \mathbb{R}$ and

$$u = \left(u^1, \cdots, u^N\right) \in \mathbb{R}^N, \quad \xi = \left(\xi_i^j\right)_{1 \leq i \leq n}^{1 \leq j \leq N} \in \mathbb{R}^{N \times n} \quad \text{and} \quad \nabla u = \left(\frac{\partial u^j}{\partial x_i}\right)_{1 \leq i \leq n}^{1 \leq j \leq N}.$$

(vi) In some cases one may be interested in an even weaker form of the Euler-Lagrange equation. More precisely, if we choose the test functions φ in (E_w) to be $C_0^\infty(\Omega)$ instead of $W_0^{1,p}(\Omega)$, then one can weaken the hypothesis (H_3) and replace it by

(H_3') there exist $p \geq 1$ and $\beta \geq 0$ so that for every $(x, u, \xi) \in \overline{\Omega} \times \mathbb{R} \times \mathbb{R}^n$

$$|f_u(x, u, \xi)|, \; |f_\xi(x, u, \xi)| \leq \beta\left(1 + |u|^p + |\xi|^p\right).$$

The proof of the theorem remains almost identical. The choice of the space, where the test function φ belongs, depends on the context. If we want to use the solution $\overline{u}$ itself as a test function, then we are obliged to choose $W_0^{1,p}(\Omega)$ as the right space (see Sections 4.3 and 4.4), while at other times (see Section 4.2) we can actually limit ourselves to the space $C_0^\infty(\Omega)$.

Proof The proof is divided into four steps.

Step 1 (Preliminary computation). From the observation that

$$f(x, u, \xi) = f(x, 0, 0) + \int_0^1 \frac{d}{dt} [f(x, tu, t\xi)] \, dt, \quad \forall (x, u, \xi) \in \overline{\Omega} \times \mathbb{R} \times \mathbb{R}^n$$

and from (H_3), we find that there exists $\gamma_1 > 0$ so that

$$|f(x, u, \xi)| \leq \gamma_1 (1 + |u|^p + |\xi|^p), \quad \forall (x, u, \xi) \in \overline{\Omega} \times \mathbb{R} \times \mathbb{R}^n. \tag{3.6}$$

In particular we deduce that

$$|I(u)| < \infty, \quad \forall u \in W^{1,p}(\Omega).$$

Step 2 (Derivative of I). We now prove that for every $u, \varphi \in W^{1,p}(\Omega)$ and every $\epsilon \in \mathbb{R}$ we have

$$\lim_{\epsilon \to 0} \frac{I(u + \epsilon \varphi) - I(u)}{\epsilon} = \int_\Omega [f_u(x, u, \nabla u) \varphi + \langle f_\xi(x, u, \nabla u); \nabla \varphi \rangle] \, dx. \tag{3.7}$$

We let

$$g(x, \epsilon) = f(x, u(x) + \epsilon \varphi(x), \nabla u(x) + \epsilon \nabla \varphi(x))$$

so that

$$I(u + \epsilon \varphi) = \int_\Omega g(x, \epsilon) \, dx.$$

Since $f \in C^1$, we consider

$$g(x, \epsilon) - g(x, 0) = \int_0^1 \frac{d}{dt} [g(x, t\epsilon)] \, dt = \epsilon \int_0^1 g_\epsilon(x, t\epsilon) \, dt$$

where

$$g_\epsilon(x, t\epsilon) = f_u(x, u + t\epsilon\varphi, \nabla u + t\epsilon\nabla\varphi) \varphi + \langle f_\xi(x, u + t\epsilon\varphi, \nabla u + t\epsilon\nabla\varphi); \nabla\varphi \rangle.$$

The hypothesis (H_3) then implies that we can find $\gamma_2 > 0$ so that, for every $\epsilon \in [-1, 1], t \in [0, 1],$

$$|g_\epsilon(x, t\epsilon)| \leq G(x) = \gamma_2 (1 + |u|^p + |\varphi|^p + |\nabla u|^p + |\nabla \varphi|^p).$$

Note that since $u, \varphi \in W^{1,p}(\Omega)$, we have $G \in L^1(\Omega)$. We now observe that, since $u, \varphi \in W^{1,p}(\Omega)$, we have from (3.6) that the functions $x \to g(x, 0)$ and $x \to g(x, \epsilon)$ are both in $L^1(\Omega)$. Summarizing the results we have that

$$\left| \frac{g(x, \epsilon) - g(x, 0)}{\epsilon} \right| \leq G(x), \quad \text{with } G \in L^1(\Omega)$$

$$\frac{g(x, \epsilon) - g(x, 0)}{\epsilon} \to g_\epsilon(x, 0) \quad \text{a.e. in } \Omega.$$

Applying Lebesgue dominated convergence theorem we deduce that (3.7) holds.

Step 3 (Derivation of (E_w) and (E)). The conclusion of the theorem follows from the preceding step. Indeed, since $\overline{u}$ is a minimizer of (P), then

$$I\left(\overline{u} + \epsilon\varphi\right) \geq I\left(\overline{u}\right), \quad \forall\varphi \in W_0^{1,p}\left(\Omega\right)$$

and thus

$$\lim_{\epsilon \to 0} \frac{I\left(\overline{u} + \epsilon\varphi\right) - I\left(\overline{u}\right)}{\epsilon} = 0$$

which combined with (3.7) implies (E_w).

To get (E) it remains to integrate by parts (using Exercise 1.4.8) to find

$$(E_w) \quad \int_\Omega \left[f_u\left(x, \overline{u}, \nabla\overline{u}\right) - \operatorname{div}\left[f_\xi\left(x, \overline{u}, \nabla\overline{u}\right)\right]\right]\varphi\, dx = 0, \quad \forall\varphi \in W_0^{1,p}\left(\Omega\right).$$

The fundamental lemma of the calculus of variations (Theorem 1.24) implies the claim.

Step 4 (Converse). Let $\overline{u}$ be a solution of (E_w) (note that any solution of (E) is necessarily a solution of (E_w)). From the convexity of f we deduce that for every $u \in u_0 + W_0^{1,p}\left(\Omega\right)$ the following holds

$$f\left(x, u, \nabla u\right) \geq f\left(x, \overline{u}, \nabla\overline{u}\right) + f_u\left(x, \overline{u}, \nabla\overline{u}\right)\left(u - \overline{u}\right)$$
$$+ \left\langle f_\xi\left(x, \overline{u}, \nabla\overline{u}\right); \left(\nabla u - \nabla\overline{u}\right)\right\rangle.$$

Integrating, using (E_w) and the fact that $u - \overline{u} \in W_0^{1,p}\left(\Omega\right)$, we immediately get that $I\left(u\right) \geq I\left(\overline{u}\right)$ and hence the theorem. ∎

3.4.2   Some examples

We now discuss some examples.

Example 3.13 In the case of Dirichlet integral we have

$$f\left(x, u, \xi\right) = f\left(\xi\right) = \frac{1}{2}\left|\xi\right|^2$$

which satisfies (H_3). The equation (E_w) is then

$$\int_\Omega \left\langle \nabla\overline{u}\left(x\right); \nabla\varphi\left(x\right)\right\rangle\, dx = 0, \quad \forall\varphi \in W_0^{1,2}\left(\Omega\right)$$

while (E) is $\Delta\overline{u} = 0$.

Example 3.14 Consider the generalization of the preceding example, where

$$f(x, u, \xi) = f(\xi) = \frac{1}{p} |\xi|^p$$

which satisfies (H_3). The equation (E) is known as the $p-$Laplace equation (so-called since when $p = 2$ it corresponds to Laplace equation)

$$\text{div}\left[|\nabla \overline{u}|^{p-2} \nabla \overline{u}\right] = 0, \quad \text{in } \Omega.$$

Example 3.15 The minimal surface problem has an integrand given by

$$f(x, u, \xi) = f(\xi) = \sqrt{1 + |\xi|^2}$$

which satisfies (H_3) with $p = 1$, since

$$|f_\xi(\xi)| = \left(\sum_{i=1}^n \frac{(\xi_i)^2}{1 + |\xi|^2}\right)^{1/2} \leq 1.$$

The equation (E) is the so-called minimal surface equation

$$\text{div} \frac{\nabla \overline{u}}{\sqrt{1 + |\nabla \overline{u}|^2}} = 0, \quad \text{in } \Omega$$

and can be rewritten as

$$\left(1 + |\nabla \overline{u}|^2\right) \Delta \overline{u} - \sum_{i,j=1}^n \overline{u}_{x_i} \overline{u}_{x_j} \overline{u}_{x_i x_j} = 0, \quad \text{in } \Omega.$$

Example 3.16 Let

$$f(x, u, \xi) = f(u, \xi) = g(u) |\xi|^2$$

with $0 \leq g(u), |g'(u)| \leq g_0$. We then have

$$|f_u(u, \xi)| = |g'(u)| |\xi|^2 \leq g_0 |\xi|^2, \quad |f_\xi(u, \xi)| = 2 |g(u)| |\xi| \leq 2g_0 |\xi|.$$

We see that if $g'(u) \neq 0$, then f does not satisfy (H_3) but only the above (H_3'). We are therefore authorized to write only

$$\int_\Omega [f_u(\overline{u}, \nabla \overline{u}) \varphi + \langle f_\xi(\overline{u}, \nabla \overline{u}); \nabla \varphi\rangle] \, dx = 0, \quad \forall \varphi \in C_0^\infty(\Omega)$$

or more generally the equation should hold for any $\varphi \in W_0^{1,2}(\Omega) \cap L^\infty(\Omega)$.

Let us now recall two examples from Section 2.2, showing that, without any hypotheses of convexity of the function f, the converse part of the theorem is false.

Example 3.17 (Poincaré-Wirtinger inequality) Let $\lambda > \pi$, $n = 1$ and

$$f(x, u, \xi) = f(u, \xi) = \frac{1}{2}(\xi^2 - \lambda^2 u^2)$$

$$(P) \quad \inf\left\{ I(u) = \int_0^1 f(u(x), u'(x))\, dx : u \in W_0^{1,2}(0,1) \right\} = m.$$

Note that $\xi \to f(u, \xi)$ is convex while $(u, \xi) \to f(u, \xi)$ is not. We have seen that $m = -\infty$ and therefore (P) has no minimizer; however the Euler-Lagrange equation

$$u'' + \lambda^2 u = 0 \quad \text{in } [0,1]$$

has $u \equiv 0$ as a solution. It is therefore not a minimizer.

Example 3.18 Let $n = 1$,

$$f(x, u, \xi) = f(\xi) = (\xi^2 - 1)^2$$

which is non-convex, and

$$(P) \quad \inf\left\{ I(u) = \int_0^1 f(u'(x))\, dx : u \in W_0^{1,4}(0,1) \right\} = m.$$

We have seen that $m = 0$. The Euler-Lagrange equation is

$$(E) \quad \frac{d}{dx}\left[\overline{u}'\left((\overline{u}')^2 - 1\right) \right] = 0$$

and its weak form is (note that f satisfies (H_3))

$$(E_w) \quad \int_0^1 \overline{u}'\left((\overline{u}')^2 - 1\right)\varphi' = 0, \quad \forall \varphi \in W_0^{1,4}(0,1).$$

It is clear that $\overline{u} \equiv 0$ is a solution of (E) and (E_w), but it is not a minimizer of (P) since $m = 0$ and $I(0) = 1$. The present example is also interesting for another reason. Indeed, the function

$$v(x) = \begin{cases} x & \text{if } x \in [0, 1/2] \\ 1 - x & \text{if } x \in (1/2, 1] \end{cases}$$

is clearly a minimizer of (P) which is not C^1; it satisfies (E_w) but not (E).

3.4.3 Exercises

Exercise 3.4.1 **(i)** Show that the theorem remains valid if we weaken the hypothesis (H_3), for example, as follows. If $1 \le p < n$, replace (H_3) by: there exist $\beta > 0$,

$$1 \le s_1 \le \frac{np - n + p}{n - p}, \quad 1 \le s_2 \le \frac{np - n + p}{n}, \quad 1 \le s_3 \le \frac{n(p - 1)}{n - p}$$

so that the following hold, for every $(x, u, \xi) \in \overline{\Omega} \times \mathbb{R} \times \mathbb{R}^n$,

$$|f_u(x, u, \xi)| \le \beta \left(1 + |u|^{s_1} + |\xi|^{s_2}\right), \quad |f_\xi(x, u, \xi)| \le \beta \left(1 + |u|^{s_3} + |\xi|^{p-1}\right).$$

(ii) Find, with the help of Sobolev imbedding theorem, other ways of weakening (H_3) and keeping the conclusions of the theorem valid.

Exercise 3.4.2 Let

$$f(x, u, \xi) = \frac{1}{p} |\xi|^p + g(x, u).$$

Find growth conditions (depending on p and n) on g that improve (H_3) and still allow us to derive, as in the preceding exercise, (E_w).

Exercise 3.4.3 Let $\Omega \subset \mathbb{R}^n$ be a bounded open set with C^1 boundary, $h \in C^\infty(\overline{\Omega})$ and $\lambda > 0$. Consider the problem

$$\inf \left\{ I(u) = \int_\Omega \left[\frac{1}{2} |\nabla u|^2 - hu\right] + \frac{\lambda}{2} \int_{\partial \Omega} |u|^2 : u \in C^2(\overline{\Omega}) \right\}.$$

Assume that there exists a minimizer $\bar{u} \in C^2(\overline{\Omega})$. Find the Euler-Lagrange equation satisfied by $\bar{u}$.

Exercise 3.4.4 Let $n = 2$, $\Omega = (0, \pi)^2$, $u = u(x, t)$, $u_t = \partial u/\partial t$, $u_x = \partial u/\partial x$ and

$$(P) \quad \inf \left\{ I(u) = \frac{1}{2} \iint_\Omega \left(u_t^2 - u_x^2\right) \, dx \, dt : u \in W_0^{1,2}(\Omega) \right\} = m.$$

(i) Show that $m = -\infty$.

(ii) Prove, formally, that the Euler-Lagrange equation associated with (P) is the wave equation $u_{tt} - u_{xx} = 0$.

3.5   The vectorial case

The problem under consideration is

$$(P) \quad \inf \left\{ I(u) = \int_\Omega f(x, u(x), \nabla u(x))\, dx : u \in u_0 + W_0^{1,p}(\Omega; \mathbb{R}^N) \right\} = m$$

where $n, N > 1$ and

- $\Omega \subset \mathbb{R}^n$ is a bounded open set;
- $f : \overline{\Omega} \times \mathbb{R}^N \times \mathbb{R}^{N \times n} \to \mathbb{R}$, $f = f(x, u, \xi)$;
- $u = \left(u^1, \cdots, u^N\right) \in \mathbb{R}^N$, $\xi = \left(\xi_i^j\right)_{\substack{1 \le j \le N \\ 1 \le i \le n}} \in \mathbb{R}^{N \times n}$ and $\nabla u = \left(\frac{\partial u^j}{\partial x_i}\right)_{\substack{1 \le j \le N \\ 1 \le i \le n}}$;
- $u \in u_0 + W_0^{1,p}(\Omega; \mathbb{R}^N)$ means that $u^j, u_0^j \in W^{1,p}(\Omega)$, $j = 1, \cdots, N$, and $u - u_0 \in W_0^{1,p}(\Omega; \mathbb{R}^N)$ (which roughly means that $u = u_0$ on $\partial\Omega$).

All the results of the preceding sections apply to the present context when $n, N > 1$. However, while for $N = 1$ (or analogously when $n = 1$) Theorem 3.3 is almost optimal, it is now far from being so. The vectorial case is intrinsically more difficult. For example, the Euler-Lagrange equations associated with (P) are then a system of partial differential equations, whose treatment is considerably harder than that of a single partial differential equation.

3.5.1   The main theorem

We present one extension of Theorem 3.3; it is not the best possible result, but it has the advantage of giving some flavour of what can be done. For the sake of clarity we essentially consider only the case $n = N = 2$; but, in a remark, we briefly mention what can be done in the higher dimensional case.

Theorem 3.19 *Let $n = N = 2$ and $\Omega \subset \mathbb{R}^2$ be a bounded open set with Lipschitz boundary. Let*

$$f : \overline{\Omega} \times \mathbb{R}^2 \times \mathbb{R}^{2 \times 2} \to \mathbb{R}, \quad f = f(x, u, \xi),$$

$$F : \overline{\Omega} \times \mathbb{R}^2 \times \mathbb{R}^{2 \times 2} \times \mathbb{R} \to \mathbb{R}, \quad F = F(x, u, \xi, \delta),$$

be continuous and satisfying

$$f(x, u, \xi) = F(x, u, \xi, \det \xi), \quad \forall (x, u, \xi) \in \overline{\Omega} \times \mathbb{R}^2 \times \mathbb{R}^{2 \times 2}$$

where $\det \xi$ denotes the determinant of the matrix ξ. Assume also that

(H_1^{vect}) *$(\xi, \delta) \to F(x, u, \xi, \delta)$ is convex for every $(x, u) \in \overline{\Omega} \times \mathbb{R}^2$;*

(H_2^{vect}) *there exist $p > \max[q, 2]$ and $\alpha_1 > 0$, $\alpha_2, \alpha_3 \in \mathbb{R}$ such that*

$$F(x, u, \xi, \delta) \ge \alpha_1 |\xi|^p + \alpha_2 |u|^q + \alpha_3, \quad \forall (x, u, \xi, \delta) \in \overline{\Omega} \times \mathbb{R}^2 \times \mathbb{R}^{2 \times 2} \times \mathbb{R}.$$

Let $u_0 \in W^{1,p}(\Omega; \mathbb{R}^2)$ be such that $I(u_0) < \infty$, then (P) has at least one solution.

Remark 3.20 (i) It is clear that, from the point of view of convexity, the theorem is more general than Theorem 3.3. Indeed, if $\xi \to f(x, u, \xi)$ is convex then choose $F(x, u, \xi, \delta) = f(x, u, \xi)$ and therefore (H_1^{vect}) and (H_1) are equivalent. However, (H_1^{vect}) is more general since, for example, a function of the form

$$f(x, u, \xi) = |\xi|^4 + 16 \left(\det \xi\right)^2$$

is non-convex (see Exercise 3.5.1), while

$$F(x, u, \xi, \delta) = |\xi|^4 + 16\delta^2$$

is obviously convex as a function of (ξ, δ).

(**ii**) The theorem is, however, slightly weaker from the point of view of coercivity. Indeed, in (H_2^{vect}) we require $p > 2$, while in (H_2) of Theorem 3.3 we only ask that $p > 1$.

(**iii**) Similar statements and proofs hold for the general case $n, N > 1$. For example, when $n = N = 3$ we ask that there exists a function

$$F : \overline{\Omega} \times \mathbb{R}^3 \times \mathbb{R}^{3\times3} \times \mathbb{R}^{3\times3} \times \mathbb{R} \to \mathbb{R}, \quad F = F(x, u, \xi, \eta, \delta)$$

such that

$$f(x, u, \xi) = F(x, u, \xi, \mathrm{adj}_2 \xi, \det \xi), \quad \forall (x, u, \xi) \in \overline{\Omega} \times \mathbb{R}^3 \times \mathbb{R}^{3\times3}$$

where $\mathrm{adj}_2 \xi$ denotes the matrix of cofactors of ξ (i.e. all the 2×2 minors of the matrix ξ). The hypotheses are then the following:

(H_1^{vect}) $(\xi, \eta, \delta) \to F(x, u, \xi, \eta, \delta)$ is convex for every $(x, u) \in \overline{\Omega} \times \mathbb{R}^3$;

(H_2^{vect}) there exist $p > \max[q, 3]$ and $\alpha_1 > 0$, $\alpha_2, \alpha_3 \in \mathbb{R}$ such that, for every $(x, u, \xi, \eta, \delta) \in \overline{\Omega} \times \mathbb{R}^3 \times \mathbb{R}^{3\times3} \times \mathbb{R}^{3\times3} \times \mathbb{R}$,

$$F(x, u, \xi, \eta, \delta) \geq \alpha_1 |\xi|^p + \alpha_2 |u|^q + \alpha_3.$$

(**iv**) When $n, N > 1$ the function f should be of the form

$$f(x, u, \xi) = F(x, u, \xi, \mathrm{adj}_2 \xi, \mathrm{adj}_3 \xi, \cdots, \mathrm{adj}_s \xi)$$

where $s = \min[N, n]$ and $\mathrm{adj}_r \xi$ denotes the matrix of all $r \times r$ minors of the matrix $\xi \in \mathbb{R}^{N\times n}$. The hypotheses are then

(H_1^{vect}) $(\xi, \eta_2, \cdots, \eta_s) \to F(x, u, \xi, \eta_2, \cdots, \eta_s)$ is convex for every $(x, u) \in \overline{\Omega} \times \mathbb{R}^N$;

(H_2^{vect}) there exist $p > \max[q, s]$ and $\alpha_1 > 0$, $\alpha_2, \alpha_3 \in \mathbb{R}$ such that

$$F(x, u, \xi, \eta_2, \cdots, \eta_s) \geq \alpha_1 |\xi|^p + \alpha_2 |u|^q + \alpha_3$$

for every $(x, u, \xi, \eta_2, \cdots, \eta_s) \in \overline{\Omega} \times \mathbb{R}^N \times \mathbb{R}^{\tau(N,n)}$, where

$$\tau(N, n) = \sum_{r=1}^{\min[N,n]} \binom{N}{r} \binom{n}{r}.$$

(v) A function f that can be written in terms of a convex function F as in the theorem is called *polyconvex*. The theorem is due to Morrey (the terminology and important applications of such results to nonlinear elasticity can be found in Ball [7]). We refer for more details to [32].

Let us now see two examples.

Example 3.21 Let $n = N = 2$, $p > 2$ and

$$f(x, u, \xi) = f(\xi) = \frac{1}{p} |\xi|^p + h(\det \xi)$$

where $h : \mathbb{R} \to \mathbb{R}$ is non-negative and convex (for example $h(\det \xi) = (\det \xi)^2$). All hypotheses of the theorem are clearly satisfied. It is also interesting to compute the associated Euler-Lagrange equation. To make them simple, consider only the case $p = 2$ and set

$$u = u(x_1, x_2) = (u^1(x_1, x_2), u^2(x_1, x_2)).$$

The system is then

$$\begin{cases} \Delta u^1 + \left[h'(\det \nabla u) u_{x_2}^2\right]_{x_1} - \left[h'(\det \nabla u) u_{x_1}^2\right]_{x_2} = 0 \\ \Delta u^2 - \left[h'(\det \nabla u) u_{x_2}^1\right]_{x_1} + \left[h'(\det \nabla u) u_{x_1}^1\right]_{x_2} = 0. \end{cases}$$

Example 3.22 Another important example coming from applications is the following. Let $n = N = 3$, $p > 3$, $q \geq 1$ and

$$f(x, u, \xi) = f(\xi) = \alpha |\xi|^p + \beta |\text{adj}_2 \xi|^q + h(\det \xi)$$

where $h : \mathbb{R} \to \mathbb{R}$ is non-negative and convex and $\alpha, \beta > 0$.

3.5.2 Weak continuity of determinants

The key ingredient in the proof of the theorem is the following lemma that is due to Morrey and Reshetnyak.

Lemma 3.23 *Let $\Omega \subset \mathbb{R}^2$ be a bounded open set with Lipschitz boundary, $p > 2$ and*

$$u^\nu = (\varphi^\nu, \psi^\nu) \rightharpoonup u = (\varphi, \psi) \quad in \ W^{1,p}(\Omega; \mathbb{R}^2) \, ;$$

then

$$\det \nabla u^\nu \rightharpoonup \det \nabla u \quad in \ L^{p/2}(\Omega) \, .$$

Remark 3.24 (i) At first glance the result is a little surprising. Indeed, we have seen in Chapter 1 (in particular Exercise 1.3.3) that if two sequences, say $(\varphi^\nu)_{x_1}$ and $(\psi^\nu)_{x_2}$, converge weakly respectively to φ_{x_1} and ψ_{x_2}, then, in general, their product $(\varphi^\nu)_{x_1} (\psi^\nu)_{x_2}$ does not converge weakly to $\varphi_{x_1} \psi_{x_2}$. Writing

$$\det \nabla u^\nu = (\varphi^\nu)_{x_1} (\psi^\nu)_{x_2} - (\varphi^\nu)_{x_2} (\psi^\nu)_{x_1}$$

we see that both terms $(\varphi^\nu)_{x_1} (\psi^\nu)_{x_2}$ and $(\varphi^\nu)_{x_2} (\psi^\nu)_{x_1}$ do not, in general, converge weakly to $\varphi_{x_1} \psi_{x_2}$ and $\varphi_{x_2} \psi_{x_1}$ but, according to the lemma, their difference, which is $\det \nabla u^\nu$, converges weakly to their difference, namely $\det \nabla u$. We therefore have a nonlinear function, the determinant, that has the property of being *weakly continuous*. This is a very rare event (see for more details [31] or Theorem 8.20 in [32, 2nd edition]).

(ii) From Hölder inequality we see that whenever $p \geq 2$ and $u \in W^{1,p}$, then $\det \nabla u \in L^{p/2}$.

(iii) The lemma is false if $1 \leq p \leq 2$ but remains partially true if $p > 4/3$; this will be seen from the proof and from Exercise 3.5.5.

(iv) The lemma generalizes to the case where $n, N > 1$ and we obtain that any minor has this property (for example when $n = N = 3$, then any 2×2 minor and the determinant are weakly continuous). Moreover they are the only nonlinear functions that have the property of weak continuity.

Proof We have to show that for every $v \in (L^{p/2})' = L^{p/(p-2)}$

$$\lim_{\nu \to \infty} \iint_\Omega \det \nabla u^\nu \, v \, dx_1 dx_2 = \iint_\Omega \det \nabla u \, v \, dx_1 dx_2 \, . \tag{3.8}$$

The proof is divided into three steps. Only the first one carries the important information, namely that the determinant has a divergence structure; the last two steps are more technical. We also draw attention to a technical fact about the exponent p. The first step can also be proved if $p > 4/3$ (see also Exercise

3.5.5). The second, in fact, requires that $p \geq 2$ and only the last one fully uses the strict inequality $p > 2$. However, in order not to burden the proof too much, we always assume that $p > 2$.

Step 1. We first prove (3.8) under the further hypotheses that $v \in C_0^\infty(\Omega)$ and $u^\nu, u \in C^2(\overline{\Omega}; \mathbb{R}^2)$.

We start by proving a preliminary result. If we let $v \in C_0^\infty(\Omega)$ and $w \in C^2(\overline{\Omega}; \mathbb{R}^2)$, $w = (\varphi, \psi)$, we always have

$$\iint_\Omega \det \nabla w \, v = - \iint_\Omega [\varphi \psi_{x_2} v_{x_1} - \varphi \psi_{x_1} v_{x_2}]. \tag{3.9}$$

Indeed, using the fact that $\varphi, \psi \in C^2$, we obtain

$$\det \nabla w = \varphi_{x_1} \psi_{x_2} - \varphi_{x_2} \psi_{x_1} = (\varphi \psi_{x_2})_{x_1} - (\varphi \psi_{x_1})_{x_2} \tag{3.10}$$

and thus

$$\iint_\Omega \det \nabla w \, v = \iint_\Omega \left[(\varphi \psi_{x_2})_{x_1} v - (\varphi \psi_{x_1})_{x_2} v \right].$$

We therefore have (3.9) after integration by parts and since $v \in C_0^\infty(\Omega)$.

The result (3.8) then easily follows. Indeed, from Rellich theorem (Theorem 1.45) we have, since $\varphi^\nu \rightharpoonup \varphi$ in $W^{1,p}$ and $p > 2$, that $\varphi^\nu \to \varphi$ in L^∞. Combining this observation with the fact that

$$\psi_{x_1}^\nu, \psi_{x_2}^\nu \rightharpoonup \psi_{x_1}, \psi_{x_2} \quad \text{in } L^p$$

we deduce (see Exercise 1.3.3) that

$$\varphi^\nu \psi_{x_1}^\nu, \varphi^\nu \psi_{x_2}^\nu \rightharpoonup \varphi \psi_{x_1}, \varphi \psi_{x_2} \quad \text{in } L^p. \tag{3.11}$$

Since $v_{x_1}, v_{x_2} \in C_0^\infty \subset L^{p'}$, we deduce from (3.9), applied to $w = u^\nu$, and from (3.11) that

$$\lim_{\nu \to \infty} \iint_\Omega \det \nabla u^\nu \, v = - \iint_\Omega [\varphi \psi_{x_2} v_{x_1} - \varphi \psi_{x_1} v_{x_2}].$$

Using again (3.9), applied to $w = u$, we have indeed obtained the claimed result (3.8).

Step 2. We now show that (3.8) still holds under the further hypothesis $v \in C_0^\infty(\Omega)$, but considering now the general case, i.e. $u^\nu, u \in W^{1,p}(\Omega; \mathbb{R}^2)$.

In fact (3.9) continues to hold under the weaker hypothesis that $v \in C_0^\infty(\Omega)$ and $w \in W^{1,p}(\Omega; \mathbb{R}^2)$; of course, the proof must be different, since this time we only know that $w \in W^{1,p}(\Omega; \mathbb{R}^2)$. Let us postpone for a moment the proof of this fact and observe that if (3.9) holds for $w \in W^{1,p}(\Omega; \mathbb{R}^2)$ then, with exactly the same argument as in the previous step, we get (3.8) under the hypotheses $v \in C_0^\infty (\Omega)$ and $u^\nu, u \in W^{1,p} (\Omega; \mathbb{R}^2)$.

We now prove the above claim and we start by regularizing $w \in W^{1,p} (\Omega; \mathbb{R}^2)$ appealing to Theorem 1.34. We therefore find for every $\epsilon > 0$ a function $w^\epsilon = (\varphi^\epsilon, \psi^\epsilon) \in C^2 (\overline{\Omega}; \mathbb{R}^2)$ so that

$$\|w - w^\epsilon\|_{W^{1,p}} \le \epsilon \quad \text{and} \quad \|w - w^\epsilon\|_{L^\infty} \le \epsilon.$$

Since $p \ge 2$ we can find (see Exercise 3.5.4) a constant α_1 (independent of ϵ but depending on w) so that

$$\|\det \nabla w - \det \nabla w^\epsilon\|_{L^{p/2}} \le \alpha_1 \epsilon. \tag{3.12}$$

It is also easy to see that we have, for α_2 a constant (independent of ϵ but depending on w),

$$\left\|\varphi \psi_{x_2} - \varphi^\epsilon \psi^\epsilon_{x_2}\right\|_{L^p} \le \alpha_2 \epsilon, \quad \left\|\varphi \psi_{x_1} - \varphi^\epsilon \psi^\epsilon_{x_1}\right\|_{L^p} \le \alpha_2 \epsilon \tag{3.13}$$

since, for example, the first inequality follows from

$$\left\|\varphi \psi_{x_2} - \varphi^\epsilon \psi^\epsilon_{x_2}\right\|_{L^p} \le \|\varphi\|_{L^\infty} \left\|\psi_{x_2} - \psi^\epsilon_{x_2}\right\|_{L^p} + \left\|\psi^\epsilon_{x_2}\right\|_{L^p} \|\varphi - \varphi^\epsilon\|_{L^\infty}.$$

Returning to (3.9) we have

$$\iint_\Omega \det \nabla w \, v + \iint_\Omega [\varphi \psi_{x_2} v_{x_1} - \varphi \psi_{x_1} v_{x_2}]$$
$$= \iint_\Omega \det \nabla w^\epsilon \, v + \iint_\Omega \left[\varphi^\epsilon \psi^\epsilon_{x_2} v_{x_1} - \varphi^\epsilon \psi^\epsilon_{x_1} v_{x_2}\right]$$
$$+ \iint_\Omega (\det \nabla w - \det \nabla w^\epsilon) \, v$$
$$+ \iint_\Omega \left[\left(\varphi \psi_{x_2} - \varphi^\epsilon \psi^\epsilon_{x_2}\right) v_{x_1} - \left(\varphi \psi_{x_1} - \varphi^\epsilon \psi^\epsilon_{x_1}\right) v_{x_2}\right].$$

Appealing to (3.9), which has already been proved to hold for $w^\epsilon = (\varphi^\epsilon, \psi^\epsilon) \in C^2$, to Hölder inequality, to (3.12) and to (3.13) we find that, α_3 being a constant independent of ϵ but depending on w,

$$\left|\iint_\Omega \det \nabla w \, v + \iint_\Omega [\varphi \psi_{x_2} v_{x_1} - \varphi \psi_{x_1} v_{x_2}]\right|$$
$$\le \alpha_3 \epsilon \left[\|v\|_{(L^{p/2})'} + \|v_{x_1}\|_{L^{p'}} + \|v_{x_2}\|_{L^{p'}}\right].$$

Since ϵ is arbitrary we have indeed obtained that (3.9) is also valid for $w \in W^{1,p} (\Omega; \mathbb{R}^2)$.

Step 3. We are finally in a position to prove the lemma, removing the last unnecessary hypothesis $(v \in C_0^\infty(\Omega))$. We want (3.8) to hold for $v \in L^{p/(p-2)}$. This is obtained by regularizing the function as in Theorem 1.13. This means, for every $\epsilon > 0$ and $v \in L^{p/(p-2)}$, that we can find $v^\epsilon \in C_0^\infty(\Omega)$ so that

$$\|v - v^\epsilon\|_{L^{p/(p-2)}} \leq \epsilon. \tag{3.14}$$

We moreover have

$$\iint_\Omega \det \nabla u^\nu \, v = \iint_\Omega \det \nabla u^\nu \, (v - v^\epsilon) + \iint_\Omega \det \nabla u^\nu \, v^\epsilon.$$

Using, once more, Hölder inequality we find

$$\left| \iint_\Omega (\det \nabla u^\nu - \det \nabla u) \, v \right|$$

$$\leq \|v - v^\epsilon\|_{L^{p/(p-2)}} \|\det \nabla u^\nu - \det \nabla u\|_{L^{p/2}} + \left| \iint_\Omega (\det \nabla u^\nu - \det \nabla u) \, v^\epsilon \right|.$$

The previous step has shown that

$$\lim_{\nu \to \infty} \left| \iint_\Omega (\det \nabla u^\nu - \det \nabla u) \, v^\epsilon \right| = 0$$

while (3.14), the fact that $u^\nu \rightharpoonup u$ in $W^{1,p}$ and Exercise 3.5.4 show that we can find $\gamma > 0$ so that

$$\|v - v^\epsilon\|_{L^{p/(p-2)}} \|\det \nabla u^\nu - \det \nabla u\|_{L^{p/2}} \leq \gamma \epsilon.$$

Since ϵ is arbitrary we have indeed obtained that (3.8) holds for $v \in L^{p/(p-2)}$ and for $u^\nu \rightharpoonup u$ in $W^{1,p}$. The lemma is therefore proved. ■

3.5.3   Proof of the main theorem

We can now proceed with the proof of Theorem 3.19.

Proof We prove the theorem under the further hypotheses (for a general proof see Theorem 8.31 in [32, 2nd edition])

$$f(x, u, \xi) = g(x, u, \xi) + h(x, \det \xi)$$

where g satisfies (H_1) and (H_2), with $p > 2$, of Theorem 3.3 and $h \in C^1(\overline{\Omega} \times \mathbb{R})$, $h \geq 0$, $\delta \to h(x, \delta)$ is convex for every $x \in \overline{\Omega}$ and there exists $\gamma > 0$ so that

$$|h_\delta(x, \delta)| \leq \gamma \left(1 + |\delta|^{(p-2)/2}\right). \tag{3.15}$$

The proof is then identical to that of Theorem 3.3, except the second step (the weak lower semicontinuity), which we discuss now. We have to prove that

$$u_\nu \rightharpoonup \overline{u} \text{ in } W^{1,p} \quad \Rightarrow \quad \liminf_{\nu \to \infty} I(u_\nu) \geq I(\overline{u})$$

where $I(u) = G(u) + H(u)$ with

$$G(u) = \int_\Omega g(x, u(x), \nabla u(x)) \, dx \quad \text{and} \quad H(u) = \int_\Omega h(x, \det \nabla u(x)) \, dx.$$

We have already proved in Theorem 3.3 that

$$\liminf_{\nu \to \infty} G(u_\nu) \geq G(\overline{u})$$

and therefore the result follows if we can show

$$\liminf_{\nu \to \infty} H(u_\nu) \geq H(\overline{u}).$$

Since h is convex and C^1 we have

$$h(x, \det \nabla u_\nu) \geq h(x, \det \nabla \overline{u}) + h_\delta(x, \det \nabla \overline{u})(\det \nabla u_\nu - \det \nabla \overline{u}). \quad (3.16)$$

We know that $\overline{u} \in W^{1,p}(\Omega; \mathbb{R}^2)$, which implies that $\det \nabla \overline{u} \in L^{p/2}(\Omega)$, and hence using (3.15) we deduce that

$$h_\delta(x, \det \nabla \overline{u}) \in L^{p/(p-2)} = L^{(p/2)'}, \quad (3.17)$$

since we can find a constant $\gamma_1 > 0$ so that

$$|h_\delta(x, \det \nabla \overline{u})|^{p/(p-2)} \leq \left[\gamma\left(1 + |\det \nabla \overline{u}|^{(p-2)/2}\right)\right]^{p/(p-2)}$$

$$\leq \gamma_1\left(1 + |\det \nabla \overline{u}|^{p/2}\right).$$

Returning to (3.16) and integrating we get

$$H(u_\nu) \geq H(\overline{u}) + \int_\Omega h_\delta(x, \det \nabla \overline{u})(\det \nabla u_\nu - \det \nabla \overline{u}) \, dx.$$

Since $u_\nu - \overline{u} \rightharpoonup 0$ in $W^{1,p}$, $p > 2$, we have from Lemma 3.23 that

$$\det \nabla u_\nu - \det \nabla \overline{u} \rightharpoonup 0 \quad \text{in } L^{p/2},$$

which, combined with (3.17) and the definition of weak convergence in $L^{p/2}$, leads to

$$\lim_{\nu \to \infty} \int_\Omega h_\delta(x, \det \nabla \overline{u})(\det \nabla u_\nu - \det \nabla \overline{u}) \, dx = 0.$$

We have therefore obtained that

$$\liminf_{\nu \to \infty} H(u_\nu) \geq H(\overline{u})$$

and the proof is complete. ∎

3.5.4   Exercises

The exercises focus on several important analytical properties of the determinant. Although we essentially deal only with the two dimensional case, most results, when properly adapted, remain valid in the higher dimensional cases.

We need in some of the exercises below the following definition.

Definition 3.25 *Let $\Omega \subset \mathbb{R}^n$ be an open set and $u_\nu, u \in L^1_{loc}(\Omega)$. We say that u_ν converges in the sense of distributions to u, and we denote it by $u_\nu \rightharpoonup u$ in $\mathcal{D}'(\Omega)$, if*

$$\lim_{\nu \to \infty} \int_\Omega u_\nu \, \varphi = \int_\Omega u \, \varphi, \quad \forall \varphi \in C_0^\infty(\Omega).$$

Remark 3.26 (i) If Ω is bounded we then have the following relations

$$u_\nu \overset{*}{\rightharpoonup} u \text{ in } L^\infty \quad \Rightarrow \quad u_\nu \rightharpoonup u \text{ in } L^1 \quad \Rightarrow \quad u_\nu \rightharpoonup u \text{ in } \mathcal{D}'.$$

(ii) The definition can be generalized to u_ν and u that are not necessarily in $L^1_{loc}(\Omega)$, but are merely what are known as "distributions", see Exercise 3.5.6.

Exercise 3.5.1 Let $\xi \in \mathbb{R}^{2 \times 2}$. Prove that the functions

$$f_1(\xi) = (\det \xi)^2 \quad \text{and} \quad f_2(\xi) = |\xi|^4 + 16(\det \xi)^2$$

are not convex.

Exercise 3.5.2 Let $\Omega \subset \mathbb{R}^2$ be a bounded open set with Lipschitz boundary, $p \geq 2$ and $u \in v + W_0^{1,p}(\Omega; \mathbb{R}^2)$.

(i) Prove that

$$\iint_\Omega \det \nabla u(x) \, dx = \iint_\Omega \det \nabla v(x) \, dx.$$

(ii) Assume, in addition, that $p > 2$. Letting $g \in C^0(\mathbb{R}^2)$, show that

$$\iint_\Omega g(u(x)) \det \nabla u(x) \, dx = \iint_\Omega g(v(x)) \det \nabla v(x) \, dx.$$

Suggestion: Prove first the results for $u, v \in C^2(\overline{\Omega}; \mathbb{R}^2)$ with $u = v$ on $\partial\Omega$.

Exercise 3.5.3 Let $\Omega \subset \mathbb{R}^2$ be a bounded open set with Lipschitz boundary, $u_0 \in W^{1,p}(\Omega; \mathbb{R}^2)$, with $p \geq 2$, and

$$(P) \quad \inf \left\{ I(u) = \iint_\Omega \det \nabla u(x) \, dx : u \in u_0 + W_0^{1,p}(\Omega; \mathbb{R}^2) \right\} = m.$$

Write the Euler-Lagrange equation associated with (P). Is the result totally surprising?

Exercise 3.5.4 Let $u, v \in W^{1,p}\left(\Omega; \mathbb{R}^2\right)$, with $p \geq 2$. Show that there exists $\gamma > 0$ (depending only on p) so that

$$\|\det \nabla u - \det \nabla v\|_{L^{p/2}} \leq \gamma \left(\|\nabla u\|_{L^p} + \|\nabla v\|_{L^p}\right) \|\nabla u - \nabla v\|_{L^p} .$$

Exercise 3.5.5 Let $\Omega \subset \mathbb{R}^2$ be a bounded open set with Lipschitz boundary. We have seen in Lemma 3.23 that, if $p > 2$, then

$$u^\nu \rightharpoonup u \text{ in } W^{1,p}\left(\Omega; \mathbb{R}^2\right) \quad \Rightarrow \quad \det \nabla u^\nu \rightharpoonup \det \nabla u \text{ in } L^{p/2}\left(\Omega\right).$$

(i) Show that the result is, in general, false if $p = 2$. To achieve this goal choose, for example, $\Omega = (0,1)^2$ and

$$u^\nu\left(x_1, x_2\right) = \frac{1}{\sqrt{\nu}}\left(1 - x_2\right)^\nu \left(\sin\left(\nu x_1\right), \cos\left(\nu x_1\right)\right).$$

(ii) Show, using Rellich theorem (Theorem 1.45), that if $u^\nu, u \in C^2\left(\overline{\Omega}; \mathbb{R}^2\right)$ and if $p > 4/3$ (so in particular for $p = 2$), then

$$u^\nu \rightharpoonup u \text{ in } W^{1,p}\left(\Omega; \mathbb{R}^2\right) \quad \Rightarrow \quad \det \nabla u^\nu \rightharpoonup \det \nabla u \text{ in } \mathcal{D}'\left(\Omega\right).$$

(iii) This last result is false if $p \leq 4/3$, see Dacorogna-Murat [35].

Exercise 3.5.6 Let $\Omega = \left\{x \in \mathbb{R}^2 : |x| < 1\right\}$ and $u\left(x\right) = x/\left|x\right|$.

(i) Recall (see Example 1.33 and Exercise 1.4.1) that $u \in W^{1,p}\left(\Omega; \mathbb{R}^2\right)$ for every $1 \leq p < 2$ (observe, however, that $u \notin W^{1,2}$ and $u \notin C^0$).

(ii) Let

$$u^\nu\left(x\right) = \frac{x}{|x| + 1/\nu} .$$

Show that $u^\nu \rightharpoonup u$ in $W^{1,p}$, for any $1 \leq p < 2$.

(iii) Let $\delta_{(0,0)}$ be the Dirac mass at $(0,0)$, which means

$$\left\langle \delta_{(0,0)}; \varphi \right\rangle = \varphi\left(0,0\right), \quad \forall \varphi \in C_0^\infty\left(\Omega\right).$$

Prove that

$$\det \nabla u^\nu \rightharpoonup \pi \delta_{(0,0)} \quad \text{in } \mathcal{D}'\left(\Omega\right).$$

Exercise 3.5.7 Let $f : \mathbb{R}^{2\times2} \to \mathbb{R}$ be polyconvex, meaning that there exists

$$F : \mathbb{R}^{2\times2} \times \mathbb{R} \to \mathbb{R}, \quad F = F\left(\xi, \delta\right),$$

convex and

$$f\left(\xi\right) = F\left(\xi, \det \xi\right), \quad \forall \xi \in \mathbb{R}^{2\times2}.$$

(i) Let $\Omega \subset \mathbb{R}^2$ be a bounded open set with Lipschitz boundary. Prove that f is *quasiconvex*, meaning that

$$\int_\Omega f\left(\xi + \nabla\varphi\left(x\right)\right) dx \geq f\left(\xi\right) \operatorname{meas} \Omega$$

for every $\xi \in \mathbb{R}^{2\times 2}$ and every $\varphi \in W_0^{1,\infty}\left(\Omega; \mathbb{R}^2\right)$.

(ii) Show that f is *rank one convex*, meaning that $\psi : \mathbb{R} \to \mathbb{R}$ defined by

$$\psi\left(t\right) = f\left(\xi + t\lambda\right)$$

is convex for every $\xi \in \mathbb{R}^{2\times 2}$ and every $\lambda \in \mathbb{R}^{2\times 2}$ with $\det\lambda = 0$.

(iii) Prove that if $f \in C^2\left(\mathbb{R}^{2\times 2}\right)$, then it satisfies the *Legendre-Hadamard condition*, namely

$$\sum_{i,j,\alpha,\beta=1}^{2} \frac{\partial^2 f\left(\xi\right)}{\partial\xi_i^\alpha \partial\xi_j^\beta} a_i a_j b^\alpha b^\beta \geq 0$$

for every $a = \left(a_1, a_2\right) \in \mathbb{R}^2$ and every $b = \left(b^1, b^2\right) \in \mathbb{R}^2$.

3.6 Relaxation theory

3.6.1 The relaxation theorem

Recall that the problem under consideration is

$$(P) \quad \inf\left\{I\left(u\right) = \int_\Omega f\left(x, u\left(x\right), \nabla u\left(x\right)\right) dx : u \in u_0 + W_0^{1,p}\left(\Omega\right)\right\} = m$$

where

- $\Omega \subset \mathbb{R}^n$ is a bounded open set with Lipschitz boundary;
- $f : \overline{\Omega} \times \mathbb{R} \times \mathbb{R}^n \to \mathbb{R}$, $f = f\left(x, u, \xi\right)$, is continuous and non-negative;
- $u_0 \in W^{1,p}\left(\Omega\right)$ with $I\left(u_0\right) < \infty$.

Before stating the main theorem, let us recall some facts from Section 1.5.

Remark 3.27 The convex envelope of f, with respect to the variable ξ, is denoted by f^{**}. It is the largest convex function (with respect to the variable ξ) that is smaller than f. In other words

$$g\left(x, u, \xi\right) \leq f^{**}\left(x, u, \xi\right) \leq f\left(x, u, \xi\right), \quad \forall\left(x, u, \xi\right) \in \overline{\Omega} \times \mathbb{R} \times \mathbb{R}^n$$

for every convex function g (more precisely, $\xi \to g\left(x, u, \xi\right)$ is convex), $g \leq f$. We have two ways of computing this function.

(i) From the duality theorem (Theorem 1.58) we have, for every $(x, u, \xi) \in \overline{\Omega} \times \mathbb{R} \times \mathbb{R}^n$,

$$f^*(x, u, \xi^*) = \sup_{\xi \in \mathbb{R}^n} \{\langle \xi; \xi^* \rangle - f(x, u, \xi)\}$$

$$f^{**}(x, u, \xi) = \sup_{\xi^* \in \mathbb{R}^n} \{\langle \xi; \xi^* \rangle - f^*(x, u, \xi^*)\}.$$

(ii) From Carathéodory theorem (Theorem 1.59) we have, for every $(x, u, \xi) \in \overline{\Omega} \times \mathbb{R} \times \mathbb{R}^n$,

$$f^{**}(x, u, \xi) = \inf \left\{ \sum_{i=1}^{n+1} \lambda_i f(x, u, \xi_i) : \xi = \sum_{i=1}^{n+1} \lambda_i \xi_i, \ \lambda_i \geq 0 \text{ and } \sum_{i=1}^{n+1} \lambda_i = 1 \right\}.$$

(iii) In general, the function $f^{**} : \overline{\Omega} \times \mathbb{R} \times \mathbb{R}^n \to \mathbb{R}$ is, however, not continuous in the variables (x, u) (see Exercise 3.6.7); of course, since f^{**} is convex in the variable ξ, it is continuous with respect to this variable. If the function f has the following structure

$$f(x, u, \xi) = f_1(x, u) f_2(\xi) + f_3(x, u)$$

with $f_1, f_3 : \overline{\Omega} \times \mathbb{R} \to \mathbb{R}$ continuous and $f_1 \geq 0$, then

$$f^{**}(x, u, \xi) = f_1(x, u) f_2^{**}(\xi) + f_3(x, u)$$

and thus $f^{**} : \overline{\Omega} \times \mathbb{R} \times \mathbb{R}^n \to \mathbb{R}$ is continuous. For more general considerations on these matters, see Marcellini-Sbordone [82] and Proposition 9.5 in [32, 2nd edition].

We have seen in Theorem 3.3 that the existence of minimizers of (P) depends strongly on the two hypotheses (H_1) and (H_2). We now briefly discuss the case where (H_1) does not hold, i.e. the function $\xi \to f(x, u, \xi)$ is no longer convex. We have seen in several examples that in general (P) has no minimizers. We present here a way of defining a "generalized" solution of (P). The main theorem (without proof) is the following.

Theorem 3.28 *Let Ω, f and u_0 be as above. Assume that f^{**} is continuous. Let $p \geq 1$ and α_1 be such that*

$$0 \leq f(x, u, \xi) \leq \alpha_1 (1 + |u|^p + |\xi|^p), \quad \forall (x, u, \xi) \in \overline{\Omega} \times \mathbb{R} \times \mathbb{R}^n.$$

Finally let

$$(\overline{P}) \quad \inf \left\{ \overline{I}(u) = \int_{\Omega} f^{**}(x, u(x), \nabla u(x)) \, dx : u \in u_0 + W_0^{1,p}(\Omega) \right\} = \overline{m}.$$

Then

(i) $\overline{m} = m$ *and for every* $u \in u_0 + W_0^{1,p}(\Omega)$, *there exists* $u_\nu \in u_0 + W_0^{1,p}(\Omega)$ *so that*

$$u_\nu \to u \ \text{in } L^p \quad \text{and} \quad I(u_\nu) \to \overline{I}(u), \ \text{as } \nu \to \infty.$$

(ii) *If, in addition,* $p > 1$ *and there exist* $\alpha_2 > 0$, $\alpha_3 \in \mathbb{R}$ *such that*

$$f(x, u, \xi) \geq \alpha_2 |\xi|^p + \alpha_3, \quad \forall (x, u, \xi) \in \overline{\Omega} \times \mathbb{R} \times \mathbb{R}^n$$

then $(\overline{P})$ *has at least one minimizer* $\overline{u} \in u_0 + W_0^{1,p}(\Omega)$ *and the sequence can be chosen so that*

$$u_\nu \rightharpoonup \overline{u} \quad \text{in } W^{1,p}.$$

Remark 3.29 (i) Theorem 3.28 (ii) allows one to define $\overline{u}$ as a generalized solution of (P), even though (P) may have no minimizer in $W^{1,p}$.

(ii) The theorem has been established by L.C. Young [112], MacShane and as stated by Ekeland (see Theorem 10.3.7 in Ekeland-Temam [45], Corollary 3.13 in Marcellini-Sbordone [82] or [32, 2nd edition]). It is false in the vectorial case (see Example 3.31 below). However the present author in [30] (see Theorems 9.1 and 9.8 in [32, 2nd edition]) has shown that a result in the same spirit can be proved.

(iii) We should emphasize that, in general (see Exercise 3.6.6), unless f satisfies some coercivity condition as in Statement (ii) of the theorem, the convergence of the sequence $\{u_\nu\}$ is of the type $u_\nu \to u$ in L^p and not $u_\nu \rightharpoonup \overline{u}$ in $W^{1,p}$.

3.6.2   Some examples

We conclude this section with two examples.

Example 3.30 Let us return to Bolza example (Example 3.10). Here we have $n = 1$,

$$f(x, u, \xi) = f(u, \xi) = \left(\xi^2 - 1\right)^2 + u^4$$

$$(P) \quad \inf \left\{ I(u) = \int_0^1 f(u(x), u'(x)) \, dx : u \in W_0^{1,4}(0, 1) \right\} = m.$$

We have already shown that $m = 0$ and that (P) has no solution. An elementary computation (see Example 1.57 (ii)) shows that

$$f^{**}(u, \xi) = \begin{cases} f(u, \xi) & \text{if } |\xi| \geq 1 \\ u^4 & \text{if } |\xi| < 1. \end{cases}$$

Therefore $\overline{u} \equiv 0$ is a solution of

$$(\overline{P}) \quad \inf\left\{ \overline{I}(u) = \int_0^1 f^{**}(u(x), u'(x))\, dx : u \in W_0^{1,4}(0,1) \right\} = \overline{m} = 0.$$

The sequence $u_\nu \in W_0^{1,4}$ ($\nu \geq 2$ being an integer) constructed in Example 3.10 satisfies the conclusions of the theorem, i.e.

$$u_\nu \rightharpoonup \overline{u} \text{ in } W^{1,4} \quad \text{and} \quad I(u_\nu) \to \overline{I}(\overline{u}) = 0, \text{ as } \nu \to \infty.$$

Example 3.31 Let $\Omega \subset \mathbb{R}^2$ be a bounded open set with Lipschitz boundary. Let $u_0 \in W^{1,4}(\Omega; \mathbb{R}^2)$ be such that

$$\iint_\Omega \det \nabla u_0(x)\, dx \neq 0.$$

Let, for $\xi \in \mathbb{R}^{2\times 2}$, $f(\xi) = (\det \xi)^2$,

$$(P) \quad \inf\left\{ I(u) = \iint_\Omega f(\nabla u(x))\, dx : u \in u_0 + W_0^{1,4}(\Omega; \mathbb{R}^2) \right\} = m$$

$$(\overline{P}) \quad \inf\left\{ \overline{I}(u) = \iint_\Omega f^{**}(\nabla u(x))\, dx : u \in u_0 + W_0^{1,4}(\Omega; \mathbb{R}^2) \right\} = \overline{m}.$$

Let us show that the conclusion of Theorem 3.28 is false in this context, by proving that $m > \overline{m}$. Indeed, it is easy to prove (see Exercise 1.5.6) that $f^{**}(\xi) \equiv 0$, which therefore implies that $\overline{m} = 0$. Let us show that $m > 0$. By Jensen inequality (see Theorem 1.54) we have, for every $u \in u_0 + W_0^{1,4}(\Omega; \mathbb{R}^2)$,

$$\iint_\Omega (\det \nabla u(x))^2\, dx \geq \text{meas}\, \Omega \left(\frac{1}{\text{meas}\, \Omega} \iint_\Omega \det \nabla u(x)\, dx \right)^2.$$

Appealing to Exercise 3.5.2 we therefore find that

$$\iint_\Omega (\det \nabla u(x))^2\, dx \geq \frac{1}{\text{meas}\, \Omega} \left(\iint_\Omega \det \nabla u_0(x)\, dx \right)^2.$$

The right-hand side being strictly positive, by hypothesis, we have indeed found that $m > 0$, which leads to the claimed counterexample.

3.6.3 Exercises

Exercise 3.6.1 Let $n = 1$ and

$$(P) \quad \inf\left\{ I(u) = \int_0^1 f(u'(x))\, dx \right\} = m$$

where $X = \{u \in W^{1,\infty}(0,1) : u(0) = \alpha,\ u(1) = \beta\}$.

(i) Assume that there exist $\lambda \in [0,1]$, $a, b \in \mathbb{R}$ such that

$$\begin{cases} \beta - \alpha = \lambda a + (1 - \lambda) b \\ f^{**}(\beta - \alpha) = \lambda f(a) + (1 - \lambda) f(b). \end{cases}$$

Show then that (P) has a solution, independently of whether f is convex or not (compare the above relations with Theorem 1.59). Of course, if f is convex, the above hypothesis is always true, it suffices to choose $\lambda = 1/2$ and $a = b = \beta - \alpha$.

(ii) Can we apply the above considerations to $f(\xi) = e^{-\xi^2}$ (see Section 2.2) and $\alpha = \beta = 0$?

(iii) What happens when $f(\xi) = \left(\xi^2 - 1\right)^2$?

Exercise 3.6.2 Let $\Omega \subset \mathbb{R}^n$ be a bounded open set with Lipschitz boundary, $f : \mathbb{R}^n \to \mathbb{R}$ be continuous, $f(x, u, \xi) = f(\xi)$, and such that

$$0 \le f(\xi) \le \alpha \left(1 + |\xi|^p\right)$$

for $\alpha > 0$ and $p \ge 1$. Let $u_0(x) = \langle \xi_0; x \rangle + a_0$, where $\xi_0 \in \mathbb{R}^n$ and $a_0 \in \mathbb{R}$. Show that

$$(P) \quad \inf \left\{ I(u) = \int_\Omega f(\nabla u(x))\, dx : u \in u_0 + W_0^{1,p}(\Omega) \right\} = f^{**}(\xi_0)\, \text{meas}\,\Omega.$$

Exercise 3.6.3 Prove Theorem 3.28 (ii).

Exercise 3.6.4 Let $\Omega \subset \mathbb{R}^2$ be a bounded open set with Lipschitz boundary,

$$f(x, u, \xi) = f(\xi) = \left((\xi_1)^2 - 1\right)^2 + (\xi_2)^4$$

where $\xi = (\xi_1, \xi_2) \in \mathbb{R}^2$ and

$$(P) \quad \inf \left\{ I(u) = \iint_\Omega f(\nabla u(x_1, x_2))\, dx_1 dx_2 : u \in W_0^{1,4}(\Omega) \right\} = m.$$

Evaluate m and show, with the help of Exercise 1.4.15, that (P) has no solution.

Exercise 3.6.5 Let

$$\Omega = \left\{ (x_1, x_2) \in \mathbb{R}^2 : |x_1 - x_2|, |x_1 + x_2| < 1 \right\}$$

$$f(x, u, \xi) = f(\xi) = \left((\xi_1)^2 - 1\right)^2 + \left((\xi_2)^2 - 1\right)^2$$

where $\xi = (\xi_1, \xi_2) \in \mathbb{R}^2$ and

$$(P) \quad \inf \left\{ I(u) = \iint_\Omega f(\nabla u(x_1, x_2)) \, dx_1 dx_2 : u \in W_0^{1,4}(\Omega) \right\} = m.$$

Compute m and prove that

$$\overline{u}(x_1, x_2) = 1 - \max \{|x_1 - x_2|, |x_1 + x_2|\}$$

is a solution of (P).

Exercise 3.6.6 Let $f(\xi) = e^{-|\xi|}$ and

$$(P) \quad \inf \left\{ I(u) = \int_0^1 f(u'(x)) \, dx : u \in W_0^{1,1}(0, 1) \right\} = m.$$

Prove, with the help of Exercise 3.6.2, that $m = 0$. Show that there is no sequence $u_\nu \in W_0^{1,1}(\Omega)$ so that

$$u_\nu \rightharpoonup u = 0 \text{ in } W^{1,1} \quad \text{and} \quad I(u_\nu) \to m = \overline{I}(0) = 0, \text{ as } \nu \to \infty$$

although there is a sequence satisfying

$$u_\nu \to u = 0 \text{ in } L^\infty \quad \text{and} \quad I(u_\nu) \to m = \overline{I}(0) = 0, \text{ as } \nu \to \infty.$$

Exercise 3.6.7 Let $\Omega \subset \mathbb{R}^n$ be an open set, $f : \overline{\Omega} \times \mathbb{R} \times \mathbb{R}^n \to \mathbb{R}$, $f = f(x, u, \xi)$, be non-negative and continuous.

(i) Prove that, if f^{**} is the convex envelope with respect to the variable ξ (as in Remark 3.27), then

$$(x, u) \to f^{**}(x, u, \xi)$$

is upper semicontinuous.

(ii) Let $n = 1$ and consider the function

$$f(u, \xi) = (|\xi| + 1)^{|u|}.$$

Show that $u \to f^{**}(u, \xi)$ is not continuous, where f^{**} is the convex envelope with respect to the variable ξ.

Chapter 4

Direct methods: regularity

4.1 Introduction

We are still considering the problem

$$(P) \quad \inf\left\{ I(u) = \int_\Omega f(x, u(x), \nabla u(x))\, dx : u \in u_0 + W_0^{1,p}(\Omega) \right\} = m$$

where

- $\Omega \subset \mathbb{R}^n$ is a bounded open set;
- $f : \overline{\Omega} \times \mathbb{R} \times \mathbb{R}^n \to \mathbb{R}$, $f = f(x, u, \xi)$;
- $u \in u_0 + W_0^{1,p}(\Omega)$ means that $u, u_0 \in W^{1,p}(\Omega)$ and $u - u_0 \in W_0^{1,p}(\Omega)$.

We have shown in Chapter 3 that, under appropriate hypotheses on f, u_0 and Ω, (P) has a minimizer $\overline{u} \in u_0 + W_0^{1,p}(\Omega)$.

We now discuss whether, in fact, the minimizer $\overline{u}$ is not more *regular*, for example $C^1(\overline{\Omega})$. More precisely, if the data f, u_0 and Ω are sufficiently regular, say C^∞, does $\overline{u} \in C^\infty$? This is one of the 23 problems of Hilbert that were mentioned in Chapter 0.

In Section 4.2 we discuss the case $n = 1$. We obtain some general results and give some counterexamples.

We then turn our attention to the higher dimensional case, which is discussed in Sections 4.3 to 4.7. This is a considerably harder problem and we want to explain why this is so using the example

$$f(x, u, \xi) = \frac{1}{2} |\xi|^2 - h(x)\, u.$$

The Euler-Lagrange equation is then $-\Delta \overline{u} = h$. When $n = 1$, the assumption $h \in C^{k,\alpha}$ or $W^{k,p}$ with $k \geq 0$ and $0 \leq \alpha \leq 1$ or $1 \leq p \leq \infty$, immediately implies $\bar{u} \in C^{k+2,\alpha}$ or $W^{k+2,p}$. When $n \geq 2$, the equation $-\Delta \bar{u} = h$ states that the trace of the Hessian matrix is in $C^{k,\alpha}$ or $W^{k,p}$ and the regularity that we want to establish is that *each* entry of the Hessian matrix is in $C^{k,\alpha}$ or $W^{k,p}$. Presented like this, the result seems unlikely. It is one of the achievements of regularity theory to prove that this is indeed the case, provided $0 < \alpha < 1$ or $1 < p < \infty$; the limit cases $\alpha = 0, 1$ or $p = 1, \infty$ turning out to be false, as seen in the exercises.

We now detail the contents of Sections 4.3 to 4.7. We proceed in several steps and restrict our attention to the case $p = 2$. The first step is to obtain one degree smoother solutions, namely to show that $\bar{u} \in W^{2,2}$. This is achieved in Section 4.3, where we obtain *interior regularity*. In Section 4.4, under some smoothness of the boundary of the set Ω, we get *regularity up to the boundary*. The main tool in both cases is the so-called *difference quotient method*.

To get even smoother solutions, we have to restrict our attention to the Dirichlet integral or more generally to quadratic integrands. This is dealt with in Section 4.5. In Section 4.6, we give a completely different proof of the regularity of solutions in the case of Dirichlet integral.

In Section 4.7 we provide, without proofs, some general theorems on higher regularity.

We should also point out that all the regularity results that we obtain here are about solutions of the Euler-Lagrange equation and therefore not only minimizers of (P).

The problem of regularity, including the closely related one concerning regularity for elliptic partial differential equations, is a difficult one that has attracted many mathematicians. We name only a few of them: Agmon, Bernstein, Calderon, De Giorgi, Douglis, E. Hopf, Leray, Lichtenstein, Morrey, Moser, Nash, Nirenberg, Rado, Schauder, Tonelli, Weyl and Zygmund.

In addition to the books that were mentioned in Chapter 3 one can consult those by Evans [47], Gilbarg-Trudinger [57] and Ladyzhenskaya-Uraltseva [77] (for Dirichlet integral see also, for example, Brézis [14], Folland [52] and John [74]).

4.2 The one dimensional case

Let us restate the problem. We consider

$$(P) \quad \inf_{u \in X} \left\{ I(u) = \int_a^b f(x, u(x), u'(x)) \, dx \right\} = m$$

where $f \in C^0([a,b] \times \mathbb{R} \times \mathbb{R})$, $f = f(x, u, \xi)$, and

$$X = \left\{ u \in W^{1,p}(a,b) : u(a) = \alpha, \ u(b) = \beta \right\}.$$

We have seen that if f satisfies

(H_1) $\xi \to f(x, u, \xi)$ is convex for every $(x, u) \in [a, b] \times \mathbb{R}$;

(H_2) there exist $p > q \geq 1$ and $\alpha_1 > 0$, $\alpha_2, \alpha_3 \in \mathbb{R}$ such that

$$f(x, u, \xi) \geq \alpha_1 |\xi|^p + \alpha_2 |u|^q + \alpha_3, \quad \forall (x, u, \xi) \in [a, b] \times \mathbb{R} \times \mathbb{R};$$

then (P) has a solution $\overline{u} \in X$.

If, furthermore, $f \in C^1([a, b] \times \mathbb{R} \times \mathbb{R})$ and verifies (see Theorem 3.11 and Remark 3.12 (vi))

(H_3') for every $R > 0$, there exists $\alpha_4 = \alpha_4(R)$ such that

$$|f_u(x, u, \xi)|, |f_\xi(x, u, \xi)| \leq \alpha_4(1 + |\xi|^p), \quad \forall (x, u, \xi) \in [a, b] \times [-R, R] \times \mathbb{R}$$

then any minimizer $\overline{u} \in X$ satisfies the weak form of the Euler-Lagrange equation

$$(E_w) \quad \int_a^b [f_u(x, \overline{u}, \overline{u}') v + f_\xi(x, \overline{u}, \overline{u}') v'] \, dx = 0, \quad \forall v \in C_0^\infty(a, b).$$

We now show that under some strengthening of the hypotheses, we have that if $f \in C^\infty$ then $\overline{u} \in C^\infty$. These results are, in part, also valid if $u : [a, b] \to \mathbb{R}^N$, for $N > 1$.

4.2.1 A simple case

We start with a very elementary result that illustrates our purpose. We consider integrands of the form

$$f(x, u, \xi) = \frac{1}{2}\xi^2 + g(x, u).$$

Proposition 4.1 *Let* $g \in C^\infty([a, b] \times \mathbb{R})$. *Then any* $\overline{u} \in W^{1,2}(a, b)$ *satisfying*

$$(E_w) \quad \int_a^b [\overline{u}' v' + g_u(x, \overline{u}) v] \, dx = 0, \quad \forall v \in C_0^\infty(a, b) \tag{4.1}$$

is, in fact, in $C^\infty([a, b])$.

Remark 4.2 The above proposition applies immediately to

$$(P) \quad \inf_{u \in X} \left\{ I(u) = \int_a^b \left[\frac{1}{2}|u'(x)|^2 + g(x, u(x)) \right] dx \right\} = m$$

where $X = \{u \in W^{1,2}(a, b) : u(a) = \alpha, u(b) = \beta\}$ and if g satisfies, in addition,

(H_2) there exist $2 > q \geq 1$ and $\alpha_2, \alpha_3 \in \mathbb{R}$ such that

$$g(x, u) \geq \alpha_2 |u|^q + \alpha_3, \quad \forall (x, u) \in [a, b] \times \mathbb{R}.$$

Indeed, under these hypotheses, (P) has at least one minimizer $\overline{u} \in W^{1,2}(a, b)$ according to Theorem 3.3. Moreover, it satisfies (4.1) in view of Theorem 3.11 (and Remark 3.12 (vi)). Finally if, in addition, $u \to g(x, u)$ is convex for every $x \in [a, b]$, then (see Exercise 3.3.4) the minimizer is unique.

Proof We start by showing that $\overline{u} \in W^{2,2}(a, b)$. This follows immediately from (4.1) and from the definition of the weak derivative. Indeed since $\overline{u} \in W^{1,2}$, we have that $\overline{u} \in L^\infty$ and thus $g_u(x, \overline{u}) \in L^2$, leading to

$$\left| \int_a^b \overline{u}' \, v' \right| \leq \| g_u(x, \overline{u}) \|_{L^2} \| v \|_{L^2}, \quad \forall v \in C_0^\infty(a, b). \tag{4.2}$$

Theorem 1.37 implies then that $\overline{u} \in W^{2,2}$. We can then integrate by parts (4.1), bearing in mind that $v(a) = v(b) = 0$, and using the fundamental lemma of the calculus of variations (see Theorem 1.24), we deduce that

$$\overline{u}''(x) = g_u(x, \overline{u}(x)), \quad \text{a.e. } x \in (a, b). \tag{4.3}$$

We are now in a position to start an iteration process. Since $\overline{u} \in W^{2,2}(a, b)$ we deduce that (see Theorem 1.44) $\overline{u} \in C^1([a, b])$ and hence the function

$$x \to g_u(x, \overline{u}(x))$$

is $C^1([a, b])$, g being C^∞. Returning to (4.3) we deduce that $\overline{u}'' \in C^1$ and hence $\overline{u} \in C^3$.

From there we can infer that $x \to g_u(x, \overline{u}(x))$ is C^3, and thus from (4.3) we obtain that $\overline{u}'' \in C^3$ and hence $\overline{u} \in C^5$.

Continuing this process we have indeed established that $\overline{u} \in C^\infty([a, b])$. $\blacksquare$

4.2.2 Two general theorems

We will make the following hypotheses. Let $f \in C^1([a, b] \times \mathbb{R} \times \mathbb{R})$, $f = f(x, u, \xi)$, satisfy

(H_1) $\xi \to f(x, u, \xi)$ is convex for every $(x, u) \in [a, b] \times \mathbb{R}$;

(H_2') there exist $p > 1$, $\alpha_1 > 0$ and, for every $R > 0$, $\alpha_2 = \alpha_2(R) \in \mathbb{R}$ such that

$$f(x, u, \xi) \geq \alpha_1 |\xi|^p + \alpha_2, \quad \forall (x, u, \xi) \in [a, b] \times [-R, R] \times \mathbb{R};$$

(H_3') for every $R > 0$, there exists $\alpha_4 = \alpha_4(R)$ such that

$$|f_u(x, u, \xi)|, |f_\xi(x, u, \xi)| \leq \alpha_4(1 + |\xi|^p), \quad \forall (x, u, \xi) \in [a, b] \times [-R, R] \times \mathbb{R}.$$

Lemma 4.3 *Let $f \in C^1\left([a,b] \times \mathbb{R} \times \mathbb{R}\right)$ satisfy (H_1), (H_2') and (H_3'). Then any solution $\overline{u} \in W^{1,p}\left(a,b\right)$ of*

$$(E_w) \quad \int_a^b \left[f_u\left(x,\overline{u},\overline{u}'\right)v + f_\xi\left(x,\overline{u},\overline{u}'\right)v'\right]dx = 0, \quad \forall v \in C_0^\infty\left(a,b\right) \tag{4.4}$$

belongs to $W^{1,\infty}\left(a,b\right)$ and the equation holds a.e., i.e.

$$\frac{d}{dx}\left[f_\xi\left(x,\overline{u},\overline{u}'\right)\right] = f_u\left(x,\overline{u},\overline{u}'\right), \quad a.e. \ x \in \left(a,b\right). \tag{4.5}$$

Proof We divide the proof of the lemma into two steps.

Step 1. Define

$$\varphi\left(x\right) = f_\xi\left(x,\overline{u}\left(x\right),\overline{u}'\left(x\right)\right) \quad \text{and} \quad \psi\left(x\right) = f_u\left(x,\overline{u}\left(x\right),\overline{u}'\left(x\right)\right).$$

We easily see that $\varphi \in W^{1,1}\left(a,b\right)$ and that $\varphi'\left(x\right) = \psi\left(x\right)$, for almost every $x \in \left(a,b\right)$, which means that (4.5) holds, namely

$$\frac{d}{dx}\left[f_\xi\left(x,\overline{u},\overline{u}'\right)\right] = f_u\left(x,\overline{u},\overline{u}'\right), \quad a.e. \ x \in \left(a,b\right).$$

Indeed, since $\overline{u} \in W^{1,p}\left(a,b\right)$, and hence $\overline{u} \in L^\infty\left(a,b\right)$, we deduce from (H_3') that $\psi \in L^1\left(a,b\right)$. We also have from (4.4) that

$$\int_a^b \psi\left(x\right)v\left(x\right)dx = -\int_a^b \varphi\left(x\right)v'\left(x\right)dx, \quad \forall v \in C_0^\infty\left(a,b\right).$$

Since $\varphi \in L^1\left(a,b\right)$ (from (H_3')), we have by the definition of weak derivatives the claim, namely $\varphi \in W^{1,1}\left(a,b\right)$ and $\varphi' = \psi$ a.e.

Step 2. Since $\varphi \in W^{1,1}\left(a,b\right)$, we have that $\varphi \in C^0\left([a,b]\right)$, which means that there exists a constant $\alpha_5 > 0$ so that

$$\left|\varphi\left(x\right)\right| = \left|f_\xi\left(x,\overline{u}\left(x\right),\overline{u}'\left(x\right)\right)\right| \le \alpha_5, \quad \forall x \in [a,b]. \tag{4.6}$$

Since $\overline{u}$ is bounded (and even continuous), we can find $R > 0$ such that $\left|\overline{u}\left(x\right)\right| \le R$ for every $x \in [a,b]$. From (H_1) we have

$$f\left(x,u,0\right) \ge f\left(x,u,\xi\right) - \xi f_\xi\left(x,u,\xi\right), \ \forall \left(x,u,\xi\right) \in [a,b] \times [-R,R] \times \mathbb{R}.$$

Combining this inequality with (H_2') we find that there exists $\alpha_6 \in \mathbb{R}$ such that, for every $\left(x,u,\xi\right) \in [a,b] \times [-R,R] \times \mathbb{R}$,

$$\xi f_\xi\left(x,u,\xi\right) \ge f\left(x,u,\xi\right) - f\left(x,u,0\right) \ge \alpha_1 \left|\xi\right|^p + \alpha_6.$$

Using (4.6) and the above inequality we find

$$\alpha_1 \left|\overline{u}'\right|^p + \alpha_6 \le \overline{u}' f_\xi\left(x,\overline{u},\overline{u}'\right) \le \left|\overline{u}'\right|\left|f_\xi\left(x,\overline{u},\overline{u}'\right)\right| \le \alpha_5 \left|\overline{u}'\right|, \quad a.e. \ x \in \left(a,b\right)$$

which implies, since $p > 1$, that $\left|\overline{u}'\right|$ is uniformly bounded. Thus the lemma. ∎

Theorem 4.4 *Let $f \in C^\infty\left([a,b] \times \mathbb{R} \times \mathbb{R}\right)$ satisfy (H_2'), (H_3') and*

$$(H_1') \quad f_{\xi\xi}\left(x,u,\xi\right) > 0, \quad \forall\left(x,u,\xi\right) \in [a,b] \times \mathbb{R} \times \mathbb{R}.$$

Then any solution $\overline{u} \in W^{1,p}\left(a,b\right)$ of

$$(E_w) \quad \int_a^b \left[f_u\left(x,\overline{u},\overline{u}'\right)v + f_\xi\left(x,\overline{u},\overline{u}'\right)v'\right]dx = 0, \quad \forall v \in C_0^\infty\left(a,b\right)$$

is in $C^\infty\left([a,b]\right)$ and satisfies

$$\frac{d}{dx}\left[f_\xi\left(x,\overline{u},\overline{u}'\right)\right] = f_u\left(x,\overline{u},\overline{u}'\right), \quad \forall x \in \left(a,b\right).$$

Remark 4.5 (i) Note that (H_1') is more restrictive than (H_1). This stronger condition is usually, but not always, as seen in Theorem 4.6 below, necessary to get higher regularity.

(ii) In order to guarantee the existence of a minimizer of

$$(P) \quad \inf_{u \in X}\left\{I\left(u\right) = \int_a^b f\left(x,u\left(x\right),u'\left(x\right)\right)dx\right\} = m$$

where $X = \left\{u \in W^{1,p}\left(a,b\right) : u\left(a\right) = \alpha,\ u\left(b\right) = \beta\right\}$, the hypothesis (H_2') must be reinforced and replaced by

(H_2) there exist $p > q \geq 1$ and $\alpha_1 > 0$, $\alpha_2, \alpha_3 \in \mathbb{R}$ such that

$$f\left(x,u,\xi\right) \geq \alpha_1\left|\xi\right|^p + \alpha_2\left|u\right|^q + \alpha_3, \quad \forall\left(x,u,\xi\right) \in [a,b] \times \mathbb{R} \times \mathbb{R}.$$

Under the hypotheses (H_1'), (H_2), (H_3'), the problem (P) admits a minimizer $\overline{u} \in W^{1,p}\left(a,b\right)$ (see Theorem 3.3) satisfying (E_w) (see Theorem 3.11 and Remark 3.12 (vi)) of the theorem. Therefore, any minimizer of (P) is necessarily C^∞.

(iii) Proposition 4.1 is, of course, a particular case of the present theorem.

(iv) The proof shows that if $f \in C^k$, $k \geq 2$, then $\overline{u} \in C^k$.

Proof We propose a different proof in Exercise 4.2.1. The present one is more direct and uses Lemma 2.9.

Step 1. We know from Lemma 4.3 that $x \to \varphi\left(x\right) = f_\xi\left(x,\overline{u}\left(x\right),\overline{u}'\left(x\right)\right)$ is in $W^{1,1}\left(a,b\right)$ and hence it is continuous. Appealing to Lemma 2.9 (and the remark following this lemma), we have that if

$$H\left(x,u,v\right) = \sup_{\xi \in \mathbb{R}}\left\{v\,\xi - f\left(x,u,\xi\right)\right\}$$

then $H \in C^{\infty}([a,b] \times \mathbb{R} \times \mathbb{R})$ and, for every $x \in [a,b]$, we have

$$\varphi(x) = f_{\xi}(x, \overline{u}(x), \overline{u}'(x)) \quad \Leftrightarrow \quad \overline{u}'(x) = H_v(x, \overline{u}(x), \varphi(x)).$$

Since H_v, $\overline{u}$ and φ are continuous, we infer that $\overline{u}'$ is continuous and hence $\overline{u} \in C^1([a,b])$. We therefore deduce that $x \to f_u(x, \overline{u}(x), \overline{u}'(x))$ is continuous, which combined with the fact that (see (4.5))

$$\frac{d}{dx}[\varphi(x)] = f_u(x, \overline{u}(x), \overline{u}'(x)), \quad \text{a.e. } x \in (a,b)$$

(or equivalently, by Lemma 2.9, $\varphi' = -H_u(x, \overline{u}, \varphi)$) leads to $\varphi \in C^1([a,b])$.

Step 2. Returning to our Hamiltonian system

$$\begin{cases} \overline{u}'(x) = H_v(x, \overline{u}(x), \varphi(x)) \\ \varphi'(x) = -H_u(x, \overline{u}(x), \varphi(x)) \end{cases}$$

we can start our iteration. Indeed, since H is C^{∞} and $\overline{u}$ and φ are C^1 we deduce from our system that, in fact, $\overline{u}$ and φ are C^2. Returning to the system we get that $\overline{u}$ and φ are C^3. Finally, we get that $\overline{u}$ is C^{∞}, as required. ∎

We conclude the section by giving an example where we can get further regularity without assuming the non-degeneracy condition $f_{\xi\xi} > 0$.

Theorem 4.6 *Let $p > 1$ and $g \in C^1([a,b] \times \mathbb{R})$. Let $\overline{u} \in W^{1,p}(a,b)$ be a solution of*

$$(E_w) \quad \int_a^b \left[|\overline{u}'|^{p-2} \overline{u}' v' + g_u(x, \overline{u}) v \right] dx = 0, \quad \forall v \in C_0^{\infty}(a,b).$$

Then $\overline{u}, |\overline{u}'|^{p-2} \overline{u}' \in C^1([a,b])$ and

$$\frac{d}{dx}\left[|\overline{u}'(x)|^{p-2} \overline{u}'(x) \right] = g_u(x, \overline{u}(x)), \quad \forall x \in [a,b].$$

Moreover, if $1 < p \leq 2$, then $\overline{u} \in C^2([a,b])$.

Remark 4.7 (i) The theorem applies to

$$(P) \quad \inf_{u \in X} \left\{ I(u) = \int_a^b f(x, u(x), u'(x)) \, dx \right\} = m$$

where $X = \{u \in W^{1,p}(a,b) : u(a) = \alpha, \ u(b) = \beta\}$ and

$$f(x, u, \xi) = \frac{1}{p} |\xi|^p + g(x, u)$$

with $g \in C^1([a,b] \times \mathbb{R})$ satisfying

(H_2) there exist $p > q \geq 1$ and $\alpha_2, \alpha_3 \in \mathbb{R}$ such that

$$g(x,u) \geq \alpha_2 |u|^q + \alpha_3, \quad \forall (x,u) \in [a,b] \times \mathbb{R}.$$

In this case (P) admits a minimizer $\overline{u} \in W^{1,p}(a,b)$ (see Theorem 3.3) satisfying (E_w) (see Theorem 3.11 and Remark 3.12 (vi)) of the theorem. Furthermore, the minimizer is unique if g is convex (see Exercise 3.3.4).

(ii) The result cannot be improved in general, see Exercises 4.2.2 and 4.2.3.

Proof According to Lemma 4.3 we know that $\overline{u} \in W^{1,\infty}(a,b)$ and, since $x \to g_u(x, \overline{u}(x))$ is continuous, we have that the equation holds everywhere, i.e.

$$\frac{d}{dx}\left[|\overline{u}'(x)|^{p-2}\overline{u}'(x)\right] = g_u(x, \overline{u}(x)), \quad x \in [a,b].$$

We thus have that $|\overline{u}'|^{p-2}\overline{u}' \in C^1([a,b])$. Call $v \equiv |\overline{u}'|^{p-2}\overline{u}'$. We may then infer that

$$\overline{u}' = |v|^{\frac{2-p}{p-1}} v.$$

Since the function $t \to |t|^{\frac{2-p}{p-1}} t$ is continuous if $p > 2$ and C^1 if $1 < p \leq 2$, we obtain, from the fact that $v \in C^1([a,b])$, the conclusions of the theorem. ∎

4.2.3   Exercises

Exercise 4.2.1 With the help of Lemma 4.3, prove Theorem 4.4 in the following manner.

(i) First show that $\overline{u} \in W^{2,\infty}(a,b)$, by proving (iii) of Theorem 1.37.

(ii) Conclude, using the fact that

$$\frac{d}{dx}[f_\xi(x, \overline{u}, \overline{u}')] = f_{\xi\xi}(x, \overline{u}, \overline{u}')\overline{u}'' + f_{u\xi}(x, \overline{u}, \overline{u}')\overline{u}' + f_{x\xi}(x, \overline{u}, \overline{u}')$$

$$= f_u(x, \overline{u}, \overline{u}').$$

Exercise 4.2.2 Let $\overline{u}(x) = \frac{7}{12}|x|^{12/7}$,

$$f_1(x,u,\xi) = \frac{1}{8}\xi^8 + 5x^4 u \quad \text{and} \quad f_2(x,u,\xi) = f_2(x,\xi) = \frac{1}{8}\xi^8 - x^5\xi$$

$$(P_i) \quad \inf_{u \in X}\left\{I_i(u) = \int_{-1}^{1} f_i(x, u(x), u'(x))\, dx\right\}, \quad i = 1, 2$$

$$X = \left\{u \in W^{1,8}(-1,1) : u(-1) = u(1) = \frac{7}{12}\right\}.$$

Prove that $\overline{u} \in C^1([-1,1])$ is the unique minimizer of (P_1) and (P_2), but $\overline{u} \notin C^2([-1,1])$, although $f_1, f_2 \in C^\infty(\mathbb{R}^3)$.

Exercise 4.2.3 Let $p > 2q > 2$ and

$$f(x, u, \xi) = f(u, \xi) = \frac{1}{p} |\xi|^p + \frac{\lambda}{q} |u|^q \quad \text{where } \lambda = \frac{qp^{q-1}(p-1)}{(p-q)^q}$$

$$\bar{u}(x) = \frac{p-q}{p} |x|^{p/(p-q)}$$

(note that if, for example, $p = 6$ and $q = 2$, then $f \in C^\infty (\mathbb{R}^2)$).

  (i) Show that $\bar{u} \in C^1 ([-1, 1])$ but $\bar{u} \notin C^2 ([-1, 1])$.

  (ii) Find some values of p and q so that

$$|\bar{u}'|^{p-2} \bar{u}', |\bar{u}|^{q-2} \bar{u} \in C^\infty ([-1, 1]),$$

although $\bar{u} \notin C^2 ([-1, 1])$.

  (iii) Show that $\bar{u}$ is the unique minimizer of

$$(P) \quad \inf_{u \in W^{1,p}(-1,1)} \left\{ I(u) = \int_{-1}^1 f(u(x), u'(x)) \, dx : u(-1) = u(1) = \frac{p-q}{p} \right\}.$$

Exercise 4.2.4 Let

$$\varphi(x) = \begin{cases} \exp(-1/x^2) & \text{if } x \neq 0 \\ 0 & \text{if } x = 0 \end{cases}$$

$$f(x, u, \xi) = f(x, \xi) = \left[\varphi(x)\xi - 2x\varphi(x)\sin\frac{\pi}{x} + \pi\varphi(x)\cos\frac{\pi}{x} \right]^2$$

and

$$(P) \quad \inf \left\{ I(u) = \int_0^1 f(x, u'(x)) \, dx : u \in W_0^{1,2}(-1, 1) \right\} = m.$$

Observe that $f \in C^\infty ([-1, 1] \times \mathbb{R})$ and $\xi \to f(x, \xi)$ is convex with $f_{\xi\xi}(x, \xi) = 2\varphi(x) > 0$ except at $x = 0$. Show that

$$\bar{u}(x) = \begin{cases} x^2 \sin(\pi/x) & \text{if } x \neq 0 \\ 0 & \text{if } x = 0 \end{cases}$$

is the unique minimizer of (P), $\bar{u} \in W_0^{1,\infty}(-1, 1)$ but $\bar{u} \notin C^1 ([-1, 1])$.

4.3 The difference quotient method: interior regularity

We now deal with the first step in getting regularity of solutions of our problem

$$(P) \quad \inf\left\{ I(u) = \int_\Omega f(x, u(x), \nabla u(x))\, dx : u \in u_0 + W_0^{1,p}(\Omega) \right\} = m.$$

In order to understand the method of difference quotient better, introduced by Nirenberg, we will first deal with the Dirichlet integral and then discuss a more general case.

4.3.1 Preliminaries

We recall (see Notation 1.36) the notion of *difference quotient*. For $\tau \in \mathbb{R}^n$, $\tau \neq 0$, we let

$$(D_\tau u)(x) = \frac{u(x+\tau) - u(x)}{|\tau|}.$$

The following properties are either elementary or in Theorem 1.37. In the following $\Omega \subset \mathbb{R}^n$ will stand for an open set.

(i) We trivially have

$$\nabla(D_\tau u) = D_\tau(\nabla u).$$

(ii) For every open set $O \subset \overline{O} \subset \Omega$, with $\overline{O}$ compact and for every $u \in W^{1,2}(\Omega)$, the following holds

$$\|D_\tau u\|_{L^2(O)} \leq \|\nabla u\|_{L^2(\Omega)} \tag{4.7}$$

where $0 \neq |\tau| < \text{dist}(O, \Omega^c)$.

(iii) If $u \in L^2(\Omega)$ satisfies

$$\|D_\tau u\|_{L^2(O)} \leq \gamma$$

for some open set $O \subset \overline{O} \subset \Omega$, with $\overline{O}$ compact, every $0 \neq |\tau| < \text{dist}(O, \Omega^c)$ and where γ denotes a constant independent of τ, then

$$u \in W^{1,2}(O) \quad \text{and} \quad \|\nabla u\|_{L^2(O)} \leq \gamma. \tag{4.8}$$

(iv) Moreover, a direct computation shows that, for $u, v \in W^{1,2}(\Omega)$ with v having compact support in Ω and $0 \neq |\tau| < \text{dist}(\text{supp}\, v, \Omega^c)$,

$$\int_\Omega \langle \nabla u; \nabla(D_{-\tau} v) \rangle = \int_\Omega \langle D_\tau(\nabla u); \nabla v \rangle. \tag{4.9}$$

We will finally use, several times, the following version of Cauchy-Schwarz inequality: given $\epsilon > 0$

$$ab \leq \epsilon a^2 + \frac{b^2}{\epsilon}$$

which leads to

$$\|uv\|_{L^1} \leq \epsilon \|u\|_{L^2}^2 + \frac{1}{\epsilon} \|v\|_{L^2}^2 .$$

4.3.2 The Dirichlet integral

We start with the model case. We recall that we have already proved in Theorem 3.1 and Exercise 3.2.1 that if $\Omega \subset \mathbb{R}^n$ is a bounded open set with Lipschitz boundary, $h \in L^2(\Omega)$ and

$$(P) \quad \inf \left\{ I(u) = \frac{1}{2} \int_\Omega |\nabla u(x)|^2 \, dx - \int_\Omega h(x) u(x) \, dx : u \in W_0^{1,2}(\Omega) \right\},$$

then there exists a unique minimizer $\overline{u} \in W_0^{1,2}(\Omega)$ satisfying the weak form of the Euler-Lagrange equation, namely

$$\int_\Omega \langle \nabla \overline{u}; \nabla \varphi \rangle = \int_\Omega h \varphi, \quad \forall \varphi \in W_0^{1,2}(\Omega). \tag{4.10}$$

We now establish that, in fact, the solution $\overline{u}$ is more regular than in $W^{1,2}$. The regularity will be obtained for solutions of (4.10) and not necessarily for the minimizer of (P).

Theorem 4.8 *Let Ω and h be as above. Let $\overline{u} \in W^{1,2}(\Omega)$ satisfy (4.10). Then $\overline{u} \in W^{1,2}(\Omega) \cap W_{loc}^{2,2}(\Omega)$. More precisely, for every open set $O \subset \overline{O} \subset \Omega$ there exists a constant $\gamma_1 = \gamma_1(O, \Omega) > 0$ such that*

$$\|\overline{u}\|_{W^{2,2}(O)} \leq \gamma_1 \left(\|\overline{u}\|_{L^2(\Omega)} + \|h\|_{L^2(\Omega)} \right) \tag{4.11}$$

and the following form of the Euler-Lagrange equation is satisfied

$$-\Delta \overline{u} = h, \quad a.e. \ in \ \Omega.$$

If, moreover, $\overline{u} \in W_0^{1,2}(\Omega)$, then there exists a constant $\gamma_2 = \gamma_2(O, \Omega) > 0$ such that

$$\|\overline{u}\|_{W^{2,2}(O)} \leq \gamma_2 \|h\|_{L^2(\Omega)} . \tag{4.12}$$

Proof We divide the proof into three steps. In the following, unless stated otherwise, all L^2 norms are understood in $L^2(\Omega)$.

Step 1. We here follow Evans [47]. We now fix an open set $O \subset \overline{O} \subset \Omega$, we find $\rho \in C_0^\infty (\Omega)$ such that

$$0 \le \rho \le 1 \quad \text{and} \quad \rho \equiv 1 \text{ in } O.$$

We then make a special choice of $\varphi \in W_0^{1,2} (\Omega)$ in the Euler-Lagrange equation (4.10), namely, for $|\tau|$ sufficiently small,

$$\varphi = D_{-\tau} \left(\rho^2 \left(D_\tau \overline{u} \right) \right).$$

Equation (4.10) then becomes

$$\int_\Omega \left\langle \nabla \overline{u}; \nabla \left(D_{-\tau} \left(\rho^2 \left(D_\tau \overline{u} \right) \right) \right) \right\rangle = \int_\Omega h \, D_{-\tau} \left(\rho^2 \left(D_\tau \overline{u} \right) \right).$$

Invoking the properties of the difference quotient, notably (4.9), we deduce that

$$\int_\Omega \left\langle D_\tau \left(\nabla \overline{u} \right); \rho^2 \left(D_\tau \left(\nabla \overline{u} \right) \right) \right\rangle + 2 \int_\Omega \left\langle D_\tau \left(\nabla \overline{u} \right); \rho \nabla \rho \left(D_\tau \overline{u} \right) \right\rangle$$
$$= \int_\Omega h D_{-\tau} \left(\rho^2 \left(D_\tau \overline{u} \right) \right)$$

and thus

$$\left\| \rho \, D_\tau \left(\nabla \overline{u} \right) \right\|_{L^2}^2 + 2 \int_\Omega \left\langle D_\tau \left(\nabla \overline{u} \right); \rho \nabla \rho \left(D_\tau \overline{u} \right) \right\rangle = \int_\Omega h D_{-\tau} \left(\rho^2 \left(D_\tau \overline{u} \right) \right). \quad (4.13)$$

We then estimate the second and third terms separately and this will lead us to the desired estimate in Step 3.

Step 2. Estimate 1. We have, for $\epsilon > 0$ fixed,

$$\int_\Omega \left\langle D_\tau \left(\nabla \overline{u} \right); \rho \nabla \rho \left(D_\tau \overline{u} \right) \right\rangle \le \int_\Omega \left| \rho \left(D_\tau \left(\nabla \overline{u} \right) \right) \right| \left| \nabla \rho \left(D_\tau \overline{u} \right) \right|$$
$$\le \epsilon \left\| \rho \left(D_\tau \left(\nabla \overline{u} \right) \right) \right\|_{L^2}^2 + \frac{1}{\epsilon} \left\| \nabla \rho \left(D_\tau \overline{u} \right) \right\|_{L^2}^2.$$

Observing that, from (4.7), we have

$$\left\| \nabla \rho \left(D_\tau \overline{u} \right) \right\|_{L^2}^2 = \int_{\text{supp}\,\rho} \left| \nabla \rho \left(D_\tau \overline{u} \right) \right|^2 \le \left\| \nabla \rho \right\|_{L^\infty}^2 \int_{\text{supp}\,\rho} \left| D_\tau \overline{u} \right|^2$$
$$\le \left\| \nabla \rho \right\|_{L^\infty}^2 \left\| \nabla \overline{u} \right\|_{L^2}^2$$

and thus

$$\int_\Omega \left\langle D_\tau \left(\nabla \overline{u} \right); \rho \nabla \rho \left(D_\tau \overline{u} \right) \right\rangle \le \epsilon \left\| \rho \left(D_\tau \left(\nabla \overline{u} \right) \right) \right\|_{L^2}^2 + \frac{1}{\epsilon} \left\| \nabla \rho \right\|_{L^\infty}^2 \left\| \nabla \overline{u} \right\|_{L^2}^2.$$

Estimate 2. It follows from (4.7) that, for $\epsilon > 0$ fixed,

$$\left| \int_\Omega h \left(D_{-\tau} \left(\rho^2 \left(D_\tau \overline{u} \right) \right) \right) \right| \le \epsilon \left\| \left(D_{-\tau} \left(\rho^2 \left(D_\tau \overline{u} \right) \right) \right) \right\|_{L^2}^2 + \frac{1}{\epsilon} \|h\|_{L^2}^2$$

$$\le \epsilon \left\| \nabla \left(\rho^2 \left(D_\tau \overline{u} \right) \right) \right\|_{L^2}^2 + \frac{1}{\epsilon} \|h\|_{L^2}^2 .$$

Since

$$\left\| \nabla \left(\rho^2 \left(D_\tau \overline{u} \right) \right) \right\|_{L^2}^2 = \int_\Omega \left| \rho^2 \left(D_\tau \left(\nabla \overline{u} \right) \right) + 2\rho \nabla \rho \left(D_\tau \overline{u} \right) \right|^2$$

$$\le 2 \int_\Omega \left| \rho^2 \left(D_\tau \left(\nabla \overline{u} \right) \right) \right|^2 + 2 \int_\Omega \left| 2\rho \nabla \rho \left(D_\tau \overline{u} \right) \right|^2$$

we obtain, as in Estimate 1 and recalling that $0 \le \rho \le 1$ and thus $\rho^2 \le \rho$,

$$\left| \int_\Omega h \left(D_{-\tau} \rho^2 \left(D_\tau \overline{u} \right) \right) \right| \le 2\epsilon \left\| \rho \left(D_\tau \left(\nabla \overline{u} \right) \right) \right\|_{L^2}^2 + 8\epsilon \left\| \nabla \rho \right\|_{L^\infty}^2 \left\| \nabla \overline{u} \right\|_{L^2}^2 + \frac{1}{\epsilon} \|h\|_{L^2}^2 .$$

Step 3. We now gather the two estimates in (4.13), to obtain

$$(1 - 4\epsilon) \left\| \rho \left(D_\tau \left(\nabla \overline{u} \right) \right) \right\|_{L^2}^2 \le \left(8\epsilon + \frac{2}{\epsilon} \right) \left\| \nabla \rho \right\|_{L^\infty}^2 \left\| \nabla \overline{u} \right\|_{L^2}^2 + \frac{1}{\epsilon} \|h\|_{L^2}^2 .$$

Therefore, choosing ϵ small, we can find $\gamma_3 = \gamma_3 \left(\epsilon, \rho \right) > 0$ such that

$$\left\| \rho \left(D_\tau \left(\nabla \overline{u} \right) \right) \right\|_{L^2}^2 \le \gamma_3 \left(\left\| \nabla \overline{u} \right\|_{L^2}^2 + \|h\|_{L^2}^2 \right) .$$

Using the definition of ρ, we infer that

$$\left\| D_\tau \left(\nabla \overline{u} \right) \right\|_{L^2(O)}^2 \le \gamma_3 \left(\left\| \nabla \overline{u} \right\|_{L^2(\Omega)}^2 + \|h\|_{L^2(\Omega)}^2 \right) .$$

From (4.8) we deduce

$$\left\| \nabla^2 \overline{u} \right\|_{L^2(O)}^2 \le \gamma_3 \left(\left\| \nabla \overline{u} \right\|_{L^2(\Omega)}^2 + \|h\|_{L^2(\Omega)}^2 \right)$$

and hence the fact that $\overline{u} \in W_{\text{loc}}^{2,2} \left(\Omega \right)$ with

$$\|\overline{u}\|_{W^{2,2}(O)}^2 \le (\gamma_3 + 1) \left(\|\overline{u}\|_{W^{1,2}(\Omega)}^2 + \|h\|_{L^2(\Omega)}^2 \right) . \tag{4.14}$$

The more precise estimates (4.11) and (4.12) are obtained in Exercise 4.3.3. Since $\overline{u} \in W_{\text{loc}}^{2,2} \left(\Omega \right)$, we can integrate the left-hand side of (4.10) and use Theorem 1.24 to find

$$-\Delta \overline{u} = h, \quad \text{a.e. in } \Omega.$$

This concludes the proof of the theorem. ∎

4.3.3   A more general case

We now generalize the previous result. We let $\Omega \subset \mathbb{R}^n$ be a bounded open set with Lipschitz boundary, $u_0 \in W^{1,2}(\Omega)$, $h \in L^2(\Omega)$ and $g \in C^1(\overline{\Omega} \times \mathbb{R}^n)$, $g = g(x, \xi)$, with

$$\xi \to g(x, \xi) \text{ convex}$$

and such that there exist $0 < \alpha_1 \leq \alpha_3$ and $\alpha_2 \in \mathbb{R}$ with

$$(G_1) \quad \alpha_1 |\xi|^2 - \alpha_2 \leq g(x, \xi) \leq \alpha_3 \left(|\xi|^2 + 1\right), \quad \forall (x, \xi) \in \overline{\Omega} \times \mathbb{R}^n.$$

Then there exists a minimizer $\overline{u} \in u_0 + W_0^{1,2}(\Omega)$ of

$$(P) \quad \inf\left\{ I(u) = \int_\Omega [g(x, \nabla u(x)) - h(x)u(x)]\, dx : u \in u_0 + W_0^{1,2}(\Omega) \right\}$$

satisfying

$$\int_\Omega \sum_{i=1}^n g_{\xi_i}(x, \nabla \overline{u}(x))\, \varphi_{x_i}(x)\, dx = \int_\Omega h(x)\varphi(x)\, dx, \quad \forall \varphi \in W_0^{1,2}(\Omega) \quad (4.15)$$

where $g_{\xi_i} = \partial g / \partial \xi_i$. The existence of a minimizer follows at once from Theorem 3.3 (and Remark 3.4 (i)). The Euler-Lagrange equation is established via Theorem 3.11 and Remark 3.12 (ii), once it has been observed that (G_1) implies (see Exercise 1.5.9) the existence of a constant $\alpha > 0$ such that

$$|g_{\xi_i}(x, \xi)| \leq \alpha(|\xi| + 1), \quad \forall (x, \xi) \in \overline{\Omega} \times \mathbb{R}^n \text{ and } \forall i = 1, \cdots, n.$$

If, moreover, g is strictly convex (in the following theorem the strict convexity of g is implied by (G_3) below), then the minimizer $\overline{u} \in u_0 + W_0^{1,2}(\Omega)$ is unique (see Exercise 3.3.4). As before, the regularity will be obtained for solutions of (4.15) and not necessarily for a minimizer of (P). In the following, as usual, partial derivatives will be denoted with sub indices, i.e.

$$g_{\xi_i} = \frac{\partial g}{\partial \xi_i}, \quad g_{\xi_i \xi_j} = \frac{\partial^2 g}{\partial \xi_i \partial \xi_j} \quad \text{and} \quad g_{\xi_i x_j} = \frac{\partial^2 g}{\partial \xi_i \partial x_j}.$$

Theorem 4.9 *Let $\Omega \subset \mathbb{R}^n$ be a bounded open set with Lipschitz boundary, $h \in L^2(\Omega)$ and $g \in C^2(\overline{\Omega} \times \mathbb{R}^n)$, $g = g(x, \xi)$, and there exist constants $\alpha_4, \alpha_5, \alpha_6 > 0$ such that*

$$(G_2) \quad \left|g_{\xi_i \xi_j}(x, \xi)\right| \leq \alpha_4, \quad \forall (x, \xi) \in \overline{\Omega} \times \mathbb{R}^n \text{ and } \forall i, j = 1, \cdots, n$$

$$(G_3) \quad \sum_{i,j=1}^n g_{\xi_i \xi_j}(x, \xi)\lambda_i \lambda_j \geq \alpha_5 |\lambda|^2, \quad \forall \xi, \lambda \in \mathbb{R}^n \text{ and } \forall x \in \overline{\Omega}$$

(G_4) $\left|g_{\xi_i x_j}(x,\xi)\right| \le \alpha_6 |\xi|, \quad \forall (x,\xi) \in \overline{\Omega} \times \mathbb{R}^n \text{ and } \forall i,j = 1,\cdots,n.$

Let $\overline{u} \in W^{1,2}(\Omega)$ be a solution of (4.15). Then $\overline{u} \in W^{2,2}_{loc}(\Omega)$ and the following form of the Euler-Lagrange equation is satisfied, namely

$$- \operatorname{div}\left[\nabla_\xi g(x, \nabla \overline{u})\right] = - \sum_{i=1}^n \left[g_{\xi_i}(x, \nabla \overline{u})\right]_{x_i} = h, \quad a.e. \text{ in } \Omega. \tag{4.16}$$

Moreover, for every open set $O \subset \overline{O} \subset \Omega$ there exists a constant $\gamma = \gamma(O, \Omega) > 0$ such that

$$\|\overline{u}\|_{W^{2,2}(O)} \le \gamma \left[\|\overline{u}\|_{W^{1,2}(\Omega)} + \|h\|_{L^2(\Omega)}\right]. \tag{4.17}$$

If, in addition,

$$\nabla_\xi g(x,0) = 0, \quad \forall x \in \overline{\Omega},$$

then the above estimate takes the more precise form

$$\|\overline{u}\|_{W^{2,2}(O)} \le \gamma \left[\|\overline{u}\|_{L^2(\Omega)} + \|h\|_{L^2(\Omega)}\right] \tag{4.18}$$

and if, further, $\overline{u} \in W^{1,2}_0(\Omega)$, the estimate is even simpler, namely

$$\|\overline{u}\|_{W^{2,2}(O)} \le \gamma \|h\|_{L^2(\Omega)}. \tag{4.19}$$

Proof We divide the proof into two steps.

Step 1. We first note that the hypothesis (G_3) implies that

$$\langle \nabla_\xi g(x,\xi) - \nabla_\xi g(x,0) ; \xi \rangle \ge \alpha_5 |\xi|^2, \quad \forall (x,\xi) \in \overline{\Omega} \times \mathbb{R}^n \tag{4.20}$$

where we denote

$$\nabla_\xi g(x,\xi) = \left(g_{\xi_1}(x,\xi), \cdots, g_{\xi_n}(x,\xi)\right).$$

This follows at once from the observation

$$g_{\xi_i}(x,\xi) - g_{\xi_i}(x,0) = \int_0^1 \frac{d}{dt}\left[g_{\xi_i}(x,t\xi)\right] dt = \int_0^1 \sum_{j=1}^n \left[g_{\xi_i \xi_j}(x,t\xi)\,\xi_j\right] dt.$$

Step 2 (difference quotient method). We then proceed exactly as in Theorem 4.8. We now fix an open set $O \subset \overline{O} \subset \Omega$, we find $\rho \in C_0^\infty(\Omega)$ such that

$$0 \le \rho \le 1 \quad \text{and} \quad \rho \equiv 1 \text{ in } O.$$

We choose $\varphi \in W^{1,2}_0(\Omega)$ in the Euler-Lagrange equation (4.15), as, for $|\tau|$ sufficiently small,

$$\varphi = D_{-\tau}\left(\rho^2 \left(D_\tau \overline{u}\right)\right).$$

Equation (4.15) then becomes

$$\int_\Omega \langle \nabla_\xi g\,(x, \nabla \overline{u})\,;\nabla \left(D_{-\tau} \left(\rho^2 \left(D_\tau \overline{u}\right)\right)\right)\rangle = \int_\Omega h D_{-\tau} \left(\rho^2 \left(D_\tau \overline{u}\right)\right).$$

Appealing to (4.9), we deduce that

$$\int_\Omega \langle D_\tau \left[\nabla_\xi g\,(x, \nabla \overline{u})\right]\,;\rho^2 \left(D_\tau \left(\nabla \overline{u}\right)\right)\rangle + 2 \int_\Omega \langle D_\tau \left[\nabla_\xi g\,(x, \nabla \overline{u})\right]\,;\rho \nabla \rho \left(D_\tau \overline{u}\right)\rangle$$

$$= \int_\Omega h D_{-\tau} \left(\rho^2 \left(D_\tau \overline{u}\right)\right)$$

$$\text{(4.21)}$$

where

$$D_\tau \left[\nabla_\xi g\,(x, \nabla u)\right] = \left(D_\tau \left[g_{\xi_1}\,(x, \nabla u)\right], \cdots, D_\tau \left[g_{\xi_n}\,(x, \nabla u)\right]\right)$$

and

$$D_\tau \left[g_{\xi_i}\,(x, \nabla u)\right] = \frac{1}{|\tau|} \left[g_{\xi_i}\,(x + \tau, \nabla u\,(x + \tau)) - g_{\xi_i}\,(x, \nabla u\,(x))\right].$$

We then estimate each one of the three terms separately and this will lead us to the desired estimate in Step 2.3; but we start with a preliminary computation.

Step 2.1. We point out some simple observations.

Observation 1. Note that

$$g_{\xi_i}\,(x + \tau, \nabla u\,(x + \tau)) - g_{\xi_i}\,(x + \tau, \nabla u\,(x))$$

$$= \int_0^1 \frac{d}{dt} \left[g_{\xi_i}\,(x + \tau, \nabla u\,(x) + t\,(\nabla u\,(x + \tau) - \nabla u\,(x)))\right] dt$$

$$= \sum_{j=1}^n \int_0^1 \left[g_{\xi_i \xi_j}\,(x + \tau, \nabla u\,(x) + t\,(\nabla u\,(x + \tau) - \nabla u\,(x)))\right.$$

$$\left. \left(u_{x_j}\,(x + \tau) - u_{x_j}\,(x)\right)\right] dt.$$

Observation 2. We next see that

$$g_{\xi_i}\,(x + \tau, \nabla u\,(x)) - g_{\xi_i}\,(x, \nabla u\,(x)) = \int_0^1 \frac{d}{dt} \left[g_{\xi_i}\,(x + t\tau, \nabla u\,(x))\right] dt$$

$$= \sum_{j=1}^n \int_0^1 g_{\xi_i x_j}\,(x + t\tau, \nabla u\,(x))\,\tau_j\,dt.$$

Observation 3. Combining the first two observations, we find that

$$D_\tau \left[g_{\xi_i}\,(x, \nabla u)\right] = \sum_{j=1}^n g_{ij} \left(D_\tau u_{x_j}\right) + \sum_{j=1}^n \widetilde{g}_{ij} \frac{\tau_j}{|\tau|}$$

where

$$g_{ij}(x) = \int_0^1 g_{\xi_i \xi_j}(x + \tau, \nabla u(x) + t(\nabla u(x + \tau) - \nabla u(x)))\, dt,$$

$$\widetilde{g}_{ij}(x) = \int_0^1 g_{\xi_i x_j}(x + t\tau, \nabla u(x))\, dt.$$

Conclusion. The following estimate is then deduced at once from the above computations and from (G_2) and (G_4)

$$|D_\tau [\nabla_\xi g(x, \nabla \overline{u})]| \le n\alpha_4 |D_\tau(\nabla \overline{u})| + n\alpha_6 |\nabla \overline{u}|. \tag{4.22}$$

Moreover, we have

$$\langle D_\tau [\nabla_\xi g(x, \nabla \overline{u})]; D_\tau(\nabla \overline{u}) \rangle = \sum_{i,j=1}^n g_{ij}\left(D_\tau \overline{u}_{x_j}\right)\left(D_\tau \overline{u}_{x_i}\right) + \sum_{i,j=1}^n \widetilde{g}_{ij} \frac{\tau_j}{|\tau|}\left(D_\tau \overline{u}_{x_i}\right)$$

and thus by (G_3) and (G_4)

$$\langle D_\tau [\nabla_\xi g(x, \nabla \overline{u})]; D_\tau(\nabla \overline{u}) \rangle \ge \alpha_5 |D_\tau(\nabla \overline{u})|^2 - n^2 \alpha_6 |\nabla \overline{u}| |D_\tau(\nabla \overline{u})|.$$

Fixing $\epsilon > 0$, we therefore get

$$\langle D_\tau [\nabla_\xi g(x, \nabla \overline{u})]; D_\tau(\nabla \overline{u}) \rangle \ge (\alpha_5 - n^2 \alpha_6 \epsilon) |D_\tau(\nabla \overline{u})|^2 - \frac{n^2 \alpha_6}{\epsilon} |\nabla \overline{u}|^2. \tag{4.23}$$

Step 2.2. We can now return to estimating the three terms in equation (4.21). In the present step, unless stated otherwise, all L^2 norms are understood in $L^2(\Omega)$.

Estimate 1. From (4.23), we get

$$(\alpha_5 - n^2 \alpha_6 \epsilon) \|\rho(D_\tau(\nabla \overline{u}))\|_{L^2}^2 - \frac{n^2 \alpha_6}{\epsilon} \|\rho \nabla \overline{u}\|_{L^2}^2$$
$$\le \int_\Omega \langle D_\tau [\nabla_\xi g(x, \nabla \overline{u})]; \rho^2(D_\tau(\nabla \overline{u})) \rangle$$

and, since $0 \le \rho \le 1$, we have

$$(\alpha_5 - n^2 \alpha_6 \epsilon) \|\rho(D_\tau(\nabla \overline{u}))\|_{L^2}^2 - \frac{n^2 \alpha_6}{\epsilon} \|\nabla \overline{u}\|_{L^2}^2$$
$$\le \int_\Omega \langle D_\tau [\nabla_\xi g(x, \nabla \overline{u})]; \rho^2(D_\tau(\nabla \overline{u})) \rangle.$$

Estimate 2. From (4.22), we have, for $\epsilon > 0$ fixed,

$$\left| 2 \int_\Omega \langle D_\tau \left[\nabla_\xi g \left(x, \nabla \overline{u} \right) \right]; \rho \nabla \rho \left(D_\tau \overline{u} \right) \rangle \right|$$

$$\leq 2n\alpha_4 \int_\Omega \left| \rho \left(D_\tau \left(\nabla \overline{u} \right) \right) \right| \left| \nabla \rho \left(D_\tau \overline{u} \right) \right| + 2n\alpha_6 \int_\Omega \left| \rho \nabla \overline{u} \right| \left| \nabla \rho \left(D_\tau \overline{u} \right) \right|$$

$$\leq 2n\alpha_4 \epsilon \left\| \rho \left(D_\tau \left(\nabla \overline{u} \right) \right) \right\|_{L^2}^2 + \left(\frac{2n\alpha_4}{\epsilon} + n\alpha_6 \right) \left\| \nabla \rho \left(D_\tau \overline{u} \right) \right\|_{L^2}^2 + n\alpha_6 \left\| \rho \nabla \overline{u} \right\|_{L^2}^2 .$$

Observing that, from (4.7), we have

$$\left\| \nabla \rho \left(D_\tau \overline{u} \right) \right\|_{L^2}^2 = \int_{\text{supp } \rho} \left| \nabla \rho \left(D_\tau \overline{u} \right) \right|^2 \leq \left\| \nabla \rho \right\|_{L^\infty}^2 \int_{\text{supp } \rho} \left| D_\tau \overline{u} \right|^2 \leq \left\| \nabla \rho \right\|_{L^\infty}^2 \left\| \nabla \overline{u} \right\|_{L^2}^2 ,$$
$$\tag{4.24}$$

we thus get, recalling that $0 \leq \rho \leq 1$,

$$\left| 2 \int_\Omega \langle D_\tau \left[\nabla_\xi g \left(x, \nabla \overline{u} \right) \right]; \rho \nabla \rho \left(D_\tau \overline{u} \right) \rangle \right|$$

$$\leq 2n\alpha_4 \epsilon \left\| \rho \left(D_\tau \left(\nabla \overline{u} \right) \right) \right\|_{L^2}^2 + \left(\frac{2n\alpha_4}{\epsilon} + n\alpha_6 \right) \left\| \nabla \rho \right\|_{L^\infty}^2 \left\| \nabla \overline{u} \right\|_{L^2}^2 + n\alpha_6 \left\| \nabla \overline{u} \right\|_{L^2}^2 .$$

Estimate 3. It follows from (4.7) that, for $\epsilon > 0$ fixed,

$$\left| \int_\Omega h \left(D_{-\tau} \rho^2 \left(D_\tau \overline{u} \right) \right) \right| \leq \epsilon \left\| \left(D_{-\tau} \rho^2 \left(D_\tau \overline{u} \right) \right) \right\|_{L^2}^2 + \frac{1}{\epsilon} \left\| h \right\|_{L^2}^2$$

$$\leq \epsilon \left\| \nabla \left(\rho^2 \left(D_\tau \overline{u} \right) \right) \right\|_{L^2}^2 + \frac{1}{\epsilon} \left\| h \right\|_{L^2}^2 .$$

Since

$$\left\| \nabla \left(\rho^2 \left(D_\tau \overline{u} \right) \right) \right\|_{L^2}^2 = \int_\Omega \left| \rho^2 \left(D_\tau \left(\nabla \overline{u} \right) \right) + 2\rho \nabla \rho \left(D_\tau \overline{u} \right) \right|^2$$

$$\leq 2 \int_\Omega \left| \rho^2 \left(D_\tau \left(\nabla \overline{u} \right) \right) \right|^2 + 2 \int_\Omega \left| 2\rho \nabla \rho \left(D_\tau \overline{u} \right) \right|^2$$

we obtain, as in (4.24) and recalling that $0 \leq \rho \leq 1$ and thus $\rho^2 \leq \rho$,

$$\left| \int_\Omega h \left(D_{-\tau} \rho^2 \left(D_\tau \overline{u} \right) \right) \right| \leq 2\epsilon \left\| \rho \left(D_\tau \left(\nabla \overline{u} \right) \right) \right\|_{L^2}^2 + 8\epsilon \left\| \nabla \rho \right\|_{L^\infty}^2 \left\| \nabla \overline{u} \right\|_{L^2}^2 + \frac{1}{\epsilon} \left\| h \right\|_{L^2}^2 .$$

Step 2.3. We now gather all these three estimates in equation (4.21), to obtain

$$\left(\alpha_5 - n^2 \alpha_6 \epsilon - 2n\alpha_4 \epsilon - 2\epsilon \right) \left\| \rho \left(D_\tau \left(\nabla \overline{u} \right) \right) \right\|_{L^2}^2$$

$$\leq \left(\frac{2n\alpha_4}{\epsilon} + n\alpha_6 + 8\epsilon \right) \left\| \nabla \rho \right\|_{L^\infty}^2 \left\| \nabla \overline{u} \right\|_{L^2}^2$$

$$+ \left(n\alpha_6 + \frac{n^2 \alpha_6}{\epsilon} \right) \left\| \nabla \overline{u} \right\|_{L^2}^2 + \frac{1}{\epsilon} \left\| h \right\|_{L^2}^2 .$$

Therefore, choosing ϵ small, we can find $\gamma_1 = \gamma_1\left(\epsilon, \rho\right) > 0$ such that

$$\left\| \rho\left(D_\tau\left(\nabla\overline{u}\right)\right) \right\|_{L^2}^2 \leq \gamma_1 \left(\left\| \nabla\overline{u} \right\|_{L^2}^2 + \left\| h \right\|_{L^2}^2 \right).$$

Using the definition of ρ, we infer that

$$\left\| \left(D_\tau\left(\nabla\overline{u}\right)\right) \right\|_{L^2(O)}^2 \leq \gamma_1 \left(\left\| \nabla\overline{u} \right\|_{L^2(\Omega)}^2 + \left\| h \right\|_{L^2(\Omega)}^2 \right).$$

From (4.8) we deduce that

$$\left\| \nabla^2\overline{u} \right\|_{L^2(O)}^2 \leq \gamma_1 \left(\left\| \nabla\overline{u} \right\|_{L^2(\Omega)}^2 + \left\| h \right\|_{L^2(\Omega)}^2 \right)$$

and hence the fact that $\overline{u} \in W_{\mathrm{loc}}^{2,2}\left(\Omega\right)$ with

$$\left\| \overline{u} \right\|_{W^{2,2}(O)}^2 \leq \left(\gamma_1 + 1\right) \left(\left\| \overline{u} \right\|_{W^{1,2}(\Omega)}^2 + \left\| h \right\|_{L^2(\Omega)}^2 \right).$$

The more refined estimates (4.18) and (4.19) are obtained in Exercise 4.3.3. We can now integrate by parts (4.15) and invoke Theorem 1.24 to get (4.16). This concludes the proof of the theorem. ∎

4.3.4 Exercises

Exercise 4.3.1 Let $a \in \mathbb{R}^n$ and $\langle .; . \rangle$ denotes the scalar product in $\mathbb{R}^n$. Find functions g of the form

$$g\left(\xi\right) = G\left(|\xi|^2\right), \quad g\left(x, \xi\right) = \sum_{i,j=1}^{n} g_{ij}\left(x\right)\xi_i\xi_j \quad \text{or} \quad g\left(\xi\right) = \frac{1}{2}|\xi|^2 + G\left(\langle a; \xi\rangle\right)$$

satisfying the hypotheses of Theorem 4.9.

Exercise 4.3.2 Let $h \in L^2\left(\Omega\right)$ and $g \in C^1\left(\overline{\Omega} \times \mathbb{R}^n\right)$, $g = g\left(x, \xi\right)$, such that there exist $\alpha, \beta > 0$ so that, for every $\left(x, \xi\right) \in \overline{\Omega} \times \mathbb{R}^n$,

$$\langle \nabla_\xi g\left(x, \xi\right); \xi \rangle \geq \alpha |\xi|^2 \quad \text{and} \quad |\nabla_\xi g\left(x, \xi\right)| \leq \beta |\xi|.$$

Let $\overline{u} \in W^{1,2}\left(\Omega\right)$ satisfy

$$\int_\Omega \langle \nabla_\xi g\left(x, \nabla\overline{u}\left(x\right)\right); \nabla\varphi\left(x\right) \rangle \, dx = \int_\Omega h\left(x\right) \varphi\left(x\right) \, dx, \quad \forall \varphi \in W_0^{1,2}\left(\Omega\right).$$

Let $O \subset \overline{O} \subset \Omega$ be an open set. Show that there exists a constant $\gamma = \gamma\left(O, \Omega\right) > 0$ such that

$$\left\| \nabla\overline{u} \right\|_{L^2(O)} \leq \gamma \left(\left\| \overline{u} \right\|_{L^2(\Omega)} + \left\| h \right\|_{L^2(\Omega)} \right).$$

Exercise 4.3.3 Assume, in addition to the hypotheses of Theorem 4.9, that

$$\nabla_\xi g(x, 0) = 0, \quad \forall x \in \overline{\Omega}.$$

(i) With the help of Exercise 4.3.2, prove that in Theorem 4.9 (or in Theorem 4.8), the estimate is not only of the form

$$\|\overline{u}\|_{W^{2,2}(O)} \leq \gamma \left(\|\overline{u}\|_{W^{1,2}(\Omega)} + \|h\|_{L^2(\Omega)} \right)$$

but also takes the more refined form (4.18) (or (4.11)), namely

$$\|\overline{u}\|_{W^{2,2}(O)} \leq \gamma \left(\|\overline{u}\|_{L^2(\Omega)} + \|h\|_{L^2(\Omega)} \right).$$

(ii) If we further assume that $u_0 = 0$, show that the estimate (4.19) (or (4.12)), has the simpler form

$$\|\overline{u}\|_{W^{2,2}(O)} \leq \gamma \|h\|_{L^2(\Omega)}.$$

4.4 The difference quotient method: boundary regularity

We now improve Theorem 4.9 in a special case, so as to get regularity up to the boundary, i.e. $\overline{u} \in W^{2,2}$, and not only in $W^{2,2}_{\text{loc}}$.

Theorem 4.10 *Let $\Omega \subset \mathbb{R}^n$ be a bounded open set with C^2 boundary, $h \in L^2(\Omega)$ and $g_{ij} = g_{ji} \in C^1(\overline{\Omega})$ such that there exists $\alpha > 0$ with*

$$(G) \quad \sum_{i,j=1}^n g_{ij}(x) \lambda_i \lambda_j \geq \alpha |\lambda|^2, \quad \forall \lambda \in \mathbb{R}^n \text{ and } \forall x \in \overline{\Omega}.$$

Then there exists a unique minimizer $\overline{u} \in W^{1,2}_0(\Omega) \cap W^{2,2}(\Omega)$ of

$$(P) \quad \inf \left\{ I(u) = \sum_{i,j=1}^n \int_\Omega g_{ij}\, u_{x_i} u_{x_j} - \int_\Omega h\, u : u \in W^{1,2}_0(\Omega) \right\}$$

and there exists a constant $\gamma > 0$ such that

$$\|\overline{u}\|_{W^{2,2}(\Omega)} \leq \gamma \|h\|_{L^2(\Omega)}. \tag{4.25}$$

Remark 4.11 **(i)** As usual, the regularity will be obtained for solutions of the Euler-Lagrange equation (4.26).

(ii) The theorem can be extended to the non-quadratic setting of Theorem 4.9.

Proof The proof is in the same spirit as that of Theorem 4.9 but technically heavier. We only outline the main steps. We refer for more details to Brézis [14], John [74] or Evans [47], which we follow here. The Euler-Lagrange equation is then

$$2 \int_\Omega \sum_{i,j=1}^n g_{ij} \bar{u}_{x_i} \varphi_{x_j} = \int_\Omega h\varphi, \quad \forall \varphi \in W_0^{1,2}(\Omega) \tag{4.26}$$

and, since $\bar{u} \in W_{loc}^{2,2}(\Omega)$, we also have, after integration by parts and using Theorem 1.24,

$$\sum_{i,j=1}^n g_{ij} \bar{u}_{x_i x_j} + \sum_{i,j=1}^n \frac{\partial}{\partial x_j}[g_{ij}]\bar{u}_{x_i} = -\frac{h}{2}, \quad \text{a.e. in } \Omega. \tag{4.27}$$

Step 1. We start with a special choice of Ω, namely

$$\Omega = Q_+ = Q \cap \{x_n > 0\}$$

where $Q = (-1,1)^n$. Let $q \subset \bar{q} \subset Q$ be open and set

$$q_+ = q \cap \{x_n > 0\}.$$

Find $\rho \in C_0^\infty(Q)$ such that

$$0 \le \rho \le 1 \quad \text{and} \quad \rho \equiv 1 \text{ in } q.$$

Note, in passing, that ρ vanishes in most of ∂Q_+ but not in the whole of ∂Q_+; more precisely, it vanishes whenever any of the $x_i = \pm 1$, $i = 1, \cdots, n-1$ and $x_n = 1$, but not necessarily when $x_n = 0$. We then proceed exactly as in Step 2 of Theorem 4.9, choosing

$$\varphi = D_{-\tau}\left(\rho^2\left(D_\tau \bar{u}\right)\right)$$

in equation (4.26), but now with

$$\tau = (\tau_1, \cdots, \tau_{n-1}, 0) \ne 0.$$

Since $\bar{u} \in W_0^{1,2}(Q_+)$ and $\rho \in C_0^\infty(Q)$, we find that $\varphi \in W_0^{1,2}(Q_+)$ for $|\tau|$ small enough. We therefore get, as in the case of interior regularity,

$$\|D_\tau(\nabla\bar{u})\|_{L^2(q_+)}^2 \le \gamma_1 \left(\|\bar{u}\|_{W^{1,2}(Q_+)}^2 + \|h\|_{L^2(Q_+)}^2\right)$$

and thus we obtain that all the $\bar{u}_{x_i x_j}$, except $\bar{u}_{x_n x_n}$ because of the fact that $\tau_n = 0$, are in $L^2(q_+)$, more precisely

$$\sum_{2 \le i+j < 2n} \|\bar{u}_{x_i x_j}\|_{L^2(q_+)}^2 \le \gamma_2 \left(\|\bar{u}\|_{W^{1,2}(Q_+)}^2 + \|h\|_{L^2(Q_+)}^2\right). \tag{4.28}$$

In order to get the full estimate we need to control $\overline{u}_{x_n x_n}$. This follows easily from (4.27) and (4.28), since

$$g_{nn}\overline{u}_{x_n x_n} = -\sum_{2 \le i+j < 2n} g_{ij}\overline{u}_{x_i x_j} - \sum_{i,j=1}^{n} \frac{\partial}{\partial x_j}[g_{ij}]\,\overline{u}_{x_i} - \frac{h}{2}$$

and since $g_{nn} \ge \alpha > 0$ by (G). We therefore have now as in Theorem 4.9, the full estimate

$$\|\overline{u}\|_{W^{2,2}(q_+)}^2 \le \gamma_3 \left(\|\overline{u}\|_{W^{1,2}(Q_+)}^2 + \|h\|_{L^2(Q_+)}^2 \right). \tag{4.29}$$

Step 2. Next we want to consider more general sets than Q_+. Recall that

$$Q = (-1,1)^n \quad \text{and} \quad Q_+ = Q \cap \{x_n > 0\}.$$

So let $W \subset \mathbb{R}^n$ be a bounded open set such that there exists a one-to-one and onto map $H : \overline{Q} \to \overline{W}$ such that

$$H \in C^2\left(\overline{Q}; \overline{W}\right), \quad H^{-1} \in C^2\left(\overline{W}; \overline{Q}\right) \quad \text{and} \quad \det \nabla H > 0.$$

Let $W_+ = H(Q_+)$, $w \subset \overline{w} \subset W$ be open and $w_+ = w \cap W_+$. We want to prove that we can find $\gamma_4 > 0$ such that

$$\|\overline{u}\|_{W^{2,2}(w_+)}^2 \le \gamma_4 \left(\|\overline{u}\|_{W^{1,2}(W_+)}^2 + \|h\|_{L^2(W_+)}^2 \right). \tag{4.30}$$

We change variables and set

$$x = H(y), \quad \overline{u}(x) = v\left(H^{-1}(x)\right), \quad \varphi(x) = \psi\left(H^{-1}(x)\right).$$

We therefore immediately have (see Exercise 4.4.1), with the change of variables $x = H(y)$,

$$\sum_{i,j=1}^{n} \int_{W_+} g_{ij}(x)\,\overline{u}_{x_i}(x)\,\varphi_{x_j}(x)\,dx = \sum_{k,l=1}^{n} \int_{Q_+} \widetilde{g}_{kl}(y)\,v_{y_k}(y)\,\psi_{y_l}(y)\,dy$$

where

$$\widetilde{g}_{kl}(y) = \sum_{i,j=1}^{n} g_{ij}(H(y))\,\frac{\partial H_k^{-1}}{\partial x_i}(H(y))\,\frac{\partial H_l^{-1}}{\partial x_j}(H(y))\,\det \nabla H(y).$$

Similarly, setting $\widetilde{h}(y) = h(H(y))\det \nabla H(y)$, we infer that

$$\int_{W_+} h(x)\,\varphi(x)\,dx = \int_{Q_+} \widetilde{h}(y)\,\psi(y)\,dy.$$

The Euler-Lagrange equation (4.26) then becomes, for every $\psi \in W_0^{1,2}(Q_+)$,

$$2 \sum_{k,l=1}^{n} \int_{Q_+} \widetilde{g}_{kl}(y) \, v_{y_k}(y) \, \psi_{y_l}(y) \, dy = \int_{Q_+} \widetilde{h}(y) \, \psi(y) \, dy . \tag{4.31}$$

Note (see Exercise 4.4.1) that the uniform ellipticity is preserved, which means that there exists a constant $\beta > 0$ such that

$$\sum_{k,l=1}^{n} \widetilde{g}_{kl}(y) \lambda_k \lambda_l \geq \beta |\lambda|^2 , \quad \forall \lambda \in \mathbb{R}^n \text{ and } \forall y \in \overline{Q}_+ .$$

We then apply Step 1 to (4.31) to get (4.29), namely

$$\|v\|_{W^{2,2}(q_+)}^2 \leq \widetilde{\gamma}_3 \left(\|v\|_{W^{1,2}(Q_+)}^2 + \left\| \widetilde{h} \right\|_{L^2(Q_+)}^2 \right)$$

where $q_+ = H^{-1}(w_+)$. Since $\overline{u} = v \circ H^{-1}$ and $h = \widetilde{h} \circ H^{-1} (\det \nabla H)^{-1}$, we have (4.30) from the above estimate.

Step 3. We then cover $\partial \Omega$ with a finite number of sets $U_1, \cdots, U_N$ as in Definition 1.42 and we then find an open set $O \subset \overline{O} \subset \Omega$ so that

$$\overline{\Omega} \subset O \cup U_1 \cup \cdots \cup U_N .$$

It follows from (4.30) applied to each U_i and from the estimate in the interior applied to O (see Theorem 4.9), that $\overline{u} \in W^{2,2}(\Omega)$ and there exists $\gamma_5 > 0$ such that

$$\|\overline{u}\|_{W^{2,2}(\Omega)}^2 \leq \gamma_5 \left(\|\overline{u}\|_{W^{1,2}(\Omega)}^2 + \|h\|_{L^2(\Omega)}^2 \right) . \tag{4.32}$$

Step 4. It remains to establish (4.25). This is proved exactly as in Exercise 4.3.3; but, for the sake of completeness, we show it again. Observe first that since $\overline{u} \in W_0^{1,2}(\Omega)$, we have, from Poincaré inequality, that there exists $\theta > 0$ such that

$$\theta \|\overline{u}\|_{W^{1,2}}^2 \leq \|\nabla \overline{u}\|_{L^2}^2 . \tag{4.33}$$

Returning to the Euler-Lagrange equation (4.26) with $\varphi = \overline{u} \in W_0^{1,2}(\Omega)$, we find

$$2 \int_{\Omega} \sum_{i,j=1}^{n} g_{ij} \overline{u}_{x_i} \overline{u}_{x_j} = \int_{\Omega} h \overline{u} .$$

Hence, appealing to (G), we find for $\epsilon > 0$ fixed

$$2\alpha \|\nabla \overline{u}\|_{L^2}^2 \leq 2 \int_{\Omega} \sum_{i,j=1}^{n} g_{ij} \overline{u}_{x_i} \overline{u}_{x_j} = \int_{\Omega} h \overline{u} \leq \epsilon \|\overline{u}\|_{W^{1,2}}^2 + \frac{1}{\epsilon} \|h\|_{L^2}^2 .$$

Choosing ϵ small and combining with (4.33), we get

$$\|\overline{u}\|^2_{W^{1,2}} \le \frac{1}{\epsilon\,(2\alpha\theta - \epsilon)} \|h\|^2_{L^2}$$

which coupled with (4.32) leads to the claim. ∎

4.4.1 Exercise

Exercise 4.4.1 Let $U \subset \mathbb{R}^n$ be a bounded open set with C^2 boundary. Assume that

$$\sum_{k,l=1}^{n} a_{kl}\,(x)\,\lambda_k\lambda_l \ge \alpha\,|\lambda|^2\,, \quad \forall\,\lambda \in \mathbb{R}^n \text{ and } \forall\,x \in \overline{U}$$

where $a_{kl} \in C\left(\overline{U}\right)$ and $\alpha > 0$. Show that a regular change of variables $H : \overline{Q} \to \overline{U}$ (in particular $\nabla H \in C\left(\overline{Q}; \mathbb{R}^{n \times n}\right)$, $\nabla H^{-1} \in C\left(\overline{U}; \mathbb{R}^{n \times n}\right)$ are invertible matrices and $\det \nabla H > 0$) and

$$x = H\,(y)\,, \quad u\,(x) = v\left(H^{-1}\,(x)\right)\,, \quad \varphi\,(x) = \psi\left(H^{-1}\,(x)\right)$$

leads to

$$\sum_{i,j=1}^{n} \int_U a_{ij}\,(x)\,u_{x_i}\,(x)\,\varphi_{x_j}\,(x)\,dx = \sum_{k,l=1}^{n} \int_Q b_{kl}\,(y)\,v_{y_k}\,(y)\,\psi_{y_l}\,(y)\,dy$$

where

$$b_{kl}\,(y) = \sum_{i,j=1}^{n} a_{ij}\,(H\,(y))\,\frac{\partial H_k^{-1}}{\partial x_i}\,(H\,(y))\,\frac{\partial H_l^{-1}}{\partial x_j}\,(H\,(y))\,\det \nabla H\,(y)\,.$$

Moreover, the following holds

$$\sum_{k,l=1}^{n} b_{kl}\,(y)\,\lambda_k\lambda_l \ge \beta\,|\lambda|^2\,, \quad \forall\,\lambda \in \mathbb{R}^n \text{ and } \forall\,y \in \overline{Q}$$

for a certain constant $\beta > 0$.

4.5 Higher regularity for the Dirichlet integral

To get higher regularity in the context of Sections 4.3 and 4.4 is a difficult task and requires new ideas which are explained in Section 4.7 (see Theorem 4.18). However, when we are considering Dirichlet integral or more generally quadratic integrands with regular coefficients (see Exercise 4.5.2) it is much easier and we discuss this matter now. Let us first express the procedure informally. Assume that h is more regular than in L^2 say in $W^{1,2}$. Once we have established the fact that $\overline{u}$ is in $W^{2,2}$ and satisfies the equation (see Theorem 4.8 or Theorem 4.10)

$$-\Delta \overline{u} = h$$

it is enough to differentiate with respect to any of the variables to have the equation

$$-\Delta \overline{u}_{x_i} = h_{x_i}$$

and restart the process to get $\overline{u}_{x_i} \in W^{2,2}$ and thus $\overline{u} \in W^{3,2}$. Iterating the process we have the maximal possible regularity.

Let us now be more precise. We recall that $\Omega \subset \mathbb{R}^n$ is a bounded open set with Lipschitz boundary, $h \in L^2(\Omega)$ and

$$(P) \quad \inf \left\{ I(u) = \frac{1}{2} \int_\Omega |\nabla u(x)|^2 \, dx - \int_\Omega h(x) u(x) \, dx : u \in W_0^{1,2}(\Omega) \right\}.$$

We have seen in Theorem 4.8 that there exists a unique minimizer $\overline{u} \in W_0^{1,2}(\Omega) \cap W_{loc}^{2,2}(\Omega)$ of (P). Higher regularity will be obtained, as in Theorem 4.8, for solutions of the Euler-Lagrange equation and not only for minimizers of (P).

Theorem 4.12 *Let $k \geq 0$ be an integer, $\Omega \subset \mathbb{R}^n$ be a bounded open set with Lipschitz boundary, $h \in W^{k,2}(\Omega)$ and let $\overline{u} \in W^{1,2}(\Omega)$ be a solution of*

$$\int_\Omega \langle \nabla \overline{u}; \nabla \varphi \rangle = \int_\Omega h \, \varphi, \quad \forall \varphi \in W_0^{1,2}(\Omega). \tag{4.34}$$

Then $\overline{u} \in W_{loc}^{k+2,2}(\Omega)$ and, for every open set $O \subset \overline{O} \subset \Omega$, there exists a constant $\gamma = \gamma(O, \Omega) > 0$ so that

$$\|\overline{u}\|_{W^{k+2,2}(O)} \leq \gamma \left(\|\overline{u}\|_{L^2(\Omega)} + \|h\|_{W^{k,2}(\Omega)} \right) \tag{4.35}$$

while if, in addition, $\overline{u} \in W_0^{1,2}(\Omega)$, then the estimate takes the simpler form

$$\|\overline{u}\|_{W^{k+2,2}(O)} \leq \gamma \|h\|_{W^{k,2}(\Omega)}.$$

Moreover, if the boundary of Ω is C^{k+2} and $\overline{u} \in W_0^{1,2}(\Omega)$, then $\overline{u} \in W^{k+2,2}(\Omega)$ and

$$\|\overline{u}\|_{W^{k+2,2}(\Omega)} \leq \gamma \|h\|_{W^{k,2}(\Omega)}. \tag{4.36}$$

In particular, if $k = \infty$, then $\overline{u} \in C^\infty(\overline{\Omega})$.

Remark 4.13 (i) If in Problem (P) we impose a non-zero boundary datum $u_0 \in W^{k+2,2}(\Omega)$, similar results hold (see Exercise 4.5.1).

(ii) An analogous result to (4.36) can be obtained in Hölder spaces (these are then known as *Schauder estimates*), under appropriate regularity hypotheses on the boundary and when $0 < a < 1$, namely

$$\|\overline{u}\|_{C^{k+2,a}} \leq \gamma \|h\|_{C^{k,a}} .$$

If $1 < p < \infty$, it can also be proved that

$$\|\overline{u}\|_{W^{k+2,p}} \leq \gamma \|h\|_{W^{k,p}} ;$$

these are then known as *Calderon-Zygmund estimates* and are considerably harder to obtain than those for $p = 2$.

(iii) Both the above results are however false if $a = 0$, $a = 1$ or $p = \infty$ (see Exercise 4.5.3) and if $p = 1$ (see Exercise 4.5.4). This is another reason why, when dealing with partial differential equations or the calculus of variations, Sobolev and Hölder spaces are more appropriate than C^k spaces.

Proof We discuss the case of interior regularity; the case of regularity up to the boundary is obtained similarly by applying, below, Theorem 4.10 instead of Theorem 4.8. We will prove the result for $k = 1$; the general case follows by repeating the argument.

Step 1. According to Theorem 4.8 (or Step 3 of Theorem 4.10), there exists a constant $\gamma_1 = \gamma_1(O, \Omega)$ such that the solution $\overline{u} \in W^{1,2}(\Omega)$ of (4.34) is in $W^{2,2}_{\text{loc}}(\Omega)$ and satisfies

$$\|\overline{u}\|_{W^{2,2}(O)} \leq \gamma_1 \left(\|\overline{u}\|_{L^2(\Omega)} + \|h\|_{L^2(\Omega)} \right) \quad \text{and} \quad -\Delta \overline{u} = h, \text{ a.e. in } \Omega. \quad (4.37)$$

Now let $h \in W^{1,2}(\Omega)$ and let us show that $\overline{u} \in W^{3,2}_{\text{loc}}(\Omega)$. The idea is simple, it consists in applying Theorem 4.8 to $\overline{u}_{x_i} = \partial \overline{u}/\partial x_i$ and observing that since $-\Delta \overline{u} = h$, then $-\Delta \overline{u}_{x_i} = h_{x_i}$. We now give some details. Let $O \subset \overline{O} \subset \Lambda \subset \overline{\Lambda} \subset \Omega$ be open sets. We know that $\overline{u}_{x_i} \in W^{1,2}(\Lambda)$, for every $i = 1, \cdots, n$, and observe that, for every $\varphi \in C_0^\infty(\Lambda)$,

$$\int_\Lambda \langle \nabla \overline{u}_{x_i}; \nabla \varphi \rangle = \int_\Lambda \langle (\nabla \overline{u})_{x_i}; \nabla \varphi \rangle = -\int_\Lambda \langle \nabla \overline{u}; (\nabla \varphi)_{x_i} \rangle$$

$$= -\int_\Lambda \langle \nabla \overline{u}; \nabla \varphi_{x_i} \rangle = -\int_\Lambda h \varphi_{x_i} = \int_\Lambda h_{x_i} \varphi .$$

Since $C_0^\infty(\Lambda)$ is dense in $W_0^{1,2}(\Lambda)$, we deduce, for every $i = 1, \cdots, n$, that

$$\int_\Lambda \langle \nabla \overline{u}_{x_i}; \nabla \varphi \rangle = \int_\Lambda h_{x_i} \varphi, \quad \forall \varphi \in W_0^{1,2}(\Lambda) .$$

Since $h \in W^{1,2}(\Lambda)$, we have that $h_{x_i} \in L^2(\Lambda)$ and hence, by Theorem 4.8 (or Step 3 of Theorem 4.10), we get that $\overline{u}_{x_i} \in W^{2,2}_{\mathrm{loc}}(\Lambda)$ and there exists a constant $\gamma_2 = \gamma_2(O, \Lambda) > 0$ such that

$$\|\overline{u}_{x_i}\|_{W^{2,2}(O)} \leq \gamma_2 \left(\|\overline{u}_{x_i}\|_{L^2(\Lambda)} + \|h_{x_i}\|_{L^2(\Lambda)} \right) \leq \gamma_2 \left(\|\overline{u}\|_{W^{1,2}(\Lambda)} + \|h\|_{W^{1,2}(\Lambda)} \right).$$

Appealing to Exercise 4.3.2, we have, since $\overline{u}$ satisfies (4.34), that there exists $\gamma_3 = \gamma_3(\Lambda, \Omega) > 0$ so that

$$\|\overline{u}\|_{W^{1,2}(\Lambda)} \leq \gamma_3 \left(\|\overline{u}\|_{L^2(\Omega)} + \|h\|_{L^2(\Omega)} \right). \tag{4.38}$$

Combining the last two inequalities, we have indeed established that (4.35), with $k = 1$, holds and thus $\overline{u} \in W^{3,2}_{\mathrm{loc}}(\Omega)$.

Step 2. When $\overline{u} \in W^{1,2}_0(\Omega)$, we proceed exactly in the same way, the only difference being that we use Exercise 4.3.3 (ii) (more precisely (7.21)) to have

$$\|\overline{u}\|_{W^{1,2}(\Omega)} \leq \gamma_3 \|h\|_{L^2(\Omega)}$$

instead of (4.38). We therefore have

$$\|\overline{u}\|_{W^{3,2}(O)} \leq \gamma \|h\|_{W^{1,2}(\Omega)}.$$

This concludes the proof of the theorem. ∎

4.5.1 Exercises

Exercise 4.5.1 Let $u_0 \in W^{k+2,2}(\Omega)$. Prove the estimates of Theorem 4.12 for

$$(P) \quad \inf \left\{ I(u) = \frac{1}{2} \int_\Omega |\nabla u|^2 - \int_\Omega h\, u : u \in u_0 + W^{1,2}_0(\Omega) \right\}.$$

Exercise 4.5.2 Let $\alpha > 0$, $h \in W^{k,2}(\Omega)$, $g_{ij} = g_{ji} \in C^\infty(\overline{\Omega})$ with

$$\sum_{i,j=1}^n g_{ij}(x) \lambda_i \lambda_j \geq \alpha |\lambda|^2, \quad \forall\, x \in \overline{\Omega} \text{ and } \forall\, \lambda \in \mathbb{R}^n.$$

Prove that the minimizer of

$$(P) \quad \inf \left\{ I(u) = \int_\Omega f(x, u(x), \nabla u(x))\, dx : u \in W^{1,2}_0(\Omega) \right\}$$

is in $W^{1,2}_0(\Omega) \cap W^{k+2,2}_{\mathrm{loc}}(\Omega)$, where

$$f(x, u, \xi) = \sum_{i,j=1}^n g_{ij}(x) \xi_i \xi_j - h(x) u.$$

Exercise 4.5.3 We show here that if $f \in C^0$, then, in general, there is no solution $u \in C^2$ of $\Delta u = f$. Let $\Omega = \left\{ x \in \mathbb{R}^2 : |x| < 1/2 \right\}$ and for $0 < \alpha < 1$, define

$$u\left(x \right) = u\left(x_1, x_2 \right) = \begin{cases} x_1 x_2 \left| \log |x| \right|^\alpha & \text{if } 0 < |x| < 1/2 \\ 0 & \text{if } x = 0. \end{cases}$$

Prove that

$$u_{x_1 x_1}, \ u_{x_2 x_2} \in C^0\left(\overline{\Omega} \right) \quad \text{and} \quad u_{x_1 x_2} \notin L^\infty\left(\Omega \right)$$

which implies that $\Delta u = u_{x_1 x_1} + u_{x_2 x_2} \in C^0$, while $u \notin C^2$; in fact u is not even in $W^{2,\infty}$.

Exercise 4.5.4 Let

$$\Omega = \left\{ x \in \mathbb{R}^2 : 0 < |x| < 1/2 \right\}$$

$$u\left(x \right) = u\left(x_1, x_2 \right) = \log \left| \log |x| \right|, \quad \text{if } x \in \Omega.$$

Show that $u \notin W^{2,1}\left(\Omega \right)$ while $\Delta u \in L^1\left(\Omega \right)$.

Exercise 4.5.5 Let $\Omega \subset \mathbb{R}^n$ be a bounded connected open set with Lipschitz boundary, $h \in W^{k,2}\left(\Omega \right)$, where $k \geq 0$ is an integer, and consider the Neumann problem

$$(N) \quad \inf \left\{ I\left(u \right) = \int_\Omega \left[\frac{1}{2} \left| \nabla u \right|^2 - h\, u \right] : u \in X \right\} = m$$

where

$$X = \left\{ u \in W^{1,2}\left(\Omega \right) : u_\Omega = 0 \right\} \quad \text{and} \quad u_\Omega = \frac{1}{\text{meas}\, \Omega} \int_\Omega u.$$

We have seen in Exercise 3.2.2 that (N) has one and only one solution $\overline{u} \in X$ that satisfies (see (7.14))

$$\int_\Omega \left[\langle \nabla \overline{u}; \nabla \psi \rangle - \left(h - h_\Omega \right) \psi \right] = 0, \quad \forall\, \psi \in W^{1,2}\left(\Omega \right).$$

Show that $\overline{u} \in W^{k+2,2}_{\text{loc}}\left(\Omega \right)$ and that, for every open set $O \subset \overline{O} \subset \Omega$, there exists a constant $\gamma = \gamma\left(O, \Omega \right) > 0$ such that

$$\left\| \overline{u} \right\|_{W^{k+2,2}(O)} \leq \gamma \left\| h - h_\Omega \right\|_{W^{k,2}(\Omega)}.$$

4.6 Weyl lemma

We now turn to a completely different approach to get interior regularity. It applies only to the case of Dirichlet integral

$$(P) \quad \inf \left\{ I(u) = \frac{1}{2} \int_{\Omega} |\nabla u(x)|^2 \, dx : u \in u_0 + W_0^{1,2}(\Omega) \right\}$$

and to its Euler-Lagrange equation

$$(E_w) \quad \int_{\Omega} \langle \nabla \overline{u}(x) ; \nabla \varphi(x) \rangle \, dx = 0, \quad \forall \varphi \in W_0^{1,2}(\Omega)$$

which in its strong form is Laplace equation

$$\Delta \overline{u} = 0 \quad \text{in } \Omega.$$

Theorem 4.14 (Weyl lemma) *Let $\Omega \subset \mathbb{R}^n$ be open and $\overline{u} \in L^1_{loc}(\Omega)$ satisfy*

$$\int_{\Omega} \overline{u}(x) \Delta \varphi(x) \, dx = 0, \quad \forall \varphi \in C_0^{\infty}(\Omega) \tag{4.39}$$

then $\overline{u} \in C^{\infty}(\Omega)$ and $\Delta \overline{u} = 0$ in Ω.

Remark 4.15 (i) The function $\overline{u}$ being defined only a.e., we have to interpret the result, as usual, up to a change of the function on a set of measure zero.

(ii) Note that a solution of the weak form of Laplace equation

$$(E_w) \quad \int_{\Omega} \langle \nabla \overline{u}(x) ; \nabla \varphi(x) \rangle \, dx = 0, \quad \forall \varphi \in W_0^{1,2}(\Omega)$$

satisfies (4.39). The converse being true if, in addition, $\overline{u} \in W^{1,2}(\Omega)$. Therefore (4.39) can be seen as a "very weak" form of Laplace equation and a solution of this equation as a "very weak" solution of $\Delta \overline{u} = 0$.

(iii) The case $n = 1$ is elementary and is discussed in Exercise 1.3.9; therefore, from now on, we assume that $n \geq 2$.

Proof The idea of the proof is to show that, up to redefining the function on a set of measure zero, $\overline{u}$ is continuous and satisfies the mean value formula

$$\overline{u}(x) = \frac{1}{\omega_n r^n} \int_{B_r(x)} \overline{u}(y) \, dy \tag{4.40}$$

for every $x \in \Omega$ and $r > 0$ sufficiently small so that

$$B_r(x) = \{y \in \mathbb{R}^n : |y - x| < r\} \subset \overline{B_r(x)} \subset \Omega.$$

and where $\omega_n = \text{meas}\,(B_1\,(0))$. A classical result (see Exercise 4.6.1) allows us to conclude that, in fact, $\overline{u} \in C^\infty\,(\Omega)$ and hence, by (4.39) and the fundamental lemma of the calculus of variations (see Theorem 1.24), $\Delta \overline{u} = 0$ in Ω, as claimed.

The proof is divided into three steps.

Step 1. We first prove that a regularization of $\overline{u}$ satisfies the mean value formula. To achieve this, we first choose an even function $\varphi \in C_0^\infty\,(B_1\,(0))$ with $\int \varphi = 1$ and define

$$\varphi_\epsilon\,(x) = \frac{1}{\epsilon^n} \varphi\left(\frac{x}{\epsilon}\right) \quad \text{and} \quad \Omega_\epsilon = \left\{x \in \Omega : \overline{B_\epsilon\,(x)} \subset \Omega\right\}.$$

Observe that since $\overline{u} \in L_{\text{loc}}^1\,(\Omega)$, then $\varphi_\epsilon * \overline{u} \in C^\infty\,(\Omega_\epsilon)$, where for $f, g \in L^1\,(\mathbb{R}^n)$ we have let

$$(f * g)\,(x) = \int_{\mathbb{R}^n} f\,(x - y)\,g\,(y)\,dy.$$

We next take any $\psi \in C_0^\infty\,(\Omega_\epsilon)$ and note that

$$\int_\Omega \Delta\,(\varphi_\epsilon * \overline{u})\,\psi = \int_\Omega (\varphi_\epsilon * \overline{u})\,\Delta\psi = \int_\Omega \overline{u}\,(\varphi_\epsilon * \Delta\psi)$$

where we have used, in the last identity, Fubini theorem and the fact that φ_ϵ is even. Since $\varphi_\epsilon * \Delta\psi = \Delta\,(\varphi_\epsilon * \psi)$, we get

$$\int_\Omega \Delta\,(\varphi_\epsilon * \overline{u})\,\psi = \int_\Omega \overline{u}\,\Delta\,(\varphi_\epsilon * \psi).$$

Appealing to (4.39) and to the fact that $\varphi_\epsilon * \psi \in C_0^\infty\,(\Omega)$, we obtain

$$\int_\Omega \Delta\,(\varphi_\epsilon * \overline{u})\,\psi = 0, \quad \forall\,\psi \in C_0^\infty\,(\Omega_\epsilon).$$

The fundamental lemma of the calculus of variations (Theorem 1.24) implies then

$$\Delta\,(\varphi_\epsilon * \overline{u}) = 0 \quad \text{in } \Omega_\epsilon.$$

Since any harmonic function satisfies the mean value formula, we have, for any $x \in \Omega_\epsilon$,

$$(\varphi_\epsilon * \overline{u})\,(x) = \frac{1}{\omega_n r^n} \int_{B_r(x)} (\varphi_\epsilon * \overline{u})\,(y)\,dy. \tag{4.41}$$

Step 2. We now prove that $\overline{u}$ itself satisfies the mean value formula. We know (see Exercise 1.3.5) that, for any bounded open set $O \subset \overline{O} \subset \Omega$ and up to a subsequence that we do not relabel,

$$\varphi_\epsilon * \overline{u} \to \overline{u} \text{ in } L^1\,(O) \quad \text{and} \quad \varphi_\epsilon * \overline{u} \to \overline{u} \text{ a.e. in } O.$$

Therefore, combining this observation with (4.41), we deduce that, for almost every $x \in O$,

$$\bar{u}(x) = \lim_{\epsilon \to 0} (\varphi_\epsilon * \bar{u})(x) = \frac{1}{\omega_n r^n} \int_{B_r(x)} \bar{u}(y)\, dy.$$

We therefore have proved, up to redefining $\bar{u}$ on a set of measure zero, that $\bar{u}$ satisfies the mean value formula, namely

$$\bar{u}(x) = \frac{1}{\omega_n r^n} \int_{B_r(x)} \bar{u}(y)\, dy$$

for every $x \in \Omega$ and $r > 0$ sufficiently small so that

$$\overline{B_r(x)} \subset \Omega.$$

Step 3. In order to conclude the proof of the theorem, from Exercise 4.6.1 and Step 2, we only need to check that $\bar{u}$ is continuous. This is easily seen as follows. Let $x, y \in \Omega$ and r be sufficiently small so that $\overline{B_r(x)} \cup \overline{B_r(y)} \subset \Omega$. We then have that

$$|\bar{u}(x) - \bar{u}(y)| = \frac{1}{\omega_n r^n} \left| \int_{B_r(x)} \bar{u}(z)\, dz - \int_{B_r(y)} \bar{u}(z)\, dz \right|$$

$$\leq \frac{1}{\omega_n r^n} \int_O |\bar{u}(z)|\, dz,$$

where $O = (B_r(x) \cup B_r(y)) \setminus (B_r(x) \cap B_r(y))$. Appealing to the fact that $\bar{u} \in L^1(B_r(x) \cup B_r(y))$ and to Exercise 1.3.10, we deduce that $\bar{u}$ is indeed continuous. ∎

4.6.1 Exercise

Exercise 4.6.1 Let $\Omega \subset \mathbb{R}^n$ be an open set. Let $u \in C^0(\Omega)$ satisfy the mean value formula, which states that

$$u(x) = \frac{1}{\omega_n r^n} \int_{B_r(x)} u(y)\, dy$$

for every $x \in \Omega$ and for every $r > 0$ sufficiently small so that

$$B_r(x) = \{y \in \mathbb{R}^n : |y - x| < r\} \subset \overline{B_r(x)} \subset \Omega.$$

Show that $u \in C^\infty(\Omega)$.

4.7   Some general results

The generalization of Section 4.5 to integrands of the form $f = f(x, u, \nabla u)$ is a difficult task. We give here, without proof, a general theorem and we refer for more results to the literature. The next theorem can be found in Morrey [86] (Theorem 1.10.4).

Theorem 4.16 *Let $\Omega \subset \mathbb{R}^n$ be a bounded open set and $f \in C^\infty (\Omega \times \mathbb{R} \times \mathbb{R}^n)$, $f = f(x, u, \xi)$. Let f satisfy, for every $(x, u, \xi) \in \Omega \times \mathbb{R} \times \mathbb{R}^n$ and $\lambda \in \mathbb{R}^n$,*

$$
(C) \quad \begin{cases}
\alpha_1 V^p - \alpha_2 \leq f(x, u, \xi) \leq \alpha_3 V^p \\[2mm]
|f_\xi|, \, |f_{x\xi}|, \, |f_u|, \, |f_{xu}| \leq \alpha_3 V^{p-1}, \, |f_{u\xi}|, \, |f_{uu}| \leq \alpha_3 V^{p-2} \\[2mm]
\displaystyle \alpha_4 V^{p-2} |\lambda|^2 \leq \sum_{i,j=1}^{n} f_{\xi_i \xi_j} (x, u, \xi) \lambda_i \lambda_j \leq \alpha_5 V^{p-2} |\lambda|^2
\end{cases}
$$

where $p \geq 2$, $V^2 = 1 + u^2 + |\xi|^2$ and $\alpha_i > 0$, $i = 1, \cdots, 5$, are constants. Then any minimizer of

$$
(P) \quad \inf \left\{ I(u) = \int_\Omega f(x, u(x), \nabla u(x)) \, dx : u \in u_0 + W_0^{1,p}(\Omega) \right\}
$$

is in $C^\infty (O)$, for every $O \subset \overline{O} \subset \Omega$.

Remark 4.17 (i) The last hypothesis in (C) implies a kind of uniform convexity of $\xi \to f(x, u, \xi)$; it guarantees the uniform ellipticity of the Euler-Lagrange equation. Functions satisfying the hypotheses of Theorem 4.9 with $h \in C^\infty$ obviously satisfy (C).

(ii) For the regularity up to the boundary, we refer to the literature.

(iii) We recall that we denote partial derivatives with sub indices, for example $f_x = (f_{x_1}, \cdots, f_{x_n})$, $f_\xi = (f_{\xi_1}, \cdots, f_{\xi_n})$, where

$$
f_{x_i} = \frac{\partial f}{\partial x_i}, \quad f_{\xi_i} = \frac{\partial f}{\partial \xi_i} \quad \text{or} \quad f_{\xi_i \xi_j} = \frac{\partial^2 f}{\partial \xi_i \partial \xi_j}.
$$

The proof of such a theorem relies on the De Giorgi-Nash-Moser theory. In the course of the proof, one transforms the nonlinear Euler-Lagrange equation into an elliptic linear equation with bounded measurable coefficients. Therefore, to obtain the desired regularity, one needs to know the regularity of the solutions of such equations and this is precisely the famous theorem that is stated below (see Giaquinta [55], Theorem 2.1 of Chapter II). It was first established by De Giorgi, then simplified by Moser and also proved, independently but at the same time, by Nash.

Theorem 4.18 *Let* $\Omega \subset \mathbb{R}^n$ *be a bounded open set and* $v \in W^{1,2}(\Omega)$ *be a solution of*

$$\sum_{i,j=1}^{n} \int_{\Omega} \left[a_{ij}(x) v_{x_i}(x) \varphi_{x_j}(x) \right] dx = 0, \quad \forall \varphi \in W_0^{1,2}(\Omega)$$

where $a_{ij} = a_{ji} \in L^\infty(\Omega)$ *and, denoting by* $\gamma > 0$ *a constant,*

$$\sum_{i,j=1}^{n} a_{ij}(x) \lambda_i \lambda_j \geq \gamma |\lambda|^2, \quad a.e. \text{ in } \Omega \text{ and } \forall \lambda \in \mathbb{R}^n.$$

Then there exists $0 < \alpha < 1$ *so that* $v \in C^{0,\alpha}(O)$, *for every* $O \subset \overline{O} \subset \Omega$.

Remark 4.19 It is interesting to try to understand, formally, the relationship between the last two theorems, for example in the case where $f = f(x, u, \xi) = f(\xi)$. The coefficients $a_{ij}(x)$ and the function v in Theorem 4.18 are, respectively, $f_{\xi_i \xi_j}(\nabla u(x))$ and u_{x_i} in Theorem 4.16. The fact that $v = u_{x_i} \in W^{1,2}$ is proved by the method of difference quotient presented in Theorem 4.9. This approach is implemented in Exercise 4.7.1.

The two preceding theorems do not generalize to the vectorial case $u : \Omega \subset \mathbb{R}^n \to \mathbb{R}^N$, with $n, N > 1$. In this case only partial regularity can, in general, be proved. We give here an example (see Giusti-Miranda [60]) of such a phenomenon.

Example 4.20 Let $n \geq 3$, $\Omega \subset \mathbb{R}^n$ be the unit ball and $u_0(x) = x$. Let $\xi = \left(\xi_i^j \right) \in \mathbb{R}^{n \times n}$ (if $\xi = \nabla u$, then $\xi_i^j = \partial u^j / \partial x_i$) and

$$f(x, u, \xi) = f(u, \xi) = \sum_{i,j=1}^{n} \left(\xi_i^j \right)^2 + \left[\sum_{i,j=1}^{n} \left(\delta_{ij} + \frac{4}{n-2} \frac{u^i u^j}{1 + |u|^2} \right) \xi_i^j \right]^2$$

where δ_{ij} is the Kronecker symbol (i.e. $\delta_{ij} = 0$ if $i \neq j$ and $\delta_{ij} = 1$ if $i = j$). Let

$$(P) \quad \inf \left\{ I(u) = \int_\Omega f(u(x), \nabla u(x)) \, dx : u \in u_0 + W_0^{1,2}(\Omega; \mathbb{R}^n) \right\}.$$

(i) It turns out (see Exercise 4.7.2) that $\overline{u}(x) = x/|x|$ is a solution of the weak form of the Euler-Lagrange equation associated with (P).

(ii) Moreover, it can be shown (see [60]) that, for n sufficiently large, $\overline{u}$ is the unique minimizer of (P).

4.7.1   Exercises

Exercise 4.7.1 Let $\Omega \subset \mathbb{R}^n$ be a bounded open set with Lipschitz boundary, $u_0 \in W^{1,2}(\Omega)$ and

$$(P) \quad \inf \left\{ I(u) = \int_\Omega f(\nabla u(x))\, dx : u \in u_0 + W_0^{1,2}(\Omega) \right\}$$

where $f \in C^2(\mathbb{R}^n)$ is convex and there exist $\gamma_1, \cdots, \gamma_5 > 0$ such that

$$\gamma_1 |\xi|^2 - \gamma_2 \le f(\xi) \le \gamma_3 \left(|\xi|^2 + 1 \right), \quad \forall\, \xi \in \mathbb{R}^n$$

$$\left| f_{\xi_i \xi_j} \right| \le \gamma_4, \quad \forall\, \xi \in \mathbb{R}^n \text{ and } \forall\, i, j = 1, \cdots, n$$

$$\sum_{i,j=1}^n f_{\xi_i \xi_j}(\xi)\, \lambda_i \lambda_j \ge \gamma_5 |\lambda|^2, \quad \forall\, \xi, \lambda \in \mathbb{R}^n.$$

With the help of Theorems 4.9 and 4.18, prove that there exists a unique minimizer $\overline{u} \in u_0 + W_0^{1,2}(\Omega)$ satisfying $\overline{u} \in W^{2,2}(O) \cap C^{1,\alpha}(O)$ for some $0 < \alpha < 1$ and for every open set $O \subset \overline{O} \subset \Omega$.

Exercise 4.7.2 Consider Example 4.20. Prove that $\overline{u}$ is a solution of the weak form of the Euler-Lagrange equation associated with (P).

Chapter 5

Minimal surfaces

5.1 Introduction

We start by explaining informally the problem under consideration. We want to find among all surfaces $\Sigma \subset \mathbb{R}^3$ (or more generally in $\mathbb{R}^{n+1}$, $n \geq 2$) with a prescribed boundary, $\partial \Sigma = \Gamma$, where (in the case $n = 2$) Γ is a simple closed curve, one that is of minimal area.

Unfortunately the formulation of the problem in more precise terms is delicate. It depends on the kind of surfaces we are considering. We consider two types of surfaces: parametric and nonparametric. The second are less general but simpler from an analytical point of view.

We start with the formulation for nonparametric (hyper)surfaces (this case is easy to generalize to $\mathbb{R}^{n+1}$). These are of the form

$$\Sigma = \left\{ v\left(x\right) = \left(x, u\left(x\right)\right) \in \mathbb{R}^{n+1} : x \in \overline{\Omega} \right\}$$

with $u : \overline{\Omega} \to \mathbb{R}$ and where $\Omega \subset \mathbb{R}^n$ is a bounded open set. The surface Σ is therefore the graph of the function u. The fact that $\partial \Sigma$ is prescribed now reads as $u = u_0$ on $\partial \Omega$, where u_0 is a given function. The area of such surfaces is given by

$$\text{Area}\left(\Sigma\right) = I\left(u\right) = \int_\Omega f\left(\nabla u\left(x\right)\right) dx$$

where, for $\xi \in \mathbb{R}^n$, we have set

$$f\left(\xi\right) = \sqrt{1 + \left|\xi\right|^2}.$$

The problem is then written in the usual form

$$(P) \quad \inf \left\{ I\left(u\right) = \int_\Omega f\left(\nabla u\left(x\right)\right) dx : u \in u_0 + W_0^{1,1}\left(\Omega\right) \right\}.$$

165

As already seen in Chapter 3, even though the function f is strictly convex and $f(\xi) \geq |\xi|^p$ with $p = 1$, we cannot use the direct methods of the calculus of variations, since we are led to work, because of the coercivity condition $f(\xi) \geq |\xi|$, in the non-reflexive space $W^{1,1}(\Omega)$. In fact, in general, there is no minimizer of (P) in $u_0 + W_0^{1,1}(\Omega)$. We therefore need a different approach to deal with this problem.

Before going further we write the associated Euler-Lagrange equation for (P)

$$(E) \quad \operatorname{div} \left[\frac{\nabla u}{\sqrt{1 + |\nabla u|^2}} \right] = \sum_{i=1}^{n} \frac{\partial}{\partial x_i} \left[\frac{u_{x_i}}{\sqrt{1 + |\nabla u|^2}} \right] = 0$$

or equivalently

$$(E) \quad Mu \equiv \left(1 + |\nabla u|^2\right) \Delta u - \sum_{i,j=1}^{n} u_{x_i} u_{x_j} u_{x_i x_j} = 0.$$

The last equation is known as the *minimal surface equation*. If $n = 2$ and $u = u(x, y)$, it reads as

$$Mu = \left(1 + u_y^2\right) u_{xx} - 2 u_x u_y u_{xy} + \left(1 + u_x^2\right) u_{yy} = 0.$$

Therefore, any $C^2(\overline{\Omega})$ minimizer of (P) should satisfy the equation (E) and conversely, since the integrand f is convex. Moreover, since f is strictly convex, the minimizer, if it exists, is unique. The equation (E) is equivalent (see Section 5.2) to the fact that the *mean curvature* of Σ, denoted by H, vanishes everywhere.

It is clear that the above problem is, geometrically, too restrictive. Indeed, if any surface can be locally represented as a graph of a function (i.e. a non-parametric surface), it is not the case globally. We are therefore led to consider the so-called *parametric surfaces*. These are sets $\Sigma \subset \mathbb{R}^{n+1}$ so that there exist a connected open set $\Omega \subset \mathbb{R}^n$ and a map $v : \overline{\Omega} \to \mathbb{R}^{n+1}$ such that

$$\Sigma = v(\overline{\Omega}) = \{v(x) : x \in \overline{\Omega}\}.$$

For example, when $n = 2$ and $v = v(x, y) \in \mathbb{R}^3$, if we denote by $v_x \times v_y$ the normal to the surface (where $a \times b$ stands for the vectorial product of $a, b \in \mathbb{R}^3$ and $v_x = \partial v / \partial x$, $v_y = \partial v / \partial y$) we find that the area is given by

$$\operatorname{Area}(\Sigma) = J(v) = \iint_{\Omega} |v_x \times v_y| \, dx \, dy.$$

More generally, if $n \geq 2$, we define (see Theorem 4.4.10 in Morrey [86])

$$g(\nabla v) = \left[\sum_{i=1}^{n+1} \left(\frac{\partial \left(v^1, \cdots, v^{i-1}, v^{i+1}, \cdots, v^{n+1}\right)}{\partial \left(x_1, \cdots, x_n\right)} \right)^2 \right]^{1/2}$$

where $\partial\left(u^1,\cdots,u^n\right)/\partial\left(x_1,\cdots,x_n\right)$ stands for the determinant of the $n\times n$ matrix $\left(\partial u^i/\partial x_j\right)_{1\leq i,j\leq n}$. In the terminology of Section 3.5 such a function g is polyconvex but not convex. The area for such a surface is given by

$$\text{Area}\left(\Sigma\right)=J\left(v\right)=\int_\Omega g\left(\nabla v\left(x\right)\right)dx.$$

The problem is then, given Γ, to find a parametric surface that minimizes

$$(Q)\quad\inf\left\{\text{Area}\left(\Sigma\right):\partial\Sigma=\Gamma\right\}.$$

It is clear that problem (Q) is more general than (P). It is, however, a more complicated problem than (P) for several reasons besides the geometrical ones. Unlike (P) it is a vectorial problem of the calculus of variations and the Euler-Lagrange equations associated with (Q) now form a system of $(n+1)$ partial differential equations. Moreover, although, as for (P), any minimizer is a solution of these equations, it is not true in general, contrary to what happens with (P), that every solution of the Euler-Lagrange equations is necessarily a minimizer of (Q). Finally, uniqueness is also lost for (Q) in contrast with what happens for (P).

We now come to the definition of *minimal surfaces*. A minimal surface is a solution of the Euler-Lagrange equations associated with (Q); it turns out that it has (see Section 5.2) zero mean curvature. We should draw the attention to the misleading terminology (this confusion is not present in the case of non-parametric surfaces): a minimal surface is not necessarily a *surface of minimal area*, while the converse is true, namely, a surface of minimal area is a minimal surface.

The problem of finding a minimal surface with a prescribed boundary is known as *Plateau problem*.

We now describe the content of the present chapter. For the most part we only consider the case $n=2$. In Section 5.2 we recall some basic facts about surfaces, mean curvature and isothermal coordinates. We then give several examples of minimal surfaces. In Section 5.3 we outline some of the main ideas of the method of Douglas, as revised by Courant and Tonelli, for solving Plateau problem. This method is valid only when $n=2$, since it strongly uses the notion and properties of conformal mappings. In Section 5.4 we briefly, and without proofs, mention some results on regularity, uniqueness and non-uniqueness of minimal surfaces. In the final section we come back to the case of nonparametric surfaces and we give some existence results.

We now briefly discuss the historical background to the problem under consideration. The problem in nonparametric form was formulated, and the equation (E) of minimal surfaces was derived, by Lagrange in 1762. It was immediately understood that the problem was a difficult one. The more general Plateau problem (named after the theoretical and experimental work of the physicist Plateau) was solved in 1930 simultaneously and independently by Douglas and Rado. One of the first two Fields medals was awarded to Douglas in 1936 for having solved the problem. Before that, many mathematicians contributed to the study of the problem: Ampère, Beltrami, Bernstein, Bonnet, Catalan, Darboux, Enneper, Haar, Korn, Legendre, Lie, Meusnier, Monge, Müntz, Riemann, H.A. Schwarz, Serret, Weierstrass, Weingarten and others. Immediately after the work of Douglas and Rado, we could cite Courant, MacShane, Morrey, Morse, Tonelli and many others more recently. It is still a very active field.

We conclude this introduction with some comments on the bibliography. We should first point out that we give many results without proofs and those that are given are only sketched. It is therefore indispensable in this chapter, even more than in the others, to refer to the bibliography. There are several excellent books but, due to the nature of the subject, they are difficult to read. The most complete to which we constantly refer are those of Dierkes-Hildebrandt-Küster-Wohlrab [41] and Nitsche [89]. By way of introduction, interesting for a general audience, one can consult Hildebrandt-Tromba [68]. We also refer to the monographs of Almgren [4], Courant [24], Dierkes-Hildebrandt-Sauvigny [42], Federer [49], Gilbarg-Trudinger [57] (for the nonparametric surfaces), Giusti [58], Morrey [86], Osserman [91] and Struwe [103].

5.2 Generalities about surfaces

5.2.1 Main definitions and some examples

We now introduce the different types of surfaces that we consider. We essentially limit ourselves to surfaces of $\mathbb{R}^3$, although in some instances we give generalizations to $\mathbb{R}^{n+1}$. Besides the references that we already mentioned, one can consult books on differential geometry, such as that of Hsiung [72].

Definition 5.1 *(i) A set $\Sigma \subset \mathbb{R}^3$ is called a* parametric surface *(or more simply a surface) if there exist a connected open set $\Omega \subset \mathbb{R}^2$ and a (non-constant) continuous map $v : \overline{\Omega} \to \mathbb{R}^3$ such that*

$$\Sigma = v\left(\overline{\Omega}\right) = \left\{v\left(x,y\right) \in \mathbb{R}^3 : (x,y) \in \overline{\Omega}\right\}.$$

(ii) We say that Σ is a nonparametric surface *if*

$$\Sigma = \left\{v\left(x,y\right) = (x, y, u\left(x,y\right)) \in \mathbb{R}^3 : (x,y) \in \overline{\Omega}\right\}$$

with $u : \overline{\Omega} \to \mathbb{R}$ continuous and where $\Omega \subset \mathbb{R}^2$ is a connected open set.

(iii) *A parametric surface is said to be* regular *of class* C^m, *(m $\geq$ 1 an integer) if, in addition,* $v \in C^m (\Omega; \mathbb{R}^3)$ *and*

$$v_x \times v_y \neq 0 \quad \text{for every } (x, y) \in \Omega$$

(where $a \times b$ *stands for the vectorial product of* $a, b \in \mathbb{R}^3$ *and* $v_x = \partial v / \partial x$, $v_y = \partial v / \partial y$). *In this case we write*

$$e_3 = \frac{v_x \times v_y}{|v_x \times v_y|}.$$

Remark 5.2 (i) In many cases we restrict our attention to the case where Ω is the unit disk and $\Sigma = v(\overline{\Omega})$ is then called a surface of the type of the disk.

(ii) In the following, we let

$$\mathcal{M}(\overline{\Omega}) = C^0(\overline{\Omega}; \mathbb{R}^3) \cap W^{1,2}(\Omega; \mathbb{R}^3).$$

(iii) For a regular surface the *area* is defined as

$$J(v) = \text{Area}(\Sigma) = \iint_\Omega |v_x \times v_y| \, dx \, dy.$$

It can be shown, following MacShane and Morrey, that if $v \in \mathcal{M}(\overline{\Omega})$, then the above formula still makes sense (see Nitsche [89] pages 195-198).

(iv) In the case of a nonparametric surface $v(x, y) = (x, y, u(x, y))$ we have

$$\text{Area}(\Sigma) = J(v) = I(u) = \iint_\Omega \sqrt{1 + u_x^2 + u_y^2} \, dx \, dy.$$

Note also that for a nonparametric surface we always have

$$|v_x \times v_y|^2 = 1 + u_x^2 + u_y^2 \neq 0.$$

We now introduce the different notions of curvature.

Definition 5.3 *Let* Σ *be a regular surface of class* C^m, *m $\geq$ 2, and let*

$$E = |v_x|^2, \quad F = \langle v_x; v_y \rangle, \quad G = |v_y|^2, \quad e_3 = \frac{v_x \times v_y}{|v_x \times v_y|},$$

$$L = \langle e_3; v_{xx} \rangle, \quad M = \langle e_3; v_{xy} \rangle, \quad N = \langle e_3; v_{yy} \rangle$$

where $\langle .; . \rangle$ *denotes the scalar product in* $\mathbb{R}^3$.

(i) The mean curvature *of* Σ, *denoted by* H, *at a point* $p \in \Sigma$ ($p = v(x, y)$) *is given by*

$$H = \frac{1}{2} \frac{E N - 2 F M + G L}{E G - F^2}.$$

(ii) The Gaussian curvature of Σ, denoted by K, at a point $p \in \Sigma$ $(p = v(x,y))$ is by definition

$$K = \frac{L\,N - M^2}{E\,G - F^2}.$$

(iii) The principal curvatures, k_1 and k_2, are defined as

$$k_1 = H + \sqrt{H^2 - K} \quad and \quad k_2 = H - \sqrt{H^2 - K}$$

so that

$$H = \frac{k_1 + k_2}{2} \quad and \quad K = k_1 k_2.$$

Remark 5.4 **(i)** We always have $H^2 \geq K$.

(ii) For a nonparametric surface $v(x,y) = (x, y, u(x,y))$, we have

$$E = 1 + u_x^2, \quad F = u_x u_y, \quad G = 1 + u_y^2, \quad EG - F^2 = 1 + u_x^2 + u_y^2$$

$$e_3 = \frac{(-u_x, -u_y, 1)}{\sqrt{1 + u_x^2 + u_y^2}}, \quad L = \frac{u_{xx}}{\sqrt{1 + u_x^2 + u_y^2}},$$

$$M = \frac{u_{xy}}{\sqrt{1 + u_x^2 + u_y^2}}, \quad N = \frac{u_{yy}}{\sqrt{1 + u_x^2 + u_y^2}}$$

and hence

$$H = \frac{(1 + u_y^2)\,u_{xx} - 2u_x u_y u_{xy} + (1 + u_x^2)\,u_{yy}}{2(1 + u_x^2 + u_y^2)^{3/2}} \quad and \quad K = \frac{u_{xx} u_{yy} - u_{xy}^2}{(1 + u_x^2 + u_y^2)^2}.$$

(iii) For a nonparametric surface in $\mathbb{R}^{n+1}$ given by $x_{n+1} = u(x_1, \cdots, x_n)$, we have that the mean curvature is defined by (see (A.14) in Gilbarg-Trudinger [57])

$$H = \frac{1}{n} \sum_{i=1}^{n} \frac{\partial}{\partial x_i} \left[\frac{u_{x_i}}{\sqrt{1 + |\nabla u|^2}} \right] = \frac{\left(1 + |\nabla u|^2\right) \Delta u - \sum_{i,j=1}^{n} u_{x_i} u_{x_j} u_{x_i x_j}}{n\left(1 + |\nabla u|^2\right)^{3/2}}.$$

In terms of the operator M defined in the introduction of the present chapter, we can write

$$Mu = n\left(1 + |\nabla u|^2\right)^{3/2} H.$$

(iv) Note that we always have (see Exercise 5.2.1)

$$|v_x \times v_y| = \sqrt{EG - F^2}.$$

We are now in a position to define the notion of minimal surface.

Definition 5.5 *A regular surface of class* C^2 *is said to be a* minimal surface *if* $H = 0$ *at every point.*

We next give several examples of minimal surfaces, starting with the non-parametric ones.

Example 5.6 The first minimal surface that comes to mind is naturally the plane, defined parametrically by $(\alpha, \beta, \gamma$ being constants)

$$\Sigma = \left\{ v\left(x, y\right) = \left(x, y, \alpha x + \beta y + \gamma\right) : \left(x, y\right) \in \mathbb{R}^2 \right\}.$$

We trivially have $H = 0$.

Example 5.7 *Scherk surface* is a minimal surface in nonparametric form given by

$$\Sigma = \left\{ v\left(x, y\right) = \left(x, y, u\left(x, y\right)\right) : \left|x\right|, \left|y\right| < \frac{\pi}{2} \right\}$$

where

$$u\left(x, y\right) = \log \cos y - \log \cos x.$$

We now turn our attention to minimal surfaces in parametric form.

Example 5.8 *Catenoids* are minimal surfaces. They are defined, for $(x, y) \in \mathbb{R}^2$, by

$$v\left(x, y\right) = \left(x, w\left(x\right) \cos y, w\left(x\right) \sin y\right) \quad \text{with} \quad w\left(x\right) = \lambda \cosh \frac{x + \mu}{\lambda},$$

where $\lambda \neq 0$ and μ are constants. We will see (in Proposition 5.11) that they are the only minimal surfaces of revolution (here around the x-axis).

Example 5.9 The *helicoid* is a minimal surface (see Exercise 5.2.2). It is given, for $(x, y) \in \mathbb{R}^2$, by

$$v\left(x, y\right) = \left(y \cos x, y \sin x, ax\right)$$

with $a \in \mathbb{R}$.

Example 5.10 *Enneper surface* is a minimal surface (see Exercise 5.2.2). It is defined, for $(x, y) \in \mathbb{R}^2$, by

$$v\left(x, y\right) = \left(x - \frac{x^3}{3} + xy^2, -y + \frac{y^3}{3} - yx^2, x^2 - y^2\right).$$

As already mentioned we have the following characterization for surfaces of revolution.

Proposition 5.11 *The only regular minimal surfaces of revolution of the form*

$$v\left(x,y\right) = \left(x, w\left(x\right)\cos y, w\left(x\right)\sin y\right),$$

are the catenoids, *i.e.*

$$w\left(x\right) = \lambda\cosh\frac{x+\mu}{\lambda}$$

where $\lambda \neq 0$ and μ are constants.

Proof We have to prove that Σ given parametrically by v is minimal if and only if

$$w\left(x\right) = \lambda\cosh\left(\left(x+\mu\right)/\lambda\right).$$

Observe first that

$$v_x = \left(1, w'\cos y, w'\sin y\right), \quad v_y = \left(0, -w\sin y, w\cos y\right)$$

$$E = 1 + \left(w'\right)^2, \quad F = 0, \quad G = w^2$$

$$v_x \times v_y = w\left(w', -\cos y, -\sin y\right), \quad e_3 = \frac{w}{|w|}\frac{\left(w', -\cos y, -\sin y\right)}{\sqrt{1+\left(w'\right)^2}}$$

$$v_{xx} = w''\left(0, \cos y, \sin y\right), \quad v_{xy} = w'\left(0, -\sin y, \cos y\right), \quad v_{yy} = -w\left(0, \cos y, \sin y\right)$$

$$L = \frac{w}{|w|}\frac{-w''}{\sqrt{1+\left(w'\right)^2}}, \quad M = 0, \quad N = \frac{|w|}{\sqrt{1+\left(w'\right)^2}}.$$

Since Σ is a regular surface, we must have $|w| > 0$ (because $|v_x \times v_y|^2 = EG - F^2 > 0$). We therefore deduce that

$$H = 0 \quad\Leftrightarrow\quad EN + GL = 0 \quad\Leftrightarrow\quad |w|\left(ww'' - \left(1+\left(w'\right)^2\right)\right) = 0$$

and thus $H = 0$ is equivalent to

$$ww'' = 1 + \left(w'\right)^2. \tag{5.1}$$

Any solution of the differential equation necessarily satisfies

$$\frac{d}{dx}\left[\frac{w\left(x\right)}{\sqrt{1+\left(w'\left(x\right)\right)^2}}\right] = 0.$$

The solution of this last differential equation (see the solution to Exercise 5.2.3) being either $w \equiv$ constant (which however does not satisfy (5.1)) or of the form

$$w\left(x\right) = \lambda \cosh \frac{x + \mu}{\lambda},$$

we have the result. ∎

5.2.2 Minimal surfaces and surfaces of minimal area

We now turn our attention to the relationship between minimal surfaces and surfaces of minimal area.

Theorem 5.12 *Let* $\Omega \subset \mathbb{R}^2$ *be a bounded connected open set with smooth boundary.*

Part 1. Let $\Sigma_0 = v\left(\overline{\Omega}\right)$ *where* $v \in C^2\left(\overline{\Omega}; \mathbb{R}^3\right)$, $v = v\left(x, y\right)$, *with* $v_x \times v_y \neq 0$ *in* $\overline{\Omega}$. *If*

$$\mathrm{Area}\left(\Sigma_0\right) \leq \mathrm{Area}\left(\Sigma\right)$$

among all regular surfaces Σ *of class* C^1 *with* $\partial\Sigma = \partial\Sigma_0$, *then* Σ_0 *is a minimal surface.*

Part 2. Let $\mathcal{S}_\Omega$ *be the set of nonparametric surfaces of the form*

$$\Sigma_u = \left\{\left(x, y, u\left(x, y\right)\right) : \left(x, y\right) \in \overline{\Omega}\right\}$$

with $u \in C^2\left(\overline{\Omega}\right)$ *and let* $\Sigma_{\overline{u}} \in \mathcal{S}_\Omega$. *The two following assertions are then equivalent.*

(i) $\Sigma_{\overline{u}}$ *is a minimal surface, which means*

$$\mathrm{M}\overline{u} = \left(1 + \overline{u}_y^2\right)\overline{u}_{xx} - 2\overline{u}_x\overline{u}_y\overline{u}_{xy} + \left(1 + \overline{u}_x^2\right)\overline{u}_{yy} = 0.$$

(ii) For every $\Sigma_u \in \mathcal{S}_\Omega$ *with* $u = \overline{u}$ *on* $\partial\Omega$

$$\mathrm{Area}\left(\Sigma_{\overline{u}}\right) \leq \mathrm{Area}\left(\Sigma_u\right) = I\left(u\right) = \iint_\Omega \sqrt{1 + u_x^2 + u_y^2}.$$

Moreover, $\Sigma_{\overline{u}}$ *is, among all surfaces of* $\mathcal{S}_\Omega$ *with* $u = \overline{u}$ *on* $\partial\Omega$, *the only one to have this property.*

Remark 5.13 (i) The converse of Part 1, namely that if Σ_0 is a minimal surface then it is of minimal area, is, in general, false. The claim of Part 2 is that the converse is true when we restrict our attention to nonparametric surfaces.

(ii) This theorem is easily extended to $\mathbb{R}^{n+1}$, $n \geq 2$.

Proof We only prove Part 2 of the theorem and we refer to Exercise 5.2.4 for Part 1. Let

$$v(x, y) = (x, y, u(x, y)), \quad (x, y) \in \overline{\Omega}$$

we then have

$$J(v) = \iint_\Omega |v_x \times v_y| = \iint_\Omega \sqrt{1 + u_x^2 + u_y^2} \equiv I(u).$$

(ii) $\Rightarrow$ **(i)** We write the associated Euler-Lagrange equation. Since $\overline{u}$ is a minimizer we have

$$I(\overline{u}) \leq I(\overline{u} + \epsilon\varphi), \quad \forall \varphi \in C_0^\infty(\Omega), \ \forall \epsilon \in \mathbb{R}$$

and hence

$$\frac{d}{d\epsilon} I(\overline{u} + \epsilon\varphi)\Big|_{\epsilon=0} = \iint_\Omega \frac{\overline{u}_x \varphi_x + \overline{u}_y \varphi_y}{\sqrt{1 + \overline{u}_x^2 + \overline{u}_y^2}} = 0, \quad \forall \varphi \in C_0^\infty(\Omega).$$

Since $\overline{u} \in C^2(\overline{\Omega})$ we have, after integration by parts and using the fundamental lemma of the calculus of variations (Theorem 1.24),

$$\frac{\partial}{\partial x}\left[\frac{\overline{u}_x}{\sqrt{1 + \overline{u}_x^2 + \overline{u}_y^2}} \right] + \frac{\partial}{\partial y}\left[\frac{\overline{u}_y}{\sqrt{1 + \overline{u}_x^2 + \overline{u}_y^2}} \right] = 0 \quad \text{in } \overline{\Omega} \qquad (5.2)$$

or equivalently

$$M\overline{u} = \left(1 + \overline{u}_y^2\right)\overline{u}_{xx} - 2\overline{u}_x \overline{u}_y \overline{u}_{xy} + \left(1 + \overline{u}_x^2\right)\overline{u}_{yy} = 0 \quad \text{in } \overline{\Omega}. \qquad (5.3)$$

This just asserts that $H = 0$ and hence

$$\Sigma_{\overline{u}} = \left\{ (x, y, \overline{u}(x, y)) : (x, y) \in \overline{\Omega} \right\}$$

is a minimal surface.

(i) $\Rightarrow$ **(ii)** We start by noting that the function

$$\xi \to f(\xi) = \sqrt{1 + |\xi|^2}$$

where $\xi \in \mathbb{R}^2$, is strictly convex. So let

$$\Sigma_{\overline{u}} = \left\{ (x, y, \overline{u}(x, y)) : (x, y) \in \overline{\Omega} \right\}$$

be a minimal surface. Since $H = 0$, we have that $\overline{u}$ satisfies (5.2) or (5.3). Let

$$\Sigma_u = \left\{ (x, y, u(x, y)) : (x, y) \in \overline{\Omega} \right\}$$

with $u \in C^2\left(\overline{\Omega}\right)$ and $u = \overline{u}$ on $\partial\Omega$. We want to show that $I\left(\overline{u}\right) \leq I\left(u\right)$. Since f is convex, we have

$$f\left(\xi\right) \geq f\left(\eta\right) + \langle \nabla f\left(\eta\right) ; \xi - \eta \rangle, \quad \forall \xi, \eta \in \mathbb{R}^2$$

and hence

$$f\left(u_x, u_y\right) \geq f\left(\overline{u}_x, \overline{u}_y\right) + \frac{1}{\sqrt{1 + \overline{u}_x^2 + \overline{u}_y^2}} \langle \left(\overline{u}_x, \overline{u}_y\right) ; \left(u_x - \overline{u}_x, u_y - \overline{u}_y\right) \rangle.$$

Integrating the above inequality and appealing to (5.2) and to the fact that $u = \overline{u}$ on $\partial\Omega$ we readily obtain the result.

The uniqueness follows from the strict convexity of f. ∎

5.2.3 Isothermal coordinates

We next introduce the notion of *isothermal coordinates* (sometimes also called *conformal parameters*). This notion will help us to understand the method of Douglas that we discuss in the next section.

Let us start with an informal presentation. Among all the parametrizations of a given curve the arc length plays a special role; for a given surface the isothermal coordinates play a similar role. They are given by

$$E = \left|v_x\right|^2 = G = \left|v_y\right|^2 \quad \text{and} \quad F = \langle v_x ; v_y \rangle = 0,$$

which means that the tangent vectors are orthogonal and have equal norms. In general and contrary to what happens for curves, we can only *locally* find such a system of coordinates (i.e. with $E = G$ and $F = 0$), according to the result of Korn, Lichtenstein and Chern [21].

Remark 5.14 (i) Note that for a nonparametric surface

$$\Sigma = \left\{v\left(x, y\right) = \left(x, y, u\left(x, y\right)\right) : \ \left(x, y\right) \in \overline{\Omega}\right\}$$

we have $E = G = 1 + u_x^2 = 1 + u_y^2$ and $F = u_x u_y = 0$ only if $u_x = u_y = 0$.

(ii) Enneper surface (Example 5.10) is globally parametrized with isothermal coordinates.

One of the remarkable aspects of minimal surfaces is that they can be globally parametrized by such coordinates as the above Enneper surface. We have the following result that we will not use explicitly. We will, in part, give some idea of the proof in the next section (for a proof, see Nitsche [89], page 175).

Theorem 5.15 *Let* $\Omega = \left\{ (x,y) \in \mathbb{R}^2 : x^2 + y^2 < 1 \right\}$. *Let* Σ *be a minimal sur-
face of the type of the disk (i.e. there exists* $\tilde{v} \in C^2 \left(\overline{\Omega}; \mathbb{R}^3 \right)$ *so that* $\Sigma = \tilde{v} \left(\overline{\Omega} \right)$)
such that $\partial \Sigma = \Gamma$ *is a simple closed curve. Then there exists a global isothermal
representation of the surface* Σ. *This means that* $\Sigma = \left\{ v \left(x, y \right) : (x,y) \in \overline{\Omega} \right\}$ *with
v satisfying*

(i) $v \in C \left(\overline{\Omega}; \mathbb{R}^3 \right) \cap C^\infty \left(\Omega; \mathbb{R}^3 \right)$ *and* $\Delta v = 0$ *in* Ω;

(ii) $E = G > 0$ *and* $F = 0$ *(i.e.* $|v_x|^2 = |v_y|^2 > 0$ *and* $\langle v_x ; v_y \rangle = 0$);

(iii) v *maps the boundary* $\partial \Omega$ *topologically onto the simple closed curve* Γ.

Remark 5.16 The second result asserts that Σ is a regular surface (i.e. $v_x \times v_y \neq 0$) since

$$|v_x \times v_y| = \sqrt{E\,G - F^2} = E = |v_x|^2 > 0.$$

To conclude we point out the deep relationship between isothermal coordinates of minimal surfaces and harmonic functions (see also Theorem 5.15), which
is one of the basic facts in the proof of Douglas.

Theorem 5.17 *Let*

$$\Sigma = \left\{ v \left(x, y \right) \in \mathbb{R}^3 : (x,y) \in \overline{\Omega} \right\}$$

be a regular surface (i.e. $v_x \times v_y \neq 0$) *of class* C^2 *globally parametrized by
isothermal coordinates; then*

$$\Sigma \text{ is a minimal surface} \quad \Leftrightarrow \quad \Delta v = 0 \text{ (i.e. } \Delta v^1 = \Delta v^2 = \Delta v^3 = 0).$$

Proof We will show that if $E = G = |v_x|^2 = |v_y|^2$ and $F = 0$, then

$$\Delta v = 2\,E\,H\,e_3 = 2\,H\,v_x \times v_y \tag{5.4}$$

where H is the mean curvature and $e_3 = (v_x \times v_y) / |v_x \times v_y|$. The result readily
follows, since $v_x \times v_y \neq 0$.
Since $E = G$ and $F = 0$, we have

$$H = \frac{L+N}{2\,E} \quad \Rightarrow \quad L + N = 2\,E\,H. \tag{5.5}$$

We next prove that $\langle v_x ; \Delta v \rangle = \langle v_y ; \Delta v \rangle = 0$. Using the equations $E = G$ and
$F = 0$, we have, after differentiation of the first one by x and the second one by
y,

$$\langle v_x ; v_{xx} \rangle = \langle v_y ; v_{xy} \rangle \quad \text{and} \quad \langle v_x ; v_{yy} \rangle + \langle v_y ; v_{xy} \rangle = 0.$$

This leads, as wished, to $\langle v_x; \Delta v \rangle = 0$ and in a similar way to $\langle v_y; \Delta v \rangle = 0$. Therefore Δv is orthogonal to v_x and v_y and thus parallel to e_3, which means that there exists $a \in \mathbb{R}$ so that $\Delta v = ae_3$. We then deduce that

$$a = \langle e_3; \Delta v \rangle = \langle e_3; v_{xx} \rangle + \langle e_3; v_{yy} \rangle = L + N. \qquad (5.6)$$

Combining (5.5) and (5.6), we immediately get (5.4) and the theorem then follows. ∎

5.2.4 Exercises

Exercise 5.2.1 (i) Let $a, b, c \in \mathbb{R}^3$. Show that

$$|a \times b|^2 = |a|^2 |b|^2 - (\langle a; b \rangle)^2$$

$$(a \times b) \times c = \langle a; c \rangle b - \langle b; c \rangle a.$$

(ii) Deduce that $|v_x \times v_y| = \sqrt{EG - F^2}$.

(iii) Show that

$$L = \langle e_3; v_{xx} \rangle = - \langle (e_3)_x ; v_x \rangle,$$

$$M = \langle e_3; v_{xy} \rangle = - \langle (e_3)_x ; v_y \rangle = - \langle (e_3)_y ; v_x \rangle,$$

$$N = \langle e_3; v_{yy} \rangle = - \langle (e_3)_y ; v_y \rangle.$$

Exercise 5.2.2 Show that the surfaces in Example 5.9 and Example 5.10 are minimal surfaces.

Exercise 5.2.3 Let Σ be a surface (of revolution) given by

$$v(x, y) = (x, w(x) \cos y, w(x) \sin y), \quad x \in (0, 1), \ y \in (0, 2\pi), \ w \geq 0.$$

(i) Show that

$$\text{Area}(\Sigma) = I(w) = 2\pi \int_0^1 w(x) \sqrt{1 + (w'(x))^2} \, dx.$$

(ii) Consider the problem (where $\alpha > 0$)

$$(P_\alpha) \quad \inf \{ I(w) : w(0) = w(1) = \alpha \}.$$

Prove that any $C^2([0, 1])$ minimizer is necessarily of the form

$$w(x) = a \cosh \frac{2x - 1}{2a} \quad \text{with} \quad a \cosh \frac{1}{2a} = \alpha.$$

Discuss the existence of such solutions, as functions of α.

Exercise 5.2.4 Prove the first part of Theorem 5.12.

5.3   The Douglas-Courant-Tonelli method

We now present the main ideas of the method of Douglas, as modified by Courant and Tonelli, for solving Plateau problem in $\mathbb{R}^3$. For a complete proof, we refer to Courant [24], Dierkes-Hildebrandt-Küster-Wohlrab [41], Nitsche [89] or for a slightly different approach to Hildebrandt-Von der Mosel [69].

Let
$$\Omega = \left\{ (x,y) \in \mathbb{R}^2 : x^2 + y^2 < 1 \right\}$$
and $\Gamma \subset \mathbb{R}^3$ be a rectifiable (i.e. of finite length) simple closed curve. Let $w_i \in \partial\Omega$ $(w_i \neq w_j)$ and $p_i \in \Gamma$ $(p_i \neq p_j)$ $i = 1, 2, 3$ be fixed. The set of admissible surfaces is then

$$
S = \left\{
\begin{array}{l}
\Sigma = v\left(\overline{\Omega}\right) \quad \text{where} \quad v : \overline{\Omega} \to \Sigma \subset \mathbb{R}^3 \text{ so that} \\
\quad (S_1) \;\; v \in \mathcal{M}\left(\overline{\Omega}\right) = C^0\left(\overline{\Omega}; \mathbb{R}^3\right) \cap W^{1,2}\left(\Omega; \mathbb{R}^3\right) \\
\quad (S_2) \;\; v : \partial\Omega \to \Gamma \text{ is weakly monotonic and onto} \\
\quad (S_3) \;\; v\left(w_i\right) = p_i, \quad i = 1, 2, 3
\end{array}
\right\}.
$$

Remark 5.18 (i) The set of admissible surfaces is then the set of parametric surfaces of the type of the disk with parametrization in $\mathcal{M}\left(\overline{\Omega}\right)$. The weakly monotonic condition in (S_2) means that we allow the map v to be constant on some parts of $\partial\Omega$; thus v is not necessarily a homeomorphism of $\partial\Omega$ onto Γ. However, the minimizer of the theorem has the property to map the boundary $\partial\Omega$ topologically onto the simple closed curve Γ. The condition (S_3) may appear a little strange, it helps us to get compactness (see the proof below).

(ii) A first natural question is to ask whether S is non-empty. If the simple closed curve Γ is rectifiable then $S \neq \emptyset$ (see for more details Dierkes-Hildebrandt-Küster-Wohlrab [41] pages 232-234 and Nitsche [89] pages 253-257).

(iii) Recall from the preceding section that for $\Sigma \in S$ we have

$$\text{Area}\left(\Sigma\right) = J\left(v\right) = \iint_\Omega \left| v_x \times v_y \right|.$$

We next define what we mean by a weak solution of Plateau problem (see [41] page 231 and [89] page 252).

Definition 5.19 *Let $\Gamma \subset \mathbb{R}^3$ be a simple closed curve. We say that*
$$\Sigma = v\left(\overline{\Omega}\right) = \left\{ v\left(x,y\right) : x^2 + y^2 \leq 1 \right\}$$
is a weak solution of Plateau problem if

(i) $v \in C^0\left(\overline{\Omega}; \mathbb{R}^3\right) \cap C^2\left(\Omega; \mathbb{R}^3\right)$,

(ii) $E = \left|v_x\right|^2 = G = \left|v_y\right|^2$, $F = \langle v_x; v_y \rangle = 0$ and $\Delta v = 0$ in Ω,

(iii) v maps the boundary $\partial\Omega$ topologically onto $\partial\Sigma = \Gamma$.

Remark 5.20 If we suppose, in addition, that the surface Σ is regular, then, in view of Theorem 5.17, any weak solution of Plateau problem is a solution of Plateau problem.

The main result of this chapter is then

Theorem 5.21 *Under the above hypotheses there exists $\Sigma_0 \in \mathcal{S}$ so that*

$$\text{Area}\,(\Sigma_0) \leq \text{Area}\,(\Sigma)\,, \quad \forall \Sigma \in \mathcal{S}.$$

Moreover, it is a weak solution of Plateau problem. More precisely, there exists $\overline{v}$ satisfying $(S_1)\,, (S_2)$ and $(S_3)\,,$ such that $\Sigma_0 = \overline{v}\,(\overline{\Omega})$ and

(i) $\overline{v} \in C^\infty\left(\Omega; \mathbb{R}^3\right)$ *with* $\Delta\overline{v} = 0$ *in* $\Omega,$

(ii) $E = |\overline{v}_x|^2 = G = |\overline{v}_y|^2$ *and* $F = \langle \overline{v}_x; \overline{v}_y \rangle = 0.$

(iii) $\overline{v}$ *maps the boundary $\partial\Omega$ topologically onto the simple closed curve Γ.*

Remark 5.22 **(i)** To solve Plateau problem completely, we must still prove that Σ_0 is a regular surface (i.e. $\overline{v}_x \times \overline{v}_y \neq 0$ everywhere). We mention in the next section some results concerning this problem. We also have a regularity result, namely that $\overline{v}$ is C^∞ and harmonic, as well as a choice of isothermal coordinates $(E = G$ and $F = 0)$.

(ii) The proof uses properties of conformal mappings in a significant way and hence cannot be generalized as such to $\mathbb{R}^{n+1}$, $n \geq 2$. The results of De Giorgi, Federer, Fleming, Morrey, Reifenberg (see Giusti [58], Morrey [86]) and others deal with such a problem.

Proof We only give the main ideas of the proof. It is divided into four steps.

Step 1. Let $\Sigma \in \mathcal{S}$ and define

$$\text{Area}\,(\Sigma) = J\,(v) = \iint_\Omega |v_x \times v_y|$$

$$D\,(v) = \frac{1}{2} \iint_\Omega \left(|v_x|^2 + |v_y|^2 \right).$$

We then trivially have

$$J\,(v) \leq D\,(v) \tag{5.7}$$

since we know that

$$|v_x \times v_y| = \sqrt{E\,G - F^2} \leq \frac{1}{2}\,(E + G) = \frac{1}{2}\left(|v_x|^2 + |v_y|^2 \right). \tag{5.8}$$

Furthermore, we have equality in (5.8) (and hence in (5.7)) if and only if $E = G$ and $F = 0$ (i.e. the parametrization is given by isothermal coordinates). We then consider the minimization problems

$$(D) \quad d = \inf \{D(v) : v \text{ satisfies } (S_1), (S_2), (S_3)\}$$

$$(A) \quad a = \inf \{\text{Area}(\Sigma) : \Sigma \in \mathcal{S}\}.$$

In Step 2, we prove that there exists a minimizer $\overline{v}$ of (D), whose components are harmonic functions, which means that $\Delta \overline{v} = 0$. Moreover, this $\overline{v}$ verifies $E = G$ and $F = 0$ (see Step 3) and hence, according to Theorem 5.17, $\Sigma_0 = \overline{v}(\overline{\Omega})$ solves Plateau problem (up to the condition $\overline{v}_x \times \overline{v}_y \neq 0$). Finally we show, in Step 4, that in fact $a = d = D(\overline{v})$ and thus, since $E = G$, $F = 0$ and (5.7) holds, we have found that Σ_0 is also of minimal area.

Step 2. We now prove that (D) has a minimizer. This does not follow from the results of the previous chapters; it would if we had chosen a fixed parametrization of the boundary Γ. Since $\mathcal{S} \neq \emptyset$, we can find a minimizing sequence $\{v_\nu\}$ so that

$$D(v_\nu) \to d. \tag{5.9}$$

Any such sequence $\{v_\nu\}$ does not, in general, converge. The idea is to replace v_ν by a harmonic function $\widetilde{v}_\nu$ such that $v_\nu = \widetilde{v}_\nu$ on $\partial\Omega$. More precisely, we define $\widetilde{v}_\nu$ as the minimizer of

$$D(\widetilde{v}_\nu) = \min \{D(v) : v = v_\nu \text{ on } \partial\Omega\}. \tag{5.10}$$

Such a $\widetilde{v}_\nu$ exists and its components are harmonic (see Chapter 3). Combining (5.9) and (5.10), we still have

$$D(\widetilde{v}_\nu) \to d.$$

Without the hypotheses $(S_2), (S_3)$, this new sequence $\{\widetilde{v}_\nu\}$ does not converge either. The condition (S_3) is important, since (see Exercise 5.3.1) Dirichlet integral is invariant under any conformal transformation from Ω onto Ω; (S_3) allows us to select a unique one. The hypothesis (S_2) and the Courant-Lebesgue lemma imply that $\{\widetilde{v}_\nu\}$ is a sequence of equicontinuous functions (see Courant [24] page 103, Dierkes-Hildebrandt-Küster-Wohlrab [41] pages 235-237 or Nitsche [89] page 257). It follows from Ascoli-Arzelà theorem (Theorem 1.3) that, up to a subsequence,

$$\widetilde{v}_{\nu_k} \to \overline{v} \quad \text{uniformly.}$$

Harnack theorem (see, for example, Gilbarg-Trudinger [57] page 21), a classical property of harmonic functions, implies that $\overline{v}$ is harmonic, satisfies $(S_1), (S_2),$ (S_3) and

$$D(\overline{v}) = d \quad \text{with} \quad \Delta \overline{v} = 0 \text{ in } \Omega. \tag{5.11}$$

Step 3. We next show that this map $\overline{v}$ also verifies $E = G$ (i.e. $|\overline{v}_x|^2 = |\overline{v}_y|^2$) and $F = 0$ (i.e. $\langle \overline{v}_x; \overline{v}_y \rangle = 0$), which in particular implies that

$$\text{Area}\left(\overline{v}\left(\Omega\right)\right) = D\left(\overline{v}\right).$$

We use, in order to establish this fact, the technique of variations of the independent variables that we have already encountered in Section 2.3, when deriving the second form of the Euler-Lagrange equation. Since the proof of this step is lengthy, we subdivide it into three substeps.

Step 3.1. Let $\lambda, \mu \in C^\infty\left(\overline{\Omega}\right)$, to be chosen later, and let $\epsilon \in \mathbb{R}$ be sufficiently small so that the map

$$\left(\begin{array}{c} x' \\ y' \end{array}\right) = \varphi^\epsilon\left(x, y\right) = \left(\begin{array}{c} \varphi_1^\epsilon\left(x, y\right) \\ \varphi_2^\epsilon\left(x, y\right) \end{array}\right) = \left(\begin{array}{c} x + \epsilon\lambda\left(x, y\right) \\ y + \epsilon\mu\left(x, y\right) \end{array}\right)$$

is a diffeomorphism from $\overline{\Omega}$ onto a simply connected domain $\overline{\Omega}^\epsilon = \varphi^\epsilon\left(\overline{\Omega}\right)$. We denote its inverse by ψ^ϵ and we find that

$$\left(\begin{array}{c} x \\ y \end{array}\right) = \psi^\epsilon\left(x', y'\right) = \left(\begin{array}{c} \psi_1^\epsilon\left(x', y'\right) \\ \psi_2^\epsilon\left(x', y'\right) \end{array}\right) = \left(\begin{array}{c} x' - \epsilon\lambda\left(x', y'\right) + o\left(\epsilon\right) \\ y' - \epsilon\mu\left(x', y'\right) + o\left(\epsilon\right) \end{array}\right)$$

where $o\left(t\right)$ stands for a function $f = f\left(t\right)$ so that $f\left(t\right)/t$ tends to 0 as t tends to 0. We therefore have

$$\varphi^\epsilon\left(\psi^\epsilon\left(x', y'\right)\right) = \left(x', y'\right) \quad \text{and} \quad \psi^\epsilon\left(\varphi^\epsilon\left(x, y\right)\right) = \left(x, y\right).$$

Moreover, the Jacobian is given by

$$\det \nabla\varphi^\epsilon\left(x, y\right) = 1 + \epsilon\left(\lambda_x\left(x, y\right) + \mu_y\left(x, y\right)\right) + o\left(\epsilon\right). \tag{5.12}$$

We now change the independent variables and write

$$u^\epsilon\left(x', y'\right) = \overline{v}\left(\psi^\epsilon\left(x', y'\right)\right).$$

We find that

$$u_{x'}^\epsilon\left(x', y'\right) = \overline{v}_x\left(\psi^\epsilon\left(x', y'\right)\right)\frac{\partial}{\partial x'}\psi_1^\epsilon\left(x', y'\right) + \overline{v}_y\left(\psi^\epsilon\left(x', y'\right)\right)\frac{\partial}{\partial x'}\psi_2^\epsilon\left(x', y'\right)$$

$$= \overline{v}_x\left(\psi^\epsilon\right) - \epsilon\left[\overline{v}_x\left(\psi^\epsilon\right)\lambda_x\left(\psi^\epsilon\right) + \overline{v}_y\left(\psi^\epsilon\right)\mu_x\left(\psi^\epsilon\right)\right] + o\left(\epsilon\right)$$

and similarly

$$u_{y'}^\epsilon\left(x', y'\right) = \overline{v}_y\left(\psi^\epsilon\right) - \epsilon\left[\overline{v}_x\left(\psi^\epsilon\right)\lambda_y\left(\psi^\epsilon\right) + \overline{v}_y\left(\psi^\epsilon\right)\mu_y\left(\psi^\epsilon\right)\right] + o\left(\epsilon\right).$$

This leads to

$$
|u^\epsilon_{x'}(x',y')|^2 + |u^\epsilon_{y'}(x',y')|^2
$$
$$
= |\overline{v}_x(\psi^\epsilon)|^2 + |\overline{v}_y(\psi^\epsilon)|^2
$$
$$
-2\epsilon\left[|\overline{v}_x(\psi^\epsilon)|^2\lambda_x(\psi^\epsilon) + |\overline{v}_y(\psi^\epsilon)|^2\mu_y(\psi^\epsilon)\right]
$$
$$
-2\epsilon\left[\langle\overline{v}_x(\psi^\epsilon);\overline{v}_y(\psi^\epsilon)\rangle(\lambda_y(\psi^\epsilon) + \mu_x(\psi^\epsilon))\right] + o(\epsilon).
$$

Integrating this identity and changing the variables, letting $(x',y') = \varphi^\epsilon(x,y)$, on the right-hand side we find (recalling (5.12)) that

$$
\iint_{\Omega^\epsilon}\left[|u^\epsilon_{x'}(x',y')|^2 + |u^\epsilon_{y'}(x',y')|^2\right]dx'dy' = \iint_\Omega\left[|\overline{v}_x(x,y)|^2 + |\overline{v}_y(x,y)|^2\right]dx\,dy
$$
$$
-\epsilon\iint_\Omega\left[\left(|\overline{v}_x|^2 - |\overline{v}_y|^2\right)(\lambda_x - \mu_y) + 2\langle\overline{v}_x;\overline{v}_y\rangle(\lambda_y + \mu_x)\right]dx\,dy + o(\epsilon).
$$
$$
(5.13)
$$

Step 3.2. We now use Riemann theorem to find a conformal mapping

$$
\alpha^\epsilon : \Omega \to \Omega^\epsilon
$$

which is also a homeomorphism from $\overline{\Omega}$ onto $\overline{\Omega}^\epsilon$. We can also impose that the mapping verifies

$$
\alpha^\epsilon(w_i) = \varphi^\epsilon(w_i)
$$

where w_i are the points appearing in the definition of $\mathcal{S}$.

Recalling that $u^\epsilon = \overline{v}\circ\psi^\epsilon$, we finally let

$$
v^\epsilon(x,y) = u^\epsilon\circ\alpha^\epsilon(x,y) = \overline{v}\circ\psi^\epsilon\circ\alpha^\epsilon(x,y).
$$

Since $\overline{v}\in\mathcal{S}$, we deduce that $v^\epsilon\in\mathcal{S}$. Therefore, using the conformal invariance of the Dirichlet integral (see Exercise 5.3.1), we find that

$$
D(v^\epsilon) = \frac{1}{2}\iint_\Omega\left[|v^\epsilon_x(x,y)|^2 + |v^\epsilon_y(x,y)|^2\right]dx\,dy
$$
$$
= \frac{1}{2}\iint_{\Omega^\epsilon}\left[|u^\epsilon_{x'}(x',y')|^2 + |u^\epsilon_{y'}(x',y')|^2\right]dx'dy'
$$

which combined with (5.13) leads to

$$
D(v^\epsilon) = D(\overline{v}) - \frac{\epsilon}{2}\iint_\Omega\left[\left(|\overline{v}_x|^2 - |\overline{v}_y|^2\right)(\lambda_x - \mu_y)\right]dx\,dy
$$
$$
-\epsilon\iint_\Omega\left[\langle\overline{v}_x;\overline{v}_y\rangle(\lambda_y + \mu_x)\right]dx\,dy + o(\epsilon).
$$

Since $v^\epsilon, \overline{v} \in \mathcal{S}$ and $\overline{v}$ is a minimizer of Dirichlet integral, we find that

$$\iint_\Omega \left[\left(|\overline{v}_x|^2 - |\overline{v}_y|^2 \right) (\lambda_x - \mu_y) + 2 \langle \overline{v}_x; \overline{v}_y \rangle (\lambda_y + \mu_x) \right] dx \, dy = 0. \qquad (5.14)$$

Step 3.3. We finally choose in an appropriate way the functions $\lambda, \mu \in C^\infty(\overline{\Omega})$ that appeared in the previous steps. We let $\sigma, \tau \in C_0^\infty(\Omega)$ be arbitrary, we then choose λ and μ so that

$$\begin{cases} \lambda_x - \mu_y = \sigma \\ \lambda_y + \mu_x = \tau \end{cases}$$

(this is always possible; find first λ satisfying

$$\Delta \lambda = \sigma_x + \tau_y$$

then choose μ such that $(\mu_x, \mu_y) = (\tau - \lambda_y, \lambda_x - \sigma))$. Returning to (5.14) we find

$$\iint_\Omega \{(E - G)\sigma + 2F\tau\} = 0, \quad \forall \sigma, \tau \in C_0^\infty(\Omega).$$

The fundamental lemma of the calculus of variations (Theorem 1.24) implies then that $E = G$ and $F = 0$. Thus, up to the condition $\overline{v}_x \times \overline{v}_y \neq 0$, Plateau problem is solved (see Theorem 5.17), since $\Delta \overline{v} = 0$, according to (5.11). We have thus found $\overline{v}$ satisfying (i) and (ii) in the statements of the theorem.

We still have to prove (iii) in the statement of the theorem. However, this follows easily from (i), (ii) and (S_2) of the theorem, see Dierkes-Hildebrandt-Küster-Wohlrab [41] page 248.

Step 4. We let $\Sigma_0 = \overline{v}(\overline{\Omega})$ where $\overline{v}$ is the element that has been found in the previous steps and satisfies in particular

$$d = \inf \{D(v) : v \text{ satisfies } (S_1), (S_2), (S_3)\} = D(\overline{v}).$$

To conclude the proof of the theorem it remains to show that

$$a = \inf \{\text{Area}(\Sigma) : \Sigma \in \mathcal{S}\} = \text{Area}(\Sigma_0) = D(\overline{v}) = d.$$

We already know, from the previous steps, that

$$a \leq \text{Area}(\Sigma_0) = D(\overline{v}) = d$$

and we therefore wish to show the reverse inequality.

A way of proving this claim is by using a result of Morrey (see Dierkes-Hildebrandt-Küster-Wohlrab [41] page 252, and for a slightly different approach

see Courant [24] and Nitsche [89]) which asserts that for any $\epsilon > 0$ and any v
satisfying $(S_1), (S_2), (S_3)$ we can find v^ϵ verifying $(S_1), (S_2), (S_3)$ so that

$$D(v^\epsilon) - \epsilon \le \text{Area}(\Sigma)$$

where $\Sigma = v(\overline{\Omega})$. Since $d \le D(v^\epsilon)$ and ϵ is arbitrary, we obtain that $d \le a$; the
other inequality being trivial, we deduce that $a = d$.

This concludes the proof of the theorem. ∎

5.3.1   Exercise

Exercise 5.3.1 Let $\Omega \subset \mathbb{R}^2$ be a bounded smooth connected open set and

$$\varphi(x,y) = (\lambda(x,y), \mu(x,y))$$

be a conformal mapping from $\overline{\Omega}$ onto $\overline{B}$. Let $v \in C^1(\overline{B}; \mathbb{R}^3)$ and $w = v \circ \varphi$; show
that

$$\iint_\Omega \left[|w_x|^2 + |w_y|^2 \right] dx\, dy = \iint_B \left[|v_\lambda|^2 + |v_\mu|^2 \right] d\lambda\, d\mu.$$

5.4   Regularity, uniqueness and non-uniqueness

We now give some results without proofs. The first concerns the regularity of
the solution found in the previous section.

In Theorem 5.21, we found a weak solution of Plateau problem with a C^∞
parametrization. We have seen, and we will see it again below, that several
minimal surfaces may exist. The next result gives a regularity result for all such
surfaces with given boundary (see Nitsche [89] page 274).

Theorem 5.23 *If* Γ *is a simple closed curve of class* $C^{k,\alpha}$ *(with* $k \ge 1$ *an integer
and* $0 < \alpha < 1$*), then every weak solution of Plateau problem (i.e. with* $\partial\Sigma = \Gamma$*)
admits a* $C^{k,\alpha}(\overline{\Omega})$ *parametrization.*

However, the most important regularity result concerns the existence of a
regular surface (i.e. with $v_x \times v_y \ne 0$) that solves Plateau problem. We have
seen in Section 5.3 that the method of Douglas does not answer this question. It
provides only a weak solution of Plateau problem. A result in this direction is the
following (see Dierkes-Hildebrandt-Küster-Wohlrab [41] page 279 and Nitsche
[89] page 334).

Theorem 5.24 *(i) If* Γ *is an analytical simple closed curve and if its total
curvature does not exceed* 4π*, then any weak solution of Plateau problem (i.e.*

with $\partial\Sigma = \Gamma$) *is a (regular) minimal surface and thus a solution of Plateau problem.*

(ii) If a weak solution of Plateau problem is of minimal area, then the result remains true without any hypothesis on the total curvature of Γ.

Remark 5.25 The second part of the theorem allows, a posteriori, to assert that the solution found in Section 5.3 is a (regular) minimal surface (i.e. $\overline{v}_x \times \overline{v}_y \neq 0$), provided Γ is analytical. Thus it is a solution of Plateau problem.

We now turn our attention to the problem of uniqueness of minimal surfaces. Recall first (Theorem 5.12) that we have uniqueness when restricted to nonparametric surfaces. For general surfaces we have the following uniqueness result (see Nitsche [89] page 351).

Theorem 5.26 *Let* Γ *be an analytical simple closed curve with total curvature not exceeding* 4π, *then Plateau problem has a unique solution.*

We now give a non-uniqueness result (for more details we refer to [41] and to [89]).

Example 5.27 (Enneper surface) (see Example 5.10). Let $r \in (1, \sqrt{3})$ and

$$\Gamma_r = \left\{ \left(r\cos\theta - \frac{r^3}{3}\cos(3\theta), -r\sin\theta - \frac{r^3}{3}\sin(3\theta), r^2\cos(2\theta) \right) : \theta \in [0, 2\pi) \right\}.$$

We have seen (Example 5.10) that

$$\Sigma_r = \left\{ \left(rx + r^3xy^2 - \frac{r^3}{3}x^3, -ry - r^3x^2y + \frac{r^3}{3}y^3, r^2(x^2 - y^2) \right) : x^2 + y^2 \leq 1 \right\}$$

is a minimal surface and that $\partial\Sigma_r = \Gamma_r$. It is possible to show (see Nitsche [89] page 338) that Σ_r is not of minimal area if $r \in (1, \sqrt{3})$; therefore, it is distinct from the one found in Theorem 5.21.

5.5 Nonparametric minimal surfaces

5.5.1 General remarks

We now discuss the case of nonparametric surfaces. Let $\Omega \subset \mathbb{R}^n$ be a bounded connected open set (in the present section we do not need to limit ourselves to the case $n = 2$). The surfaces that we consider are of the form

$$\Sigma = \left\{ (x, u(x)) = (x_1, \cdots, x_n, u(x_1, \cdots, x_n)) : x \in \overline{\Omega} \right\}.$$

The area of such a surface is given by

$$I\left(u\right) = \int_{\Omega} \sqrt{1 + \left|\nabla u\left(x\right)\right|^2}\, dx.$$

As already seen in Theorem 5.12, we have that any $C^2\left(\overline{\Omega}\right)$ minimizer of

$$(P) \quad \inf\left\{I\left(u\right) : u = u_0 \text{ on } \partial\Omega\right\}$$

satisfies the minimal surface equation

$$(E) \quad Mu \equiv \left(1 + \left|\nabla u\right|^2\right)\Delta u - \sum_{i,j=1}^{n} u_{x_i} u_{x_j} u_{x_i x_j} = 0$$

$$\Leftrightarrow \quad \sum_{i=1}^{n} \frac{\partial}{\partial x_i}\left[\frac{u_{x_i}}{\sqrt{1 + \left|\nabla u\right|^2}}\right] = 0$$

and hence Σ has mean curvature that vanishes everywhere. The converse is also true; moreover, we have uniqueness of such solutions. However, the existence of a minimizer still needs to be proved because Theorem 5.21 does not deal with this case. The techniques for solving such a problem are much more analytical than the previous ones.

Before proceeding with the existence theorems we would like to mention a famous related problem known as *Bernstein problem*. The problem is posed in the whole space $\mathbb{R}^n$ (i.e. $\Omega = \mathbb{R}^n$) and we seek for $C^2\left(\mathbb{R}^n\right)$ solutions of the minimal surface equation

$$(E) \quad \sum_{i=1}^{n} \frac{\partial}{\partial x_i}\left(\frac{u_{x_i}}{\sqrt{1 + \left|\nabla u\right|^2}}\right) = 0 \quad \text{in } \mathbb{R}^n$$

or of its equivalent form $Mu = 0$. In terms of regular surfaces, we are searching for a nonparametric surface (defined over the whole of $\mathbb{R}^n$) in $\mathbb{R}^{n+1}$ that has vanishing mean curvature. Obviously the function

$$u\left(x\right) = \langle a; x\rangle + b$$

with $a \in \mathbb{R}^n$ and $b \in \mathbb{R}$, which in geometrical terms represents a hyperplane, is a solution of the equation. The question is whether this is the only solution.

In the case $n = 2$, Bernstein has shown that, indeed, this is the only C^2 solution (the result is known as Bernstein theorem). Since then several authors have found different proofs of this theorem. The extension to higher dimensions is however much harder. De Giorgi extended the result to the case $n = 3$,

Almgren to the case $n = 4$ and Simons to $n = 5, 6, 7$. In 1969, Bombieri, De Giorgi and Giusti proved that when $n \geq 8$, there exists a nonlinear $u \in C^2(\mathbb{R}^n)$ (and hence the surface is not a hyperplane) satisfying equation (E). For more details on Bernstein problem, see Giusti [58] Chapter 17 and Nitsche [89] pages 123-124 and 429-430.

We now return to our problem in a bounded connected open set. We start by quoting a result of Jenkins and Serrin; for a proof see Gilbarg-Trudinger [57] page 297.

Theorem 5.28 (Jenkins-Serrin) *Let $\Omega \subset \mathbb{R}^n$ be a bounded connected open set with $C^{2,\alpha}$, $0 < \alpha < 1$, boundary and let $u_0 \in C^{2,\alpha}(\overline{\Omega})$. The problem*

$$\begin{cases} Mu = 0 & in \ \Omega \\ u = u_0 & on \ \partial\Omega \end{cases}$$

has a solution for every u_0 if and only if the mean curvature of $\partial\Omega$ is everywhere non-negative.

Remark 5.29 (i) We now briefly mention a related result due to Finn and Osserman. It roughly says that if Ω is a non-convex connected open set, there exists a continuous u_0 so that the problem

$$\begin{cases} Mu = 0 & in \ \Omega \\ u = u_0 & on \ \partial\Omega \end{cases}$$

has no C^2 solution. Such a u_0 can even have arbitrarily small norm $\|u_0\|_{C^0}$.

(ii) The above theorem follows several earlier works that started with Bernstein (see Nitsche [89] pages 352-358).

5.5.2 Korn-Müntz theorem

We end the present chapter with a simple theorem of which ideas contained in the proof are used in several different problems of partial differential equations.

Theorem 5.30 (Korn-Müntz) *Let $\Omega \subset \mathbb{R}^n$ be a bounded connected open set with $C^{2,\alpha}$, $0 < \alpha < 1$, boundary and consider the problem*

$$\begin{cases} Mu = \left(1 + |\nabla u|^2\right)\Delta u - \sum_{i,j=1}^{n} u_{x_i} u_{x_j} u_{x_i x_j} = 0 & in \ \Omega \\ u = u_0 & on \ \partial\Omega. \end{cases}$$

Then there exists $\epsilon > 0$ so that, for every $u_0 \in C^{2,\alpha}(\overline{\Omega})$ with $\|u_0\|_{C^{2,\alpha}} \leq \epsilon$, the above problem has a (unique) $C^{2,\alpha}(\overline{\Omega})$ solution.

Remark 5.31 It is interesting to compare this theorem with the preceding one. We see that we have not made any assumption on the mean curvature of $\partial\Omega$; but we require that the $C^{2,\alpha}$ norm of u_0 be small. We should also observe that the above-mentioned result of Finn and Osserman (see Remark 5.29 (i)) shows that if Ω is non-convex and if $\|u_0\|_{C^0} \leq \epsilon$ this is, in general, not sufficient to get the existence of solutions. Therefore we cannot, in general, replace the condition $\|u_0\|_{C^{2,\alpha}}$ small, by $\|u_0\|_{C^0}$ small.

Proof The proof is divided into four steps. We write

$$Mu = 0 \quad \Leftrightarrow \quad \Delta u = N(u) \equiv \sum_{i,j=1}^{n} u_{x_i} u_{x_j} u_{x_i x_j} - |\nabla u|^2 \Delta u. \qquad (5.15)$$

From estimates of the linearized equation $\Delta u = f$ (Step 1) and estimates of the nonlinear part $N(u)$ (Step 2), we will be able to conclude (Step 3).

Step 1. Let us recall the classical Schauder estimates concerning Poisson equation (see Theorem 6.6 and page 103 in Gilbarg-Trudinger [57]). If $\Omega \subset \mathbb{R}^n$ is a bounded connected open set of $\mathbb{R}^n$ with $C^{2,\alpha}$ boundary and if

$$\begin{cases} \Delta u = f & \text{in } \Omega \\ u = \varphi & \text{on } \partial\Omega \end{cases} \qquad (5.16)$$

we can find a constant $\gamma = \gamma(\Omega) > 0$ so that the (unique) solution u of (5.16) satisfies

$$\|u\|_{C^{2,\alpha}} \leq \gamma \left(\|f\|_{C^{0,\alpha}} + \|\varphi\|_{C^{2,\alpha}} \right). \qquad (5.17)$$

Step 2. We now estimate the nonlinear term N. We will show that we can find a constant, still denoted by $\gamma = \gamma(\Omega) > 0$, so that for every $u, v \in C^{2,\alpha}(\overline{\Omega})$ we have

$$\|N(u) - N(v)\|_{C^{0,\alpha}} \leq \gamma \left(\|u\|_{C^{2,\alpha}} + \|v\|_{C^{2,\alpha}} \right)^2 \|u - v\|_{C^{2,\alpha}}. \qquad (5.18)$$

From

$$N(u) - N(v) = \sum_{i,j=1}^{n} u_{x_i} u_{x_j} \left(u_{x_i x_j} - v_{x_i x_j} \right)$$

$$+ \sum_{i,j=1}^{n} v_{x_i x_j} \left(u_{x_i} u_{x_j} - v_{x_i} v_{x_j} \right)$$

$$+ |\nabla u|^2 (\Delta v - \Delta u) + \Delta v \left(|\nabla v|^2 - |\nabla u|^2 \right)$$

we deduce that (5.18) holds from Proposition 1.9 and its proof (see Exercise 1.2.2).

Step 3. We are now in a position to show the theorem. We define a sequence $\{u_\nu\}_{\nu=1}^\infty$ of $C^{2,\alpha}\left(\overline{\Omega}\right)$ functions in the following way

$$\begin{cases} \Delta u_1 = 0 & \text{in } \Omega \\ u_1 = u_0 & \text{on } \partial\Omega \end{cases} \tag{5.19}$$

and by induction

$$\begin{cases} \Delta u_{\nu+1} = N\left(u_\nu\right) & \text{in } \Omega \\ u_{\nu+1} = u_0 & \text{on } \partial\Omega. \end{cases} \tag{5.20}$$

The previous estimates will allow us (see Step 4) to deduce that for $\|u_0\|_{C^{2,\alpha}} \le \epsilon$, ϵ to be determined, we have

$$\|u_{\nu+1} - u_\nu\|_{C^{2,\alpha}} \le K \|u_\nu - u_{\nu-1}\|_{C^{2,\alpha}} \tag{5.21}$$

for some $K < 1$. If (5.21) is true, we obtain

$$\|u_{\nu+1} - u_\nu\|_{C^{2,\alpha}} \le K^\nu \|u_1 - u_0\|_{C^{2,\alpha}}.$$

Therefore if $\nu > \mu$, we have

$$\|u_\nu - u_\mu\|_{C^{2,\alpha}} \le \left(\sum_{s=\mu}^{\nu-1} K^s\right) \|u_1 - u_0\|_{C^{2,\alpha}} \le \frac{K^\mu}{1-K} \|u_1 - u_0\|_{C^{2,\alpha}}.$$

This readily implies that $\{u_\nu\}$ is a Cauchy sequence in $C^{2,\alpha}$ and thus it converges to a certain u in $C^{2,\alpha}$. We moreover have

$$\begin{cases} \Delta u = N\left(u\right) & \text{in } \Omega \\ u = u_0 & \text{on } \partial\Omega \end{cases}$$

which is the claimed result.

Step 4. We now establish (5.21), which amounts to find the appropriate $\epsilon > 0$. We start by choosing $0 < K < 1$ and we then choose $\epsilon > 0$ sufficiently small so that

$$2\gamma^2 \epsilon \left(1 + \frac{\gamma^4 \epsilon^2}{1-K}\right) \le \sqrt{K} \tag{5.22}$$

where γ is the constant appearing in Step 1 and Step 2 (we can consider, without loss of generality, that they are the same).

We therefore only need to show that if $\|u_0\|_{C^{2,\alpha}} \le \epsilon$, we have indeed (5.21). Note that for every $\nu \ge 2$ we have

$$\begin{cases} \Delta \left(u_{\nu+1} - u_\nu\right) = N\left(u_\nu\right) - N\left(u_{\nu-1}\right) & \text{in } \Omega \\ u_{\nu+1} - u_\nu = 0 & \text{on } \partial\Omega. \end{cases}$$

From Step 1 and Step 2, we find that, for every $\nu \geq 2$,

$$\|u_{\nu+1} - u_\nu\|_{C^{2,\alpha}} \leq \gamma \|N(u_\nu) - N(u_{\nu-1})\|_{C^{0,\alpha}}$$

and thus

$$\|u_{\nu+1} - u_\nu\|_{C^{2,\alpha}} \leq \gamma^2 \left(\|u_\nu\|_{C^{2,\alpha}} + \|u_{\nu-1}\|_{C^{2,\alpha}}\right)^2 \|u_\nu - u_{\nu-1}\|_{C^{2,\alpha}}. \qquad (5.23)$$

Similarly for $\nu = 1$, we have

$$\|u_2 - u_1\|_{C^{2,\alpha}} \leq \gamma \|N(u_1)\|_{C^{0,\alpha}} \leq \gamma^2 \|u_1\|_{C^{2,\alpha}}^3. \qquad (5.24)$$

From now on, since all norms will be $C^{2,\alpha}$ norms, we will denote them simply by $\|\cdot\|$. From (5.23), we deduce that it is enough to show

$$\gamma^2 \left(\|u_\nu\| + \|u_{\nu-1}\|\right)^2 \leq K, \quad \nu \geq 2 \qquad (5.25)$$

to obtain (5.21) and thus the theorem. To prove (5.25), it is sufficient to show that

$$\|u_\nu\| \leq \gamma\epsilon \left(1 + \frac{\gamma^4\epsilon^2}{1-K}\right), \quad \nu \geq 1. \qquad (5.26)$$

The inequality (5.21) then follows from the choice of ϵ in (5.22). We prove (5.26) by induction. Observe that by Step 1 and from (5.19), we have

$$\|u_1\| \leq \gamma \|u_0\| \leq \gamma\epsilon \left(\leq \gamma\epsilon \left(1 + \frac{\gamma^4\epsilon^2}{1-K}\right)\right). \qquad (5.27)$$

We now show (5.26) for $\nu = 2$. We have, from (5.24) and from (5.27), that

$$\|u_2\| \leq \|u_1\| + \|u_2 - u_1\| \leq \|u_1\| \left(1 + \gamma^2 \|u_1\|^2\right) \leq \gamma\epsilon \left(1 + \gamma^4\epsilon^2\right)$$

and hence, since $K \in (0,1)$, we deduce the inequality (5.26). Suppose now, from the hypothesis of induction, that (5.26) is valid up to order ν and let us prove that it holds true for $\nu + 1$. We trivially have that

$$\|u_{\nu+1}\| \leq \|u_1\| + \sum_{j=1}^{\nu} \|u_{j+1} - u_j\|. \qquad (5.28)$$

We deduce from (5.23) and (5.25) (since (5.26) holds for every $1 \leq j \leq \nu$) that

$$\|u_{j+1} - u_j\| \leq K \|u_j - u_{j-1}\| \leq K^{j-1} \|u_2 - u_1\|.$$

Returning to (5.28), we find

$$\|u_{\nu+1}\| \leq \|u_1\| + \|u_2 - u_1\| \sum_{j=1}^{\nu} K^{j-1} \leq \|u_1\| + \frac{1}{1-K} \|u_2 - u_1\|.$$

Appealing to (5.24) and then to (5.27), we obtain

$$\|u_{\nu+1}\| \leq \|u_1\| \left(1 + \frac{\gamma^2 \|u_1\|^2}{1 - K}\right) \leq \gamma\epsilon \left(1 + \frac{\gamma^4\epsilon^2}{1 - K}\right),$$

which is exactly (5.26). The theorem then follows. ∎

5.5.3 Exercise

Exercise 5.5.1 Let $u \in C^2\left(\mathbb{R}^2\right)$, $u = u\left(x, y\right)$, be a solution of the minimal surface equation

$$Mu = \left(1 + u_y^2\right) u_{xx} - 2\, u_x\, u_y\, u_{xy} + \left(1 + u_x^2\right) u_{yy} = 0.$$

Show that there exists a convex function $\varphi \in C^2\left(\mathbb{R}^2\right)$, so that

$$\varphi_{xx} = \frac{1 + u_x^2}{\sqrt{1 + u_x^2 + u_y^2}}, \quad \varphi_{xy} = \frac{u_x u_y}{\sqrt{1 + u_x^2 + u_y^2}}, \quad \varphi_{yy} = \frac{1 + u_y^2}{\sqrt{1 + u_x^2 + u_y^2}}.$$

Deduce that

$$\varphi_{xx}\varphi_{yy} - \varphi_{xy}^2 = 1.$$

Chapter 6

Isoperimetric inequality

6.1 Introduction

Let $A \subset \mathbb{R}^2$ be a bounded open set whose boundary, ∂A, is a sufficiently regular, simple closed curve. Denote by $L(\partial A)$ the length of the boundary and by $M(A)$ the measure (the area) of A. The isoperimetric inequality states that

$$[L(\partial A)]^2 - 4\pi M(A) \geq 0.$$

Furthermore, equality holds if and only if A is a disk (i.e. ∂A is a circle).

This is one of the oldest problems in mathematics. A variant of this inequality is known as Dido problem (Dido is said to have been a Phoenician princess). Several more or less rigorous proofs were known at the time of the Ancient Greeks; the most notable attempt at proving the inequality is due to Zenodorus, who proved the inequality for polygons. There are also significant contributions by Archimedes and Pappus. Coming closer to us, one can mention, among many, Euler, Galileo, Legendre, L'Huilier, Riccati and Simpson. A special tribute should be paid to Steiner, who derived necessary conditions through a clever argument of symmetrization. The first proof that agrees with modern standards is due to Weierstrass. Since then, many proofs have been given, notably by Blaschke, Bonnesen, Carathéodory, Edler, Frobenius, Hurwitz, Lebesgue, Liebmann, Minkowski, H.A. Schwarz, Sturm and Tonelli, among others. We refer to Porter [97] for an interesting article on the history of the inequality.

We give here the proof of Hurwitz as modified by H. Lewy and Hardy-Littlewood-Polya [65]. In particular, we show that the isoperimetric inequality is equivalent to Wirtinger inequality which we have already encountered in a weaker form (see Poincaré-Wirtinger inequality). This inequality reads as

$$\int_{-1}^{1} (u')^2 \geq \pi^2 \int_{-1}^{1} u^2, \quad \forall u \in X$$

where

$$X = \left\{ u \in W^{1,2}(-1,1) : u(-1) = u(1) \text{ and } \int_{-1}^{1} u = 0 \right\}.$$

It also states that equality holds if and only if $u(x) = \alpha \cos(\pi x) + \beta \sin(\pi x)$, for any $\alpha, \beta \in \mathbb{R}$.

In Section 6.3 we discuss the generalization to $\mathbb{R}^n$, $n \geq 3$, of the isoperimetric inequality. It reads as follows

$$[L(\partial A)]^n - n^n \omega_n [M(A)]^{n-1} \geq 0$$

for every bounded open set $A \subset \mathbb{R}^n$ with sufficiently regular boundary, ∂A; and where ω_n is the measure of the unit ball of $\mathbb{R}^n$, $M(A)$ stands for the measure of A and $L(\partial A)$ for the $(n-1)$ measure of ∂A. Moreover, if A is sufficiently regular (for example, convex), there is equality if and only if A is a ball.

The inequality in higher dimensions is considerably harder to prove; we briefly discuss, in Section 6.3, the main ideas of one of the proofs. When $n = 3$, the first complete proof is that of H.A. Schwarz. Soon after there were generalizations to higher dimensions and other proofs notably by A. Aleksandrov, Blaschke, Bonnesen, H. Hopf, Liebmann, Minkowski and E. Schmidt; or more recently by De Giorgi [40] and Gromov [62].

In recent years considerable efforts have been made on the stability of the isoperimetric inequality, see, for example, Fusco-Maggi-Pratelli [53].

Finally, numerous generalizations of this inequality have been studied in relation to problems of mathematical physics, see Bandle [9], Payne [94] and Polya-Szegö [96] for more references.

There are several articles and books devoted to the subject; we recommend the review article of Osserman [92] and the books by Berger [10], Blaschke [11], Federer [49], Hardy-Littlewood-Polya [65] (for the two dimensional case) and Webster [108]. The book of Hildebrandt-Tromba [68] also has a chapter on this matter.

6.2   The case of dimension 2

6.2.1   Wirtinger inequality

We start with the key result for proving the isoperimetric inequality; but before that we introduce the following notation, for any $p \geq 1$,

$$W_{\text{per}}^{1,p}(a,b) = \left\{ u \in W^{1,p}(a,b) : u(a) = u(b) \right\}$$

and
$$C^1_{per}\left([a,b]\right) = \left\{u \in C^1\left([a,b]\right) : u\left(a\right) = u\left(b\right)\right\}.$$

Theorem 6.1 (Wirtinger inequality) *Let*
$$X = \left\{u \in W^{1,2}_{per}\left(-1,1\right) : \int_{-1}^{1} u = 0\right\}$$

then
$$\int_{-1}^{1}\left(u'\right)^2 \geq \pi^2 \int_{-1}^{1} u^2, \quad \forall\, u \in X.$$

Furthermore, the equality holds if and only if $u\left(x\right) = \alpha \cos\left(\pi x\right) + \beta \sin\left(\pi x\right),$ *for any* $\alpha, \beta \in \mathbb{R}$.

Remark 6.2 (i) We will show, in Exercise 6.2.4, that for smooth functions Wirtinger inequality is implied by the isoperimetric inequality and is thus equivalent.

(ii) More generally we have if
$$X = \left\{u \in W^{1,2}_{per}\left(a,b\right) : \int_a^b u = 0\right\}$$

that
$$\int_a^b \left(u'\right)^2 \geq \left(\frac{2\pi}{b-a}\right)^2 \int_a^b u^2, \quad \forall\, u \in X.$$

(iii) The inequality can also be generalized (see Croce-Dacorogna [28]) to
$$\left(\int_a^b |u'|^p\right)^{1/p} \geq \alpha\left(p,q,r\right)\left(\int_a^b |u|^q\right)^{1/q}, \quad \forall\, u \in X$$

for some appropriate $\alpha\left(p,q,r\right)$ (in particular $\alpha\left(2,2,2\right) = 2\pi/\left(b-a\right)$) and where
$$X = \left\{u \in W^{1,p}_{per}\left(a,b\right) : \int_a^b |u|^{r-2} u = 0\right\}.$$

(iv) We have seen in Example 2.24 a weaker form of the inequality, known as Poincaré-Wirtinger inequality, namely
$$\int_0^1 \left(u'\right)^2 \geq \pi^2 \int_0^1 u^2, \quad \forall\, u \in W^{1,2}_0\left(0,1\right).$$

This inequality can be inferred from the theorem by setting
$$u\left(x\right) = -u\left(-x\right) \quad \text{if } x \in \left(-1,0\right).$$

Proof An alternative proof, more in the spirit of Example 2.24, is proposed in Exercise 6.2.1. The proof given here is, essentially, the classical proof of Hurwitz. We divide the proof into two steps.

Step 1. We start by proving the theorem under the further restriction that $u \in X \cap C^2[-1,1]$. We express u in Fourier series

$$u(x) = \sum_{n=1}^{+\infty} [a_n \cos(n\pi x) + b_n \sin(n\pi x)].$$

Note that there is no constant term since $\int_{-1}^{1} u = 0$. We know at the same time that

$$u'(x) = \pi \sum_{n=1}^{+\infty} [-na_n \sin(n\pi x) + nb_n \cos(n\pi x)].$$

We can now invoke *Parseval formula* to get

$$\int_{-1}^{1} u^2 = \sum_{n=1}^{+\infty} (a_n^2 + b_n^2) \quad \text{and} \quad \int_{-1}^{1} (u')^2 = \pi^2 \sum_{n=1}^{+\infty} (a_n^2 + b_n^2) n^2.$$

The desired inequality follows at once

$$\int_{-1}^{1} (u')^2 \geq \pi^2 \int_{-1}^{1} u^2, \quad \forall u \in X \cap C^2.$$

Moreover, the equality holds if and only if $a_n = b_n = 0$, for every $n \geq 2$. This implies that the equality holds if and only if $u(x) = \alpha \cos(\pi x) + \beta \sin(\pi x)$, for any $\alpha, \beta \in \mathbb{R}$, as claimed.

Step 2. We now show that we can remove the restriction $u \in X \cap C^2[-1,1]$. By the usual density argument we can find for every $u \in X$ a sequence $u_\nu \in X \cap C^2[-1,1]$ so that

$$u_\nu \to u \quad \text{in } W^{1,2}(-1,1).$$

Therefore, for every $\epsilon > 0$, we can find ν sufficiently large so that

$$\int_{-1}^{1} (u')^2 \geq \int_{-1}^{1} (u_\nu')^2 - \epsilon \quad \text{and} \quad \int_{-1}^{1} u_\nu^2 \geq \int_{-1}^{1} u^2 - \epsilon.$$

Combining these inequalities with Step 1 we find

$$\int_{-1}^{1} (u')^2 \geq \pi^2 \int_{-1}^{1} u^2 - (\pi^2 + 1)\epsilon.$$

Letting $\epsilon \to 0$ we have indeed obtained the inequality.

We still need to see that the equality in X holds if and only if $u(x) = \alpha \cos(\pi x) + \beta \sin(\pi x)$, for any $\alpha, \beta \in \mathbb{R}$. This has been proved in Step 1 only if $u \in X \cap C^2[-1,1]$. This property is established in Exercise 6.2.2. ∎

We get as a direct consequence of the theorem

Corollary 6.3 *The following inequality holds*

$$\int_{-1}^{1} \left((u')^2 + (v')^2 \right) \geq 2\pi \int_{-1}^{1} u\, v', \quad \forall u, v \in W_{per}^{1,2}(-1,1).$$

Furthermore, the equality holds if and only if

$$(u(x) - r_1)^2 + (v(x) - r_2)^2 = r_3^2, \quad \forall x \in [-1,1]$$

where $r_1, r_2, r_3 \in \mathbb{R}$ are constants.

Proof Let

$$r_1 = \frac{1}{2}\int_{-1}^{1} u, \quad r_2 = \frac{1}{2}\int_{-1}^{1} v, \quad f = u - r_1, \quad g = v - r_2.$$

Note that

$$f, g \in X = \left\{ u \in W_{per}^{1,2}(-1,1) : \int_{-1}^{1} u = 0 \right\}$$

$$\int_{-1}^{1} \left((f')^2 + (g')^2 \right) = \int_{-1}^{1} \left((u')^2 + (v')^2 \right) \quad \text{and} \quad \int_{-1}^{1} f\, g' = \int_{-1}^{1} u\, v'.$$

We can therefore rewrite our problem as

$$\int_{-1}^{1} \left((u')^2 + (v')^2 - 2\pi\, u\, v' \right) = \int_{-1}^{1} \left((f')^2 + (g')^2 - 2\pi\, f\, g' \right)$$

$$= \int_{-1}^{1} (g' - \pi f)^2 + \int_{-1}^{1} \left((f')^2 - \pi^2 f^2 \right).$$

From Theorem 6.1, we deduce that the second term in the above inequality is non-negative, while the first one is trivially non-negative; thus the inequality is established.

We now discuss the equality case. If the equality holds we should have

$$g' = \pi f \quad \text{and} \quad \int_{-1}^{1} \left((f')^2 - \pi^2 f^2 \right) = 0.$$

This implies, according to Theorem 6.1, that

$$f(x) = \alpha \cos(\pi x) + \beta \sin(\pi x) \quad \text{and} \quad g(x) = \alpha \sin(\pi x) - \beta \cos(\pi x)$$

and thus
$$f^2 + g^2 = \alpha^2 + \beta^2.$$

We therefore have, setting $r_3 = \sqrt{\alpha^2 + \beta^2}$, that

$$(u(x) - r_1)^2 + (v(x) - r_2)^2 = r_3^2, \quad \forall x \in [-1, 1]$$

as wished. ∎

6.2.2   The isoperimetric inequality

We are now in a position to prove the isoperimetric inequality in its analytic form; we postpone the discussion of its geometric meaning for later.

Theorem 6.4 (Isoperimetric inequality) *For $u, v \in W_{per}^{1,1}(a, b)$, let*

$$L(u, v) = \int_a^b \sqrt{(u')^2 + (v')^2}$$

$$M(u, v) = \frac{1}{2} \int_a^b (u\, v' - v\, u') = \int_a^b u\, v'.$$

Then

$$[L(u, v)]^2 - 4\pi M(u, v) \geq 0, \quad \forall u, v \in W_{per}^{1,1}(a, b).$$

Moreover, among all $u, v \in C_{per}^1([a, b])$, the equality holds if and only if

$$(u(x) - r_1)^2 + (v(x) - r_2)^2 = r_3^2, \quad \forall x \in [a, b]$$

where $r_1, r_2, r_3 \in \mathbb{R}$ are constants.   .

Remark 6.5 The uniqueness holds under weaker regularity hypotheses, but we do not discuss this here. However, we point out that the very same proof for the uniqueness is valid for piecewise C_{per}^1 functions u and v.

Proof We divide the proof into three steps.

Step 1. We first prove the theorem under the further restrictions that $u, v \in C_{per}^1([a, b])$ and

$$(u'(x))^2 + (v'(x))^2 > 0, \quad \forall x \in [a, b].$$

We start by reparametrizing the curve by a multiple of its arc length, namely

$$\begin{cases} y = \eta(x) = -1 + \frac{2}{L(u,v)} \int_a^x \sqrt{(u')^2 + (v')^2} \\ \varphi(y) = u(\eta^{-1}(y)) \quad \text{and} \quad \psi(y) = v(\eta^{-1}(y)). \end{cases}$$

It is easy to see that $\varphi, \psi \in C^1_{per}([a,b])$ and

$$\sqrt{(\varphi'(y))^2 + (\psi'(y))^2} = \frac{L(u,v)}{2}, \quad \forall\, y \in [-1,1].$$

We therefore have

$$L(u,v) = \int_{-1}^1 \sqrt{(\varphi'(y))^2 + (\psi'(y))^2}\, dy = \left(2 \int_{-1}^1 \left[(\varphi'(y))^2 + (\psi'(y))^2\right] dy\right)^{1/2}$$

$$M(u,v) = \int_{-1}^1 \varphi(y)\, \psi'(y)\, dy.$$

We, however, know from Corollary 6.3 that

$$\int_{-1}^1 \left((\varphi')^2 + (\psi')^2\right) \geq 2\pi \int_{-1}^1 \varphi \psi', \quad \forall\, \varphi, \psi \in W^{1,2}_{per}(-1,1)$$

which implies the claim

$$[L(u,v)]^2 - 4\pi M(u,v) \geq 0, \quad \forall\, u, v \in C^1_{per}([a,b]).$$

The uniqueness in the equality case follows also from the corresponding one in Corollary 6.3.

Step 2. We next remove the hypothesis $(u')^2 + (v')^2 > 0$, still keeping the fact that $u, v \in C^1_{per}([a,b])$. The details are discussed in Exercise 6.2.3.

Step 3. We now remove the hypothesis $u, v \in C^1_{per}([a,b])$. As before, given $u, v \in W^{1,1}_{per}(a,b)$, we can find $u_\nu, v_\nu \in C^1_{per}([a,b])$ so that

$$u_\nu, v_\nu \to u, v \quad \text{in } W^{1,1}(a,b) \cap L^\infty(a,b).$$

Therefore, for every $\epsilon > 0$, we can find ν sufficiently large so that

$$[L(u,v)]^2 \geq [L(u_\nu,v_\nu)]^2 - \epsilon \quad \text{and} \quad M(u_\nu,v_\nu) \geq M(u,v) - \epsilon$$

and hence, combining these inequalities with Step 2, we get

$$[L(u,v)]^2 - 4\pi M(u,v) \geq [L(u_\nu,v_\nu)]^2 - 4\pi M(u_\nu,v_\nu) - (1+4\pi)\epsilon$$
$$\geq -(1+4\pi)\epsilon.$$

Since ϵ is arbitrary, we have indeed obtained the inequality. ∎

We now briefly discuss the geometrical meaning of the inequality obtained in Theorem 6.4. Any bounded open set A, whose boundary ∂A is a closed curve that possesses a parametrization $u, v \in W^{1,1}_{\text{per}}(a, b)$ so that its length and area are given by

$$L(\partial A) = L(u, v) = \int_a^b \sqrt{(u')^2 + (v')^2}$$

$$M(A) = M(u, v) = \frac{1}{2} \int_a^b (u\,v' - v\,u') = \int_a^b u\,v'$$

will therefore satisfy the isoperimetric inequality

$$[L(\partial A)]^2 - 4\pi M(A) \geq 0.$$

This is, of course, the case for any simple closed smooth curve, whose interior is A.

One should also note that very wild sets A can be allowed. Indeed sets A that can be approximated by sets A_ν that satisfy the isoperimetric inequality and are such that

$$L(\partial A_\nu) \to L(\partial A) \quad \text{and} \quad M(A_\nu) \to M(A), \text{ as } \nu \to \infty$$

also verify the inequality.

6.2.3   Exercises

Exercise 6.2.1 Prove Theorem 6.1 in a manner analogous to that of Example 2.24.

Exercise 6.2.2 Let

$$(P) \inf \left\{ I(u) = \int_{-1}^1 \left((u')^2 - \pi^2 u^2 \right) : u \in X \right\} = m$$

where

$$X = \left\{ u \in W^{1,2}_{\text{per}}(-1, 1) : \int_{-1}^1 u = 0 \right\}.$$

Prove that, for any $\alpha, \beta \in \mathbb{R}$,

$$u(x) = \alpha \cos(\pi x) + \beta \sin(\pi x)$$

are the only minimizers in X.

Suggestion. We have seen in Theorem 6.1 that $m = 0$ and the minimum is attained in $X \cap C^2[-1, 1]$ if and only if $u(x) = \alpha \cos(\pi x) + \beta \sin(\pi x)$, for any $\alpha, \beta \in \mathbb{R}$. Show that any minimizer of (P) is $C^2[-1, 1]$. Conclude.

Exercise 6.2.3 Prove Step 2 of Theorem 6.4 for any $u, v \in C^1_{\text{per}}([a, b])$.

Exercise 6.2.4 Deduce Wirtinger inequality, for C^1 functions, from the isoperimetric inequality (Theorem 6.4).

6.3   The case of dimension n

The above proof does not generalize to $\mathbb{R}^n$, $n \geq 3$. A completely different and harder proof is necessary to deal with this case. We will present one of the classical proofs, based on Brunn-Minkowski theorem. Several others are available, notably one using Steiner symmetrization (see Berger [10] Section 12.11.2 or De Giorgi [40] for a more general setting) or the one by Gromov [62] (see also Berger [10] Section 12.11.4). Another more geometrical proof is described now. The inequality

$$L^n - n^n \, \omega_n \, M^{n-1} \geq 0$$

($L = L(\partial A)$ and $M = M(A)$) is equivalent to the minimization of L for fixed M together with showing that the minimizers are given by spheres. We can then write the associated Euler-Lagrange equation, with a Lagrange multiplier corresponding to the constraint that M is fixed (see Exercise 6.3.2). We then obtain that for ∂A to be a minimizer it must have constant mean curvature (we recall that a minimal surface is a surface with vanishing mean curvature, see Chapter 5). The question is then to show that the sphere is, among all compact surfaces with constant mean curvature, the only one to have this property. This is the result proved by Aleksandrov, Hopf, Liebmann, Reilly and others (see Hsiung [72] page 280, for a proof). We immediately see that this result only partially answers the problem. Indeed, we have only found a necessary condition that the minimizer should satisfy. Moreover, this method requires a strong regularity on the minimizer.

6.3.1   Minkowski-Steiner formula

We now introduce some definitions and give some intermediate results.

Definition 6.6 *(i) For $A, B \subset \mathbb{R}^n$, $n \geq 1$, we define*

$$A + B = \{a + b : a \in A, \ b \in B\}.$$

(ii) For $x \in \mathbb{R}^n$ and $A \subset \mathbb{R}^n$, we let

$$d(x, A) = \inf \{|x - a| : a \in A\}.$$

Example 6.7 (i) If $n = 1$, $A = [a, b]$, $B = [c, d]$, we have

$$A + B = [a + c, \ b + d].$$

(ii) If we let $B_R = \{x \in \mathbb{R}^n : |x| < R\}$, we get

$$B_R + B_S = B_{R+S}.$$

Proposition 6.8 *Let $A \subset \mathbb{R}^n$, $n \geq 1$, be compact and*

$$B_R = \{x \in \mathbb{R}^n : |x| < R\}.$$

The following properties then hold.

(i) $A + \overline{B_R} = \{x \in \mathbb{R}^n : d(x, A) \leq R\}$.

(ii) If A is convex, then $A + B_R$ is also convex.

Proof (i) Let $x \in A + \overline{B_R}$ and

$$X = \{x \in \mathbb{R}^n : d(x, A) \leq R\}.$$

We then have that $x = a + b$ for some $a \in A$ and $b \in \overline{B_R}$, and hence

$$|x - a| = |b| \leq R$$

which implies that $x \in X$. Conversely, since A is compact, we can find, for every $x \in X$, an element $a \in A$ so that $|x - a| \leq R$. Letting $b = x - a$, we have indeed found that $x \in A + \overline{B_R}$.

(ii) This is trivial. ∎

We now examine the meaning of the proposition in a simple example.

Example 6.9 If A is a rectangle in $\mathbb{R}^2$, we find that $A + \overline{B_R}$ is given by the figure below. Anticipating, a little, the following results, we see that we have

$$M(A + \overline{B_R}) = M(A) + RL(\partial A) + R^2 \pi$$

where $L(\partial A)$ is the perimeter of ∂A.

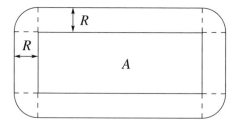

Figure 6.1: $A + B_R$

We now define the meaning (see Berger [10] Section 12.10.9.3) of $L(\partial A)$ and $M(A)$.

Definition 6.10 *Let* $n \geq 2$ *and* $A \subset \mathbb{R}^n$ *be a compact set. We define*

   (i) $M(A)$ *as the Lebesgue measure of* A;

   (ii) $L(\partial A)$ *as given by the* Minkowski-Steiner *formula*

$$L(\partial A) = \liminf_{\epsilon \to 0} \frac{M(A + \overline{B_\epsilon}) - M(A)}{\epsilon}$$

where $B_\epsilon = \{x \in \mathbb{R}^n : |x| < \epsilon\}$.

Remark 6.11 **(i)** The right-hand side in the above formula is called the (lower) *Minkowski content.* The first natural question that comes to mind is to ask if this definition of $L(\partial A)$ corresponds to the usual notion of the $(n-1)$ measure of ∂A. This is the case if A is "sufficiently regular". This is a deep result that we will not prove and that we will not even formulate precisely (see Federer [49] for a thorough discussion on this matter and the remark below when A is convex).

   (ii) When $A \subset \mathbb{R}^n$ is convex, the above limit is a true limit and we can show (see Berger [10] Sections 12.10.6 and 9.12.4.6) that

$$M(A + \overline{B_\epsilon}) = M(A) + L(\partial A)\epsilon + \sum_{i=2}^{n-1} L_i(A)\epsilon^i + \omega_n \epsilon^n$$

where $L_i(A)$ are some (continuous) functions of A and ω_n is the measure of the unit ball in $\mathbb{R}^n$ given by

$$\omega_n = \frac{2\,\pi^{n/2}}{n\,\Gamma(n/2)} = \begin{cases} \pi^k/k! & \text{if } n = 2k \\ 2^{k+1}\pi^k/1.3.5\cdots(2k+1) & \text{if } n = 2k+1. \end{cases}$$

Example 6.12 If $A = \overline{B_R}$, we find the well known formula for the area of the sphere $S_R = \partial B_R$

$$\begin{aligned} L(S_R) &= \lim_{\epsilon \to 0} \frac{M(\overline{B_R} + \overline{B_\epsilon}) - M(\overline{B_R})}{\epsilon} \\ &= \lim_{\epsilon \to 0} \frac{[(R+\epsilon)^n - R^n]\,\omega_n}{\epsilon} = nR^{n-1}\omega_n \end{aligned}$$

where ω_n is as above.

6.3.2   Brunn-Minkowski theorem

We now state the theorem (for a proof see Subsection 6.3.4) that plays a central role in the proof of the isoperimetric inequality.

Theorem 6.13 (Brunn-Minkowski theorem) *Let $A, B \subset \mathbb{R}^n$, $n \geq 1$, be compact, then the following inequality holds*

$$[M(A+B)]^{1/n} \geq [M(A)]^{1/n} + [M(B)]^{1/n}.$$

Remark 6.14 (i) The same proof establishes that the function $A \to (M(A))^{1/n}$ is concave. We thus have

$$[M(\lambda A + (1-\lambda)B)]^{1/n} \geq \lambda[M(A)]^{1/n} + (1-\lambda)[M(B)]^{1/n}$$

for every compact $A, B \subset \mathbb{R}^n$ and for every $\lambda \in [0,1]$.

(ii) One can even show that the function is strictly concave if and only if A and B are not homothetic.

Example 6.15 Let $n = 1$.

(i) If $A = [a, b]$, $B = [c, d]$, we have $A + B = [a + c, \ b + d]$ and

$$M(A+B) = M(A) + M(B).$$

(ii) If $A = [0, 1]$, $B = [0, 1] \cup [2, 3]$, we find $A + B = [0, 4]$ and hence

$$M(A+B) = 4 > M(A) + M(B) = 3.$$

6.3.3 Proof of the isoperimetric inequality

We are now in a position to state and prove the isoperimetric inequality.

Theorem 6.16 (Isoperimetric inequality) *Let $A \subset \mathbb{R}^n$, $n \geq 2$, be a compact set, $L = L(\partial A)$, $M = M(A)$ and ω_n be as above, then the following inequality holds*

$$L^n - n^n \omega_n M^{n-1} \geq 0.$$

Furthermore, the equality holds, among all convex sets, if and only if A is a ball.

Remark 6.17 (i) The proof that we give is also valid in the case $n = 2$. However, it is unduly complicated and less precise than the one given in the preceding section.

(ii) Concerning the uniqueness that we do not prove below (see Berger [10] Section 12.11), we should point out that it is uniqueness only among convex sets. In dimension 2, we did not need this restriction; since for a non-convex set A, its convex hull has larger area and smaller perimeter. In higher dimensions this is not true. In the case $n \geq 3$, one can still obtain uniqueness by assuming some regularity of the boundary ∂A, in order to avoid "hairy" spheres (i.e. sets that have zero n and $(n-1)$ measures but non-zero lower dimensional measures).

Proof (Theorem 6.16). Let $A \subset \mathbb{R}^n$ be compact, we have from the definition of L (see Minkowski-Steiner formula) and from Theorem 6.13 that

$$L(\partial A) = \liminf_{\epsilon \to 0} \frac{M\left(A + \overline{B_\epsilon}\right) - M(A)}{\epsilon}$$

$$\geq \liminf_{\epsilon \to 0} \left[\frac{\left[(M(A))^{1/n} + (M(B_\epsilon))^{1/n} \right]^n - M(A)}{\epsilon} \right].$$

Since $M(B_\varepsilon) = \epsilon^n \omega_n$, we get

$$L(\partial A) \geq M(A) \liminf_{\epsilon \to 0} \frac{\left[1 + \epsilon \left(\frac{\omega_n}{M(A)} \right)^{1/n} \right]^n - 1}{\epsilon}$$

$$= M(A) \cdot n \left(\frac{\omega_n}{M(A)} \right)^{1/n}$$

and the isoperimetric inequality follows. ∎

6.3.4 Proof of Brunn-Minkowski theorem

We conclude with some idea of the proof of Brunn-Minkowski theorem (for more details see Berger [10] Section 11.8.8, Federer [49] page 277 or Webster [108] Theorem 6.5.7). In Exercise 6.3.1 we propose a proof of the theorem valid in the case $n = 1$. Another proof in the case of $\mathbb{R}^n$ can be found in Pisier [95].

Proof (Theorem 6.13). The proof is divided into four steps.

Step 1. We first prove an elementary inequality. Let $u_i > 0$, $\lambda_i \geq 0$ with $\sum_{i=1}^n \lambda_i = 1$, then

$$\prod_{i=1}^n u_i^{\lambda_i} \leq \sum_{i=1}^n \lambda_i u_i. \tag{6.1}$$

This is a direct consequence of the fact that the logarithmic function is concave and hence

$$\log \left(\sum_{i=1}^n \lambda_i u_i \right) \geq \sum_{i=1}^n \lambda_i \log u_i = \log \left(\prod_{i=1}^n u_i^{\lambda_i} \right).$$

Step 2. Let $\mathcal{F}$ be the family of all open sets A of the form

$$A = \prod_{i=1}^n (a_i, b_i).$$

We next prove the theorem for $A, B \in \mathcal{F}$. We even show that for every $\lambda \in [0, 1]$, $A, B \in \mathcal{F}$ we have

$$[M (\lambda A + (1 - \lambda) B)]^{1/n} \geq \lambda [M (A)]^{1/n} + (1 - \lambda) [M (B)]^{1/n} . \qquad (6.2)$$

The theorem follows from (6.2) by setting $\lambda = 1/2$. If we let

$$A = \prod_{i=1}^{n} (a_i, b_i) \quad \text{and} \quad B = \prod_{i=1}^{n} (c_i, d_i)$$

we obtain

$$\lambda A + (1 - \lambda) B = \prod_{i=1}^{n} (\lambda a_i + (1 - \lambda) c_i, \lambda b_i + (1 - \lambda) d_i) .$$

Setting, for $1 \leq i \leq n$,

$$u_i = \frac{b_i - a_i}{\lambda (b_i - a_i) + (1 - \lambda) (d_i - c_i)} , \quad v_i = \frac{d_i - c_i}{\lambda (b_i - a_i) + (1 - \lambda) (d_i - c_i)} \qquad (6.3)$$

we find that

$$\lambda u_i + (1 - \lambda) v_i = 1, \quad 1 \leq i \leq n, \qquad (6.4)$$

$$\frac{M (A)}{M (\lambda A + (1 - \lambda) B)} = \prod_{i=1}^{n} u_i \quad \text{and} \quad \frac{M (B)}{M (\lambda A + (1 - \lambda) B)} = \prod_{i=1}^{n} v_i . \qquad (6.5)$$

We now combine (6.1), (6.4) and (6.5) to deduce that

$$\begin{aligned}
\frac{\lambda [M (A)]^{1/n} + (1 - \lambda) [M (B)]^{1/n}}{[M (\lambda A + (1 - \lambda) B)]^{1/n}} &= \lambda \prod_{i=1}^{n} u_i^{1/n} + (1 - \lambda) \prod_{i=1}^{n} v_i^{1/n} \\
&\leq \lambda \sum_{i=1}^{n} \frac{u_i}{n} + (1 - \lambda) \sum_{i=1}^{n} \frac{v_i}{n} \\
&= \frac{1}{n} \sum_{i=1}^{n} (\lambda u_i + (1 - \lambda) v_i) = 1
\end{aligned}$$

and hence the result.

Step 3. We now prove (6.2) for any A and B of the form

$$A = \bigcup_{\mu=1}^{M} A_\mu \quad \text{and} \quad B = \bigcup_{\nu=1}^{N} B_\nu$$

where A_μ, $B_\nu \in \mathcal{F}$, $A_\nu \cap A_\mu = B_\nu \cap B_\mu = \emptyset$ if $\mu \neq \nu$. The proof is achieved through induction on $M + N$. Step 2 has established the result when $M = N = 1$.

We now assume that $M > 1$. We then choose $i \in \{1, \cdots, n\}$ and $a \in \mathbb{R}$ such that if

$$A^+ = A \cap \{x \in \mathbb{R}^n : x_i > a\} \quad \text{and} \quad A^- = A \cap \{x \in \mathbb{R}^n : x_i < a\}$$

then A^+ and A^- contain at least one of the A_μ, $1 \leq \mu \leq M$, i.e. the hyperplane $\{x_i = a\}$ separates at least two of the A_μ (see Figure 6.2).

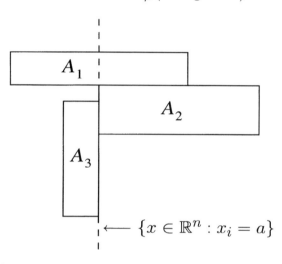

Figure 6.2: separating hyperplanes

We clearly have

$$M\left(A^+\right) + M\left(A^-\right) = M\left(A\right). \tag{6.6}$$

We next choose $b \in \mathbb{R}$ (such a b exists by an argument of continuity) so that if

$$B^+ = B \cap \{x \in \mathbb{R}^n : x_i > b\} \quad \text{and} \quad B^- = B \cap \{x \in \mathbb{R}^n : x_i < b\}$$

then

$$\frac{M\left(A^+\right)}{M\left(A\right)} = \frac{M\left(B^+\right)}{M\left(B\right)} \quad \text{and} \quad \frac{M\left(A^-\right)}{M\left(A\right)} = \frac{M\left(B^-\right)}{M\left(B\right)}. \tag{6.7}$$

We let

$$A_\mu^\pm = A^\pm \cap A_\mu \quad \text{and} \quad B_\nu^\pm = B^\pm \cap B_\nu$$

provided these intersections are non-empty; and we deduce that

$$A^\pm = \bigcup_{\mu=1}^{M^\pm} A_\mu^\pm \quad \text{and} \quad B^\pm = \bigcup_{\nu=1}^{N^\pm} B_\nu^\pm.$$

By construction we have $M^+ < M$ and $M^- < M$, while N^+, $N^- \leq N$. If $\lambda \in [0,1]$, we see that

$$\lambda A^+ + (1-\lambda) B^+ \quad \text{and} \quad \lambda A^- + (1-\lambda) B^-$$

are separated by $\{x : x_i = \lambda a + (1-\lambda) b\}$ and thus

$$M(\lambda A + (1-\lambda) B) = M(\lambda A^+ + (1-\lambda) B^+) + M(\lambda A^- + (1-\lambda) B^-).$$

Applying the hypothesis of induction to A^+, B^+ and A^-, B^-, we deduce that

$$M(\lambda A + (1-\lambda) B) \geq \left[\lambda \left[M(A^+) \right]^{1/n} + (1-\lambda) \left[M(B^+) \right]^{1/n} \right]^n$$
$$+ \left[\lambda \left[M(A^-) \right]^{1/n} + (1-\lambda) \left[M(B^-) \right]^{1/n} \right]^n.$$

Using (6.7) we obtain

$$M(\lambda A + (1-\lambda) B) \geq \frac{M(A^+)}{M(A)} \left[\lambda \left[M(A) \right]^{1/n} + (1-\lambda) \left[M(B) \right]^{1/n} \right]^n$$
$$+ \frac{M(A^-)}{M(A)} \left[\lambda \left[M(A) \right]^{1/n} + (1-\lambda) \left[M(B) \right]^{1/n} \right]^n.$$

The identity (6.6) and the above inequality then imply (6.2).

Step 4. We finally show (6.2) for any compact set, concluding thus the proof of the theorem. Letting $\epsilon > 0$, we can then approximate the compact sets A and B, by A_ϵ and B_ϵ as in Step 3, so that

$$|M(A) - M(A_\epsilon)|, \quad |M(B) - M(B_\epsilon)| \leq \epsilon, \tag{6.8}$$

$$|M(\lambda A + (1-\lambda) B) - M(\lambda A_\epsilon + (1-\lambda) B_\epsilon)| \leq \epsilon. \tag{6.9}$$

Applying (6.2) to A_ϵ, B_ϵ, using (6.8) and (6.9), we obtain, after passing to the limit as $\epsilon \to 0$, the claim

$$[M(\lambda A + (1-\lambda) B)]^{1/n} \geq \lambda [M(A)]^{1/n} + (1-\lambda) [M(B)]^{1/n}.$$

This achieves the proof of the theorem. ∎

6.3.5   Exercises

Exercise 6.3.1 Let $A, B \subset \mathbb{R}$ be compact,

$$\bar{a} = \min\{a : a \in A\} \quad \text{and} \quad \bar{b} = \max\{b : b \in B\}.$$

Prove that

$$(\bar{a} + B) \cup (\bar{b} + A) \subset A + B$$

and deduce that

$$M(A) + M(B) \leq M(A + B).$$

Exercise 6.3.2 Denote by $\mathcal{A}_1$ (respectively $\mathcal{A}_2$) the set of bounded open sets $A \subset \mathbb{R}^3$ whose boundary ∂A is the image of a bounded smooth connected open set $\Omega \subset \mathbb{R}^2$ by a $C^1(\overline{\Omega}; \mathbb{R}^3)$ (respectively C^2) map v, $v = v(x, y)$, with $v_x \times v_y \neq 0$ in $\overline{\Omega}$. Denote by $L(\partial A)$ and $M(A)$ the area of the boundary ∂A and the volume of A respectively.

Show that if there exists $A_0 \in \mathcal{A}_2$ so that

$$L(\partial A_0) = \inf_{A \in \mathcal{A}_1} \{ L(\partial A) : M(A) = M(A_0) \}$$

then ∂A_0 has constant mean curvature.

Exercise 6.3.3 Let $\Omega \subset \mathbb{R}^n$ be a bounded open set. Recall that the Sobolev imbedding theorem (see Theorem 1.44 and Remark 1.46 (v)), coupled with Poincaré inequality, states that there exists a constant $\gamma = \gamma(\Omega)$ so that, for every $u \in W_0^{1,1}(\Omega)$,

$$\|u\|_{L^{\frac{n}{n-1}}} \leq \gamma \|\nabla u\|_{L^1} .$$

Prove that if $A \subset \Omega$ is a compact set, then the following inequality holds

$$[L(\partial A)]^n - \frac{1}{\gamma^n} [M(A)]^{n-1} \geq 0.$$

Chapter 7

Solutions to the exercises

7.1 Chapter 1. Preliminaries

7.1.1 Continuous and Hölder continuous functions

Exercise 1.2.1. (i) We start by observing that, since $\alpha \in (0, 1]$,

$$g(t) = 1 - t^\alpha - (1 - t)^\alpha \leq 0 \quad \text{for every } t \in [0, 1].$$

This is easily seen by noticing that g is convex in $[0, 1]$ and $g(0) = g(1) = 0$. We then get that

$$[u_\alpha]_{C^{0,\alpha}} = \sup_{\substack{x \neq y \\ x, y \in [0,1]}} \left\{ \frac{|x^\alpha - y^\alpha|}{|x - y|^\alpha} \right\} = \sup_{t \in (0,1)} \left\{ \frac{1 - t^\alpha}{(1 - t)^\alpha} \right\} = 1$$

and hence $u_\alpha \in C^{0,\alpha}([0, 1])$.

(ii) It is obvious that the function u is continuous. If it were Hölder continuous, we should have for a certain $\alpha \in (0, 1]$ a constant $\gamma_\alpha > 0$ so that, for every $x \in (0, 1/2]$,

$$\frac{-1}{\log x} = |u(x) - u(0)| \leq \gamma_\alpha \, x^\alpha$$

and this is clearly impossible.

(iii) Observe that

$$
\begin{aligned}
|\cos(2^n x) - \cos(2^n y)| &= 2 \left| \sin(2^{n-1}(x - y)) \right| \left| \sin(2^{n-1}(x + y)) \right| \\
&\leq 2 \left| \sin(2^{n-1}(x - y)) \right| \leq 2 \left| \sin(2^{n-1}(x - y)) \right|^\alpha \\
&\leq 2 \left(2^{n-1} |x - y| \right)^\alpha \leq 2^{n\alpha+1} |x - y|^\alpha.
\end{aligned}
$$

We therefore obtain that, for $0 < \alpha < \lambda \le 1$,

$$|u_\lambda(x) - u_\lambda(y)| \le \frac{2}{2^{\lambda-\alpha} - 1} |x - y|^\alpha .$$

Note that the function u_λ, which is essentially the classical Weierstrass example, is nowhere differentiable for any $\lambda \in (0, 1]$ (see for example [102] page 118). ■

Exercise 1.2.2. (i) We have

$$\|uv\|_{C^{0,\alpha}} = \|uv\|_{C^0} + [uv]_{C^{0,\alpha}} .$$

Since

$$[uv]_{C^{0,\alpha}} \le \sup \frac{|u(x)v(x) - u(y)v(y)|}{|x - y|^\alpha}$$

$$\le \|u\|_{C^0} \sup \frac{|v(x) - v(y)|}{|x - y|^\alpha} + \|v\|_{C^0} \sup \frac{|u(x) - u(y)|}{|x - y|^\alpha}$$

we deduce that

$$\|uv\|_{C^{0,\alpha}} \le \|u\|_{C^0} \|v\|_{C^0} + \|u\|_{C^0} [v]_{C^{0,\alpha}} + \|v\|_{C^0} [u]_{C^{0,\alpha}}$$

$$\le 2 \|u\|_{C^{0,\alpha}} \|v\|_{C^{0,\alpha}} .$$

(ii) The inclusion $C^{0,\alpha} \subset C^0$ is obvious. Let us show that $C^{0,\beta} \subset C^{0,\alpha}$. Observe that

$$\sup_{\substack{x,y\in\overline{\Omega} \\ 0<|x-y|<1}} \left\{ \frac{|u(x) - u(y)|}{|x - y|^\alpha} \right\} \le \sup_{\substack{x,y\in\overline{\Omega} \\ 0<|x-y|<1}} \left\{ \frac{|u(x) - u(y)|}{|x - y|^\beta} \right\} \le [u]_{C^{0,\beta}} .$$

Since

$$\sup_{\substack{x,y\in\overline{\Omega} \\ |x-y|\ge 1}} \left\{ \frac{|u(x) - u(y)|}{|x - y|^\alpha} \right\} \le \sup_{x,y\in\overline{\Omega}} \{|u(x) - u(y)|\} \le 2 \|u\|_{C^0}$$

we get

$$\|u\|_{C^{0,\alpha}} = \|u\|_{C^0} + [u]_{C^{0,\alpha}}$$

$$\le \|u\|_{C^0} + \max\{2 \|u\|_{C^0}, [u]_{C^{0,\beta}}\} \le 3 \|u\|_{C^{0,\beta}} .$$

(iii) We now assume that Ω is bounded and convex, and show that $C^1(\overline{\Omega}) \subset C^{0,1}(\overline{\Omega})$. Let $x, y \in \overline{\Omega}$. Since $\overline{\Omega}$ is convex, we have that $[x, y] \subset \overline{\Omega}$ (where $[x, y]$ denotes the segment joining x to y). We can therefore write

$$u(x) - u(y) = \int_0^1 \frac{d}{dt} u(y + t(x - y)) \, dt = \int_0^1 \langle \nabla u(y + t(x - y)); x - y \rangle \, dt$$

and thus

$$|u(x) - u(y)| \leq \max_{z \in \overline{\Omega}} |\nabla u(z)| \, |x - y| \leq \|u\|_{C^1} |x - y| . \quad \blacksquare$$

Exercise 1.2.3. We easily see that

$$\operatorname{grad} u(x_1, x_2) = \begin{cases} (0, 2\beta x_2^{2\beta-1}) & \text{if } (x_1, x_2) \in \Omega \text{ and } x_1, x_2 > 0 \\ (0, -2\beta x_2^{2\beta-1}) & \text{if } (x_1, x_2) \in \Omega \text{ and } x_1 < 0 < x_2 \\ (0, 0) & \text{if } (x_1, x_2) \in \Omega \text{ and } x_2 \leq 0. \end{cases}$$

Since $\beta > 1/2$, we get

$$\lim_{(x_1, x_2) \to (0,0)} \operatorname{grad} u(x_1, x_2) = (0, 0)$$

and hence $u \in C^1(\overline{\Omega})$. Let us now show that $u \notin C^{0,\alpha}(\overline{\Omega})$, for every $\frac{1}{2} < \beta < \alpha \leq 1$. We proceed by contradiction. If $u \in C^{0,\alpha}(\overline{\Omega})$, we should then have, for a certain constant γ,

$$|u(x_1, x_2) - u(y_1, y_2)| \leq \gamma |(x_1, x_2) - (y_1, y_2)|^\alpha, \quad \forall \, (x_1, x_2), (y_1, y_2) \in \overline{\Omega}.$$

Choosing $(x_1, x_2) = (a, \sqrt{a})$ and $(y_1, y_2) = (-a, \sqrt{a})$ for $0 < a < a + a^2 < 1$, we get

$$2a^\beta = |u(a, \sqrt{a}) - u(-a, \sqrt{a})| \leq \gamma |(a, \sqrt{a}) - (-a, \sqrt{a})|^\alpha = \gamma \, 2^\alpha a^\alpha$$

and hence

$$a^{\beta-\alpha} \leq \gamma \, 2^{\alpha-1}.$$

Letting $a \to 0$, we obtain the desired contradiction, since $\beta < \alpha$. $\quad \blacksquare$

Exercise 1.2.4. (i) Let $x, y \in \overline{\Omega}$ with $u(x) \neq u(y)$, then

$$\frac{|v(u(x)) - v(u(y))|}{|x - y|^{\alpha\beta}} = \frac{|v(u(x)) - v(u(y))|}{|u(x) - u(y)|^\beta} \left(\frac{|u(x) - u(y)|}{|x - y|^\alpha} \right)^\beta$$

$$\leq [v]_{C^{0,\beta}} [u]_{C^{0,\alpha}}^\beta .$$

Since this inequality is trivially true for $x \neq y$ and $u(x) = u(y)$, we have

$$[v \circ u]_{C^{0,\alpha\beta}} \leq [v]_{C^{0,\beta}} [u]_{C^{0,\alpha}}^\beta$$

which implies that $v \circ u \in C^{0,\alpha\beta}$.

The above result is sharp when $0 < \alpha, \beta < 1$. Indeed, let $n = 1$, $u(x) = x^\alpha$ and $v(y) = y^\beta$. Then $u \in C^{0,\alpha}([0,1])$ and $v \in C^{0,\beta}([0,1])$; however $v \circ u(x) = x^{\alpha\beta}$ and thus $v \circ u \in C^{0,\alpha\beta}$ and in no better space.

(ii) We now discuss the case $k = 1$; for higher derivatives $(k \geq 2)$ we proceed with a straightforward induction. Let $u \in C^{1,\alpha}(\overline{\Omega}; \mathbb{R}^n)$, $u = u(x) = (u^1, \cdots, u^n)$, and $v \in C^{1,\alpha}(u(\overline{\Omega}))$, $v = v(y)$. Let $i = 1, \cdots, n$ and consider

$$(v \circ u)_{x_i} = \sum_{j=1}^n \left(v_{y_j} \circ u\right) u^j_{x_i}.$$

Note that it follows from Proposition 1.9 (iii) that $u \in C^{0,1}(\overline{\Omega}; \mathbb{R}^n)$ and hence, by (i) above, that $v_{y_j} \circ u \in C^{0,\alpha}(\overline{\Omega})$. Appealing to Proposition 1.9 (i), we deduce that

$$\left(v_{y_j} \circ u\right) u^j_{x_i} \in C^{0,\alpha}(\overline{\Omega})$$

which in turn implies $v \circ u \in C^{1,\alpha}(\overline{\Omega})$. ∎

Exercise 1.2.5. (i) We discuss the case of u_+, the other case being handled similarly.

1) Let us first check that indeed u_+ is an extension of u. Letting $x \in \Omega$, we therefore get

$$u(x) \leq u(y) + \gamma |x - y|^\alpha \text{ for every } y \in \overline{\Omega} \quad \Rightarrow \quad u(x) \leq u_+(x).$$

Now, clearly, choosing $y = x$ in the definition of u_+ leads to $u_+(x) \leq u(x)$. Thus u_+ is indeed an extension of u.

2) Let $x, z \in \mathbb{R}^n$. Assume, without loss of generality, that $u_+(z) \leq u_+(x)$. For every $\epsilon > 0$, we can find $y_z \in \overline{\Omega}$ such that

$$-\epsilon + u(y_z) + \gamma |z - y_z|^\alpha \leq u_+(z) \leq u(y_z) + \gamma |z - y_z|^\alpha.$$

We hence obtain

$$\begin{aligned}
|u_+(x) - u_+(z)| &= u_+(x) - u_+(z) \\
&\leq u(y_z) + \gamma |x - y_z|^\alpha + \epsilon - u(y_z) - \gamma |z - y_z|^\alpha \\
&\leq \epsilon + \gamma |x - z|^\alpha.
\end{aligned}$$

Letting $\epsilon \to 0$ we have the claim.

(ii) Let v be such that $[v]_{C^{0,\alpha}(\mathbb{R}^n)} = \gamma$. We therefore have for $x \in \mathbb{R}^n$ and for every $y \in \overline{\Omega}$ (and thus $v(y) = u(y)$)

$$-\gamma |x - y|^\alpha \leq v(x) - v(y) = v(x) - u(y) \leq \gamma |x - y|^\alpha.$$

This leads to

$$u(y) - \gamma |x - y|^\alpha \leq v(x) \leq u(y) + \gamma |x - y|^\alpha$$

and hence $u_-(x) \leq v(x) \leq u_+(x)$ as required. ∎

7.1.2 L^p spaces

Exercise 1.3.1. (i) Hölder inequality. Let $a, b > 0$ and $1/p + 1/p' = 1$, with $1 < p < \infty$. Since the function $f(x) = \log x$ is concave, we have that

$$\log\left(\frac{1}{p} a^p + \frac{1}{p'} b^{p'}\right) \geqslant \frac{1}{p} \log a^p + \frac{1}{p'} \log b^{p'} = \log(ab)$$

and hence

$$\frac{1}{p} a^p + \frac{1}{p'} b^{p'} \geq ab.$$

Assume that $\|u\|_{L^p}$, $\|v\|_{L^{p'}} \neq 0$, otherwise Hölder inequality is trivial. Choose then

$$a = \frac{|u|}{\|u\|_{L^p}}, \quad b = \frac{|v|}{\|v\|_{L^{p'}}}$$

and integrate to get the inequality for $1 < p < \infty$. The cases $p = 1$ and $p = \infty$ are trivial.

Minkowski inequality. The cases $p = 1$ and $p = \infty$ follow from the triangle inequality. We therefore assume that $1 < p < \infty$.

Step 1. Let us first show that L^p is a vector space. Observe that for every $x \in [0, 1]$, we have

$$1 \leq 1 + x \leq 2 \leq 2(1 + x^p)^{1/p}$$

and thus

$$|u + v|^p \leq (|u| + |v|)^p \leq 2^p (|u|^p + |v|^p).$$

Therefore, if $u, v \in L^p$, then $u + v \in L^p$.

Step 2. Use Hölder inequality to get

$$\|u + v\|_{L^p}^p = \int_\Omega |u + v|^p \leq \int_\Omega |u| |u + v|^{p-1} + \int_\Omega |v| |u + v|^{p-1}$$

$$\leq \|u\|_{L^p} \left\||u + v|^{p-1}\right\|_{L^{p'}} + \|v\|_{L^p} \left\||u + v|^{p-1}\right\|_{L^{p'}}.$$

The result then follows, since

$$\left\||u + v|^{p-1}\right\|_{L^{p'}} = \|u + v\|_{L^p}^{p-1}.$$

(ii) Use Hölder inequality with $\alpha = (p + q)/q$ and hence $\alpha' = (p + q)/p$ to obtain

$$\int_\Omega |uv|^{pq/p+q} \leq \left(\int_\Omega |u|^{pq\alpha/(p+q)}\right)^{1/\alpha} \left(\int_\Omega |v|^{pq\alpha'/(p+q)}\right)^{1/\alpha'}$$

$$\leq \left(\int_\Omega |u|^p\right)^{q/(p+q)} \left(\int_\Omega |v|^q\right)^{p/(p+q)}.$$

(iii) The inclusion $L^\infty(\Omega) \subset L^p(\Omega)$ is trivial. The other inclusions follow from Hölder inequality. Indeed we have

$$\int_\Omega |u|^q = \int_\Omega \left(|u|^q \cdot 1\right) \leq \left(\int_\Omega |u|^{q \cdot p/q}\right)^{q/p} \left(\int_\Omega 1^{p/(p-q)}\right)^{(p-q)/p}$$

$$\leq (\text{meas } \Omega)^{(p-q)/p} \left(\int_\Omega |u|^p\right)^{q/p}$$

and hence

$$\|u\|_{L^q} \leq (\text{meas } \Omega)^{(p-q)/pq} \|u\|_{L^p}$$

which gives the desired inclusion.

If, however, the measure is not finite, the result is not valid as the simple example $\Omega = (1, \infty)$, $u(x) = 1/x$ shows; indeed, we have $u \in L^2$ but $u \notin L^1$. ∎

Exercise 1.3.2. A direct computation leads to

$$\|u_\nu\|_{L^p}^p = \int_0^1 |u_\nu(x)|^p \, dx = \int_0^{1/\nu} \nu^{\alpha p} dx = \nu^{\alpha p - 1}.$$

We therefore have that $u_\nu \to 0$ in L^p provided $\alpha p - 1 < 0$. If $\alpha = 1/p$, let us show that $u_\nu \rightharpoonup 0$ in L^p. We have to prove that for every $\varphi \in L^{p'}(0, 1)$, the following convergence holds

$$\lim_{\nu \to \infty} \int_0^1 u_\nu \, \varphi = 0.$$

By a density argument, and since $\|u_\nu\|_{L^p} = 1$, it is sufficient to prove the result when φ is a step function, which means that there exist $0 = a_0 < a_1 < \cdots < a_I = 1$ so that $\varphi(x) = \alpha_i$ whenever $x \in (a_{i-1}, a_i)$, $1 \leq i \leq I$. We hence find, for ν sufficiently large, that

$$\int_0^1 u_\nu(x) \, \varphi(x) \, dx = \alpha_1 \int_{a_0}^{1/\nu} \nu^{1/p} dx = \alpha_1 \nu^{(1/p)-1} \to 0. \quad ∎$$

Exercise 1.3.3. **(i)** We have to show that for every $\varphi \in L^\infty$, then

$$\lim_{\nu \to \infty} \int_\Omega (u_\nu v_\nu - uv) \, \varphi = 0.$$

Rewriting the integral we have

$$\int_\Omega (u_\nu v_\nu - uv) \, \varphi = \int_\Omega u_\nu (v_\nu - v) \, \varphi + \int_\Omega (u_\nu - u) \, v\varphi.$$

Since $u_\nu - u \rightharpoonup 0$ in L^p, we can find γ such that $\|u_\nu\|_{L^p} \leq \gamma$. Furthermore, since

$$(v_\nu - v) \to 0 \quad \text{in } L^{p'}$$

we deduce from Hölder inequality that the first integral tends to 0. The second one also tends to 0, since

$$u_\nu - u \rightharpoonup 0 \text{ in } L^p \quad \text{and} \quad v\varphi \in L^{p'}$$

(the last fact follows from the hypotheses $v \in L^{p'}$ and $\varphi \in L^\infty$).

Let us show that the result is, in general, false if $v_\nu \rightharpoonup v$ in $L^{p'}$ (instead of $v_\nu \to v$ in $L^{p'}$). Choose $p = p' = 2$,

$$u_\nu(x) = v_\nu(x) = \sin(\nu x) \quad \text{and} \quad \Omega = (0, 2\pi).$$

We have that $u_\nu, v_\nu \rightharpoonup 0$ in L^2 but the product $u_\nu v_\nu$ does not tend to 0 weakly in L^2, since

$$u_\nu(x) v_\nu(x) = \sin^2(\nu x) \rightharpoonup 1/2 \neq 0 \quad \text{in } L^2.$$

(ii) We want to prove that $\|u_\nu - u\|_{L^2} \to 0$. We write

$$\int_\Omega |u_\nu - u|^2 = \int_\Omega u_\nu^2 - 2 \int_\Omega u u_\nu + \int_\Omega u^2.$$

The first integral tends to $\int u^2$ since $u_\nu^2 \rightharpoonup u^2$ in L^1 (choosing $\varphi(x) \equiv 1 \in L^\infty$ in the definition of weak convergence). The second one tends to $-2 \int u^2$ since $u_\nu \rightharpoonup u$ in L^2 and $u \in L^2$. The claim then follows. ∎

Exercise 1.3.4. It follows from Example 1.53 that

$$|u_\nu|^p \geq |u|^p + x^*(u)(u_\nu - u)$$

where

$$x^*(u) = \begin{cases} p|u|^{p-2} u & \text{if } 1 < p < \infty \\ +1 & \text{if } p = 1 \text{ and } u \geq 0 \\ -1 & \text{if } p = 1 \text{ and } u < 0. \end{cases}$$

Integrating, we get that

$$\int_\Omega |u_\nu|^p \geq \int_\Omega |u|^p + \int_\Omega x^*(u)(u_\nu - u).$$

Noting that $x^*(u) \in L^{p'}(\Omega)$, we have, passing to the limit and invoking that $u_\nu \rightharpoonup u$ in L^p,

$$\liminf_{\nu \to \infty} \int_\Omega |u_\nu|^p \geq \int_\Omega |u|^p$$

which is exactly what had to be proved. It should be noted that the above argument is at the root of the direct methods presented in Chapter 3. ∎

Exercise 1.3.5. (i) The case $p = \infty$ is trivial, since

$$\int_{\mathbb{R}^n} \varphi_\nu(x)\, dx = \int_{\mathbb{R}^n} \varphi(x)\, dx = 1.$$

So assume that $1 \leq p < \infty$. We next compute

$$u_\nu(x) = \int_{\mathbb{R}^n} \varphi_\nu(x - y)\, u(y)\, dy = \int_{\mathbb{R}^n} \varphi_\nu(y)\, u(x - y)\, dy$$

$$= \nu^n \int_{\mathbb{R}^n} \varphi(\nu y)\, u(x - y)\, dy = \int_{\mathbb{R}^n} \varphi(z)\, u\left(x - \frac{z}{\nu}\right) dz.$$

We therefore find

$$|u_\nu(x)| \leq \int_{\mathbb{R}^n} \varphi(z)\left| u\left(x - \frac{z}{\nu}\right)\right| dz$$

$$= \int_{\mathbb{R}^n} |\varphi(z)|^{1/p'} \left[|\varphi(z)|^{1/p} \left| u\left(x - \frac{z}{\nu}\right)\right| \right] dz.$$

Hölder inequality leads to

$$|u_\nu(x)| \leq \left(\int_{\mathbb{R}^n} \varphi(z)\, dz \right)^{1/p'} \left(\int_{\mathbb{R}^n} \varphi(z) \left| u\left(x - \frac{z}{\nu}\right)\right|^p dz \right)^{1/p}.$$

Since $\int \varphi = 1$, we have, after interchanging the order of integration,

$$\|u_\nu\|_{L^p}^p = \int_{\mathbb{R}^n} |u_\nu(x)|^p\, dx \leq \int_{\mathbb{R}^n} \int_{\mathbb{R}^n} \left\{ \varphi(z) \left| u\left(x - \frac{z}{\nu}\right)\right|^p dz \right\} dx$$

$$= \int_{\mathbb{R}^n} \left\{ \varphi(z) \int_{\mathbb{R}^n} \left| u\left(x - \frac{z}{\nu}\right)\right|^p dx \right\} dz \leq \|u\|_{L^p}^p.$$

(ii) The result follows, since φ is C^∞ and

$$D^a u_\nu(x) = \int_{\mathbb{R}^n} D^a \varphi_\nu(x - y)\, u(y)\, dy.$$

(iii) Let $K \subset \mathbb{R}^n$ be a fixed compact. Since u is continuous, we have that for every $\epsilon > 0$ there exists $\delta = \delta(\epsilon, K) > 0$ so that

$$|y| \leq \delta \quad \Rightarrow \quad |u(x - y) - u(x)| \leq \epsilon,\ \forall\, x \in K.$$

Since $\varphi = 0$ if $|x| > 1$, $\int \varphi = 1$, and hence $\int \varphi_\nu = 1$, we find that

$$u_\nu(x) - u(x) = \int_{\mathbb{R}^n} [u(x-y) - u(x)] \varphi_\nu(y) \, dy$$
$$= \int_{(-1/\nu, 1/\nu)^n} [u(x-y) - u(x)] \varphi_\nu(y) \, dy.$$

Taking $x \in K$ and $\nu > 1/\delta$, we deduce that $|u_\nu(x) - u(x)| \leq \epsilon$, and thus the claim.

(iv) Since $u \in L^p(\mathbb{R}^n)$ and $1 \leq p < \infty$, we deduce (see Theorem 1.13 (vi)) that for every $\epsilon > 0$, there exists $\overline{u} \in C_0(\mathbb{R}^n)$ so that

$$\|u - \overline{u}\|_{L^p} \leq \epsilon. \tag{7.1}$$

Define then

$$\overline{u}_\nu(x) = (\varphi_\nu * \overline{u})(x) = \int_{\mathbb{R}^n} \varphi_\nu(x-y) \overline{u}(y) \, dy.$$

Since $u - \overline{u} \in L^p$, it follows from (i) that

$$\|u_\nu - \overline{u}_\nu\|_{L^p} \leq \|u - \overline{u}\|_{L^p} \leq \epsilon. \tag{7.2}$$

Moreover, since $\operatorname{supp} \overline{u}$ is compact and $\varphi = 0$ if $|x| > 1$, we find that there exists a compact set K so that $\operatorname{supp} \overline{u}, \operatorname{supp} \overline{u}_\nu \subset K$ (for every ν). From (iii) we then get that $\|\overline{u}_\nu - \overline{u}\|_{L^p} \to 0$. Combining (7.1) and (7.2), we deduce that

$$\|u_\nu - u\|_{L^p} \leq \|u_\nu - \overline{u}_\nu\|_{L^p} + \|\overline{u}_\nu - \overline{u}\|_{L^p} + \|\overline{u} - u\|_{L^p}$$
$$\leq 2\epsilon + \|\overline{u}_\nu - \overline{u}\|_{L^p}.$$

Letting $\nu \to \infty$, we get the claim, since ϵ is arbitrary. ∎

Exercise 1.3.6. We adopt the same hypotheses and notations of Theorem 1.22. Step 1 remains unchanged and we modify Step 2 as follows. We define

$$v_\nu(x) = \int_0^x u_\nu(t) \, dt = \int_0^x u(\nu t) \, dt = \frac{1}{\nu} \int_0^{\nu x} u(s) \, ds$$
$$= \frac{1}{\nu} \int_{[\nu x]}^{\nu x} u(s) \, ds = \frac{1}{\nu} \int_0^{\nu x - [\nu x]} u(s) \, ds$$

where $[a]$ stands for the integer part of $a \geq 0$ and where we have used the periodicity of u and the fact that $\overline{u} = \int_0^1 u = 0$. We therefore find that

$$\|v_\nu\|_{L^\infty} \leq \frac{1}{\nu} \max_{x \in [0,1]} \int_0^{\nu x - [\nu x]} |u| \, ds = \frac{1}{\nu} \int_0^1 |u| \, ds = \frac{1}{\nu} \|u\|_{L^1} \leq \frac{1}{\nu} \|u\|_{L^p} \tag{7.3}$$

and, according to (1.11) and (1.12) in Lemma 1.39, we also have

$$\int_0^1 u_\nu(x)\psi(x)\,dx = -\int_0^1 v_\nu(x)\psi'(x)\,dx, \quad \text{for every } \psi \in C_0^\infty(0,1). \quad (7.4)$$

Recall that we have to show that

$$\lim_{\nu\to\infty} \int_0^1 u_\nu(x)\varphi(x)\,dx = 0, \quad \forall \varphi \in L^{p'}(0,1). \quad (7.5)$$

Let $\epsilon > 0$ be arbitrary. Since $\varphi \in L^{p'}(0,1)$ and $1 < p \leq \infty$, which implies $1 \leq p' < \infty$ (i.e. $p' \neq \infty$), we have from Theorem 1.13 that there exists $\psi \in C_0^\infty(0,1)$ so that

$$\|\varphi - \psi\|_{L^{p'}} \leq \epsilon. \quad (7.6)$$

We now appeal to (7.4) to get that

$$\int_0^1 u_\nu(x)\varphi(x)\,dx = \int_0^1 u_\nu(x)[\varphi(x) - \psi(x)]\,dx + \int_0^1 u_\nu(x)\psi(x)\,dx$$

$$= \int_0^1 u_\nu(x)[\varphi(x) - \psi(x)]\,dx - \int_0^1 v_\nu(x)\psi'(x)\,dx.$$

Using Hölder inequality, (1.1), (7.3) and (7.6), we obtain that

$$\left| \int_0^1 u_\nu(x)\varphi(x)\,dx \right| \leq \epsilon\|u\|_{L^p} + \|v_\nu\|_{L^\infty}\|\psi'\|_{L^1} \leq \epsilon\|u\|_{L^p} + \frac{1}{\nu}\|u\|_{L^p}\|\psi'\|_{L^1}.$$

Letting $\nu \to \infty$, we hence deduce that

$$0 \leq \limsup_{\nu\to\infty} \left| \int_0^1 u_\nu\,\varphi\,dx \right| \leq \epsilon\|u\|_{L^p}.$$

Since ϵ is arbitrary, we immediately have (7.5) and thus the result. ∎

Exercise 1.3.7. (i) Let $f \in C_0^\infty(\Omega)$, with $\int_\Omega f(x)\,dx = 1$, be a fixed function. Let $w \in C_0^\infty(\Omega)$ be arbitrary and

$$\psi(x) = w(x) - \left[\int_\Omega w(y)\,dy\right] f(x).$$

We therefore have $\psi \in C_0^\infty(\Omega)$ and $\int \psi = 0$, which leads to

$$0 = \int_\Omega u(x)\psi(x)\,dx = \int_\Omega u(x)w(x)\,dx - \int_\Omega f(x)u(x)\,dx \cdot \int_\Omega w(y)\,dy$$

$$= \int_\Omega \left[u(x) - \int_\Omega u(y)f(y)\,dy\right] w(x)\,dx.$$

Appealing to Theorem 1.24, we deduce that $u(x) = \int u(y) f(y) \, dy = \text{constant}$ a.e.

(ii) Let $\psi \in C_0^\infty(a, b)$, with $\int_a^b \psi = 0$, be arbitrary and define

$$\varphi(x) = \int_a^x \psi(t) \, dt \, .$$

Note that $\psi = \varphi'$ and $\varphi \in C_0^\infty(a, b)$. We may thus apply (i) and get the result. ∎

Exercise 1.3.8. We define $(N + 1)$ linear functionals on $C_0^\infty(\Omega)$ by

$$\Lambda(\psi) = \int_\Omega u(x) \psi(x) \, dx \quad \text{and} \quad \Lambda_i(\psi) = \int_\Omega \alpha_i(x) \psi(x) \, dx, \ i = 1, \cdots, N.$$

Note that our hypothesis guarantees that

$$\Lambda(\psi) = 0, \quad \text{for every } \psi \in C_0^\infty(\Omega) \text{ with } \Lambda_i(\psi) = 0, \ i = 1, \cdots, N.$$

Lemma 3.9 page 62 in [100] then implies that there exist constants $a_1, \cdots, a_N \in \mathbb{R}$ such that

$$\Lambda = \sum_{i=1}^N a_i \Lambda_i \, .$$

This means that

$$\int_\Omega \left[u(x) - \sum_{i=1}^N a_i \alpha_i(x) \right] \psi(x) \, dx = 0, \quad \forall \psi \in C_0^\infty(\Omega) \, .$$

Theorem 1.24 then implies the result, namely

$$u(x) = \sum_{i=1}^N a_i \alpha_i(x) \quad \text{a.e. } x \in \Omega. \quad ∎$$

Exercise 1.3.9. Let

$$X = \left\{ \psi \in C_0^\infty(a, b) : \int_a^b x^{k-1} \psi(x) \, dx = 0, \ k = 1, \cdots, n \right\}.$$

Let $\psi \in X$ be arbitrary and observe that

$$\varphi(x) = \int_a^x \frac{(x - t)^{n-1}}{(n - 1)!} \psi(t) \, dt$$

is such that $\varphi \in C_0^\infty(a, b)$ and

$$\varphi^{(n)}(x) = \psi(x) \, .$$

We therefore have

$$\int_a^b u(x)\,\psi(x)\,dx = 0, \quad \forall \psi \in X.$$

Appealing to Exercise 1.3.8 we have the result. ∎

Exercise 1.3.10. (i) Let, for $\nu \in \mathbb{N}$,

$$u_\nu(x) = \min\{|u(x)|, \nu\}.$$

The monotone convergence theorem implies that, for every $\epsilon > 0$, we can find ν sufficiently large so that

$$\int_\Omega |u| \leq \frac{\epsilon}{2} + \int_\Omega u_\nu.$$

Choose $\delta = \epsilon/2\nu$. We therefore deduce that, if meas $E \leq \delta$, then

$$\int_E |u| = \int_E u_\nu + \int_E [|u| - u_\nu] \leq \nu\,\text{meas}\,E + \frac{\epsilon}{2} \leq \epsilon.$$

For a more general setting see, for example, Theorem 5.18 in De Barra [38].

(ii) If $u \in L^p(\Omega)$ with $1 < p < \infty$, use Hölder inequality to get

$$\int_E |u| \leq \left(\int_E |u|^p\right)^{1/p}\left(\int_E 1^{p'}\right)^{1/p'} \leq \|u\|_{L^p(\Omega)}\,(\text{meas}\,E)^{1/p'}.$$

For $p = \infty$ and hence $p' = 1$, the result is straightforward. ∎

7.1.3 Sobolev spaces

Exercise 1.4.1. Let $\sigma_{n-1} = \text{meas}(\partial B_1(0))$ (i.e. $\sigma_1 = 2\pi$, $\sigma_2 = 4\pi, \cdots$).

(i) The result follows from the following observation

$$\|u\|_{L^p}^p = \int_{B_R} |u(x)|^p\,dx = \sigma_{n-1}\int_0^R r^{n-1}|f(r)|^p\,dr.$$

(ii) We find, if $x \neq 0$, that

$$u_{x_i} = f'(|x|)\frac{x_i}{|x|} \quad \Rightarrow \quad |\nabla u(x)| = |f'(|x|)|.$$

Assume, for a moment, that we have already proved that u is weakly differentiable in B_R, then

$$\|\nabla u\|_{L^p}^p = \sigma_{n-1}\int_0^R r^{n-1}|f'(r)|^p\,dr,$$

which is the claim.

Let us now show that u_{x_i}, as above, is indeed the weak derivative (with respect to x_i) of u. We have to prove that, for every $\varphi \in C_0^\infty (B_R)$,

$$\int_{B_R} u \, \varphi_{x_i} = - \int_{B_R} \varphi \, u_{x_i} . \tag{7.7}$$

Let $\epsilon > 0$ be sufficiently small and observe that (recall that $\varphi = 0$ on ∂B_R)

$$\int_{B_R} u \, \varphi_{x_i} \, dx = \int_{B_R \setminus B_\epsilon} u \, \varphi_{x_i} \, dx + \int_{B_\epsilon} u \, \varphi_{x_i} \, dx$$

$$= - \int_{B_R \setminus B_\epsilon} \varphi \, u_{x_i} \, dx - \int_{\partial B_\epsilon} u \, \varphi \, \frac{x_i}{|x|} \, d\sigma + \int_{B_\epsilon} u \, \varphi_{x_i} \, dx$$

$$= - \int_{B_R} \varphi \, u_{x_i} \, dx + \int_{B_\epsilon} \varphi \, u_{x_i} \, dx + \int_{B_\epsilon} u \, \varphi_{x_i} \, dx - \int_{\partial B_\epsilon} u \, \varphi \, \frac{x_i}{|x|} \, d\sigma.$$

Since the elements $\varphi \, u_{x_i}$ and $u \, \varphi_{x_i}$ are both in $L^1 (B_R)$, we deduce (see Exercise 1.3.10) that

$$\lim_{\epsilon \to 0} \int_{B_\epsilon} \varphi \, u_{x_i} = \lim_{\epsilon \to 0} \int_{B_\epsilon} u \, \varphi_{x_i} = 0.$$

Moreover, by hypothesis, we have the claim (i.e. (7.7)), since

$$\left| \int_{\partial B_\epsilon} u \, \varphi \, \frac{x_i}{|x|} \, d\sigma \right| \le \sigma_{n-1} \, \|\varphi\|_{L^\infty} \, \epsilon^{n-1} \, |f(\epsilon)| \to 0, \quad \text{as } \epsilon \to 0.$$

(iii) 1) The first example follows at once and gives

$$\psi \in L^p \Leftrightarrow sp < n \quad \text{and} \quad \psi \in W^{1,p} \Leftrightarrow (s+1) \, p < n.$$

2) We find, for every $0 < s < 1/2$ and $p \ge 1$, that

$$\int_0^{1/2} r \, |\log r|^{sp} \, dr < \infty, \qquad \int_0^{1/2} r^{-1} \, |\log r|^{2(s-1)} \, dr = \frac{|\log 2|^{2s-1}}{1 - 2s} < \infty.$$

The first guarantees that $\psi \in L^p$ and the second that $\psi \in W^{1,2}$. The fact that $\psi \notin L^\infty$ is obvious.

3) Clearly $u \in L^\infty$. We moreover have, denoting by δ_{ij} the Kronecker symbol, that

$$u_{x_j}^i = \frac{\partial u^i}{\partial x_j} = \frac{\delta_{ij} \, |x|^2 - x_i x_j}{|x|^3} \quad \Rightarrow \quad |\nabla u|^2 = \frac{n-1}{|x|^2} .$$

It is easy to see, as above, that $u_{x_j}^i$ is indeed the weak derivative (with respect to x_j) of u^i. We moreover find

$$\int_\Omega |\nabla u (x)|^p \, dx = (n-1)^{p/2} \, \sigma_{n-1} \int_0^1 r^{n-1-p} dr.$$

This quantity is finite if and only if $p \in [1, n)$. ∎

Exercise 1.4.2. The inclusion $AC\left([a, b]\right) \subset C\left([a, b]\right)$ is easy. Indeed, by definition any function in $AC\left([a, b]\right)$ is uniformly continuous in (a, b) and therefore can be continuously extended to $[a, b]$.

Let us now discuss the second inclusion, namely $W^{1,1}\left(a, b\right) \subset AC\left([a, b]\right)$. Let $u \in W^{1,1}\left(a, b\right)$. We know from Lemma 1.39 that

$$u\left(b_k\right) - u\left(a_k\right) = \int_{a_k}^{b_k} u'\left(t\right) dt.$$

We therefore find

$$\sum_k \left|u\left(b_k\right) - u\left(a_k\right)\right| \leq \sum_k \int_{a_k}^{b_k} \left|u'\left(t\right)\right| dt.$$

Let $E = \bigcup_k \left(a_k, b_k\right)$. A classical property of Lebesgue integral (see Exercise 1.3.10) asserts that if $u' \in L^1$, then, for every $\epsilon > 0$, there exists $\delta > 0$ so that

$$\mathrm{meas}\, E = \sum_k \left|b_k - a_k\right| < \delta \quad \Rightarrow \quad \int_E \left|u'\right| < \epsilon.$$

The claim then follows. ∎

Exercise 1.4.3. This follows from Hölder inequality, since

$$\left|u\left(x\right) - u\left(y\right)\right| \leq \int_y^x \left|u'\left(t\right)\right| dt \leq \left(\int_y^x \left|u'\left(t\right)\right|^p dt\right)^{1/p} \left(\int_y^x 1^{p'} dt\right)^{1/p'}$$

$$\leq \left(\int_y^x \left|u'\left(t\right)\right|^p dt\right)^{1/p} \left|x - y\right|^{1/p'}$$

and by the properties of Lebesgue integrals (see Exercise 1.3.10) the quantity $\left(\int_y^x \left|u'\left(t\right)\right|^p dt\right)^{1/p}$ tends to 0 as $\left|x - y\right|$ tends to 0. ∎

Exercise 1.4.4. Observe first that if $v \in W^{1,p}\left(a, b\right), p > 1$ and $y < x$, then

$$\left|v\left(x\right) - v\left(y\right)\right| = \left|\int_y^x v'\left(z\right) dz\right| \leq \left(\int_y^x \left|v'\left(z\right)\right|^p dz\right)^{1/p} \left(\int_y^x dz\right)^{1/p'}$$

and thus

$$\left|v\left(x\right) - v\left(y\right)\right| \leq \left\|v'\right\|_{L^p} \left|x - y\right|^{1/p'}. \tag{7.8}$$

Let us now show that if $u_\nu \rightharpoonup u$ in $W^{1,p}$, then $u_\nu \to u$ in L^∞. Without loss of generality, we can take $u \equiv 0$. Assume, for the sake of contradiction, that $u_\nu \not\to 0$ in L^∞. We can therefore find $\epsilon > 0$, $\{\nu_i\}$ so that

$$\|u_{\nu_i}\|_{L^\infty} \geq \epsilon, \quad \nu_i \to \infty. \tag{7.9}$$

From (7.8) we have that the subsequence $\{u_{\nu_i}\}$ is equicontinuous (note also that by Theorem 1.44 and Theorem 1.20 (iii) we have $\|u_{\nu_i}\|_{L^\infty} \leq c' \|u_{\nu_i}\|_{W^{1,p}} \leq c$) and hence from Ascoli-Arzelà theorem, we find, up to a subsequence,

$$u_{\nu_{i_j}} \to v \quad \text{in } L^\infty. \tag{7.10}$$

However, we must have $v = 0$ since (7.10) implies $u_{\nu_{i_j}} \rightharpoonup v$ in L^p and by uniqueness of the limits (we already know that $u_{\nu_{i_j}} \rightharpoonup u = 0$ in L^p) we deduce that $v = 0$ a.e., which contradicts (7.9). ∎

Exercise 1.4.5. From Theorem 1.20 we have that there exist $u, v_k \in L^p(\Omega)$, $k = 1, \cdots, n$, and a subsequence such that

$$u_{\nu_i} \rightharpoonup u \text{ in } L^p \quad \text{and} \quad (u_{\nu_i})_{x_k} \rightharpoonup v_k \text{ in } L^p.$$

Moreover $v_k = u_{x_k}$, since for every $\varphi \in C_0^\infty(\Omega)$

$$\int v_k\, \varphi = \lim_{\nu_i \to \infty} \int (u_{\nu_i})_{x_k}\, \varphi = -\lim_{\nu_i \to \infty} \int u_{\nu_i}\, \varphi_{x_k} = -\int u\, \varphi_{x_k}. \quad \blacksquare$$

Exercise 1.4.6. It is enough to prove the result when $u = 0$. Let $\varphi \in L^{p'}(\Omega)$ $(1 \leq p' < \infty)$ and $\epsilon > 0$. We can find $\psi \in C_0^\infty(\Omega)$ such that

$$\|\varphi - \psi\|_{L^{p'}} \leq \epsilon.$$

Note that, since $u_\nu \in W^{1,p}$ and $\psi \in C_0^\infty(\Omega)$, we can write, for every $i = 1, \cdots, n$,

$$\int (u_\nu)_{x_i}\, \varphi = \int (u_\nu)_{x_i}\, \psi + \int (u_\nu)_{x_i}\, (\varphi - \psi)$$

$$= -\int u_\nu\, \psi_{x_i} + \int (u_\nu)_{x_i}\, (\varphi - \psi).$$

We therefore have, since $\|\nabla u_\nu\|_{L^p} \leq \gamma$,

$$\left| \int (u_\nu)_{x_i}\, \varphi \right| \leq \left| \int u_\nu\, \psi_{x_i} \right| + \|\nabla u_\nu\|_{L^p} \|\varphi - \psi\|_{L^{p'}} \leq \left| \int u_\nu\, \psi_{x_i} \right| + \gamma\epsilon.$$

Passing to the limit, bearing in mind that $u_\nu \rightharpoonup 0$ in L^p, we get

$$\limsup_{\nu \to \infty} \left| \int (u_\nu)_{x_i}\, \varphi \right| \leq \lim_{\nu \to \infty} \left| \int u_\nu\, \psi_{x_i} \right| + \gamma\epsilon = \gamma\epsilon.$$

Since ϵ is arbitrary, we have the claim, namely

$$\lim_{\nu \to \infty} \int (u_\nu)_{x_i}\, \varphi = 0, \quad \text{for every } \varphi \in L^{p'} \text{ and } i = 1, \cdots, n. \quad \blacksquare$$

Exercise 1.4.7. It is clear that $u_\nu \to 0$ in L^∞. We also find

$$\frac{\partial u_\nu}{\partial x_1} = \sqrt{\nu}\,(1 - x_2)^\nu \cos(\nu x_1), \quad \frac{\partial u_\nu}{\partial x_2} = -\sqrt{\nu}\,(1 - x_2)^{\nu-1} \sin(\nu x_1)$$

which implies that there exists a constant $\gamma > 0$ independent of ν, such that

$$\iint_\Omega |\nabla u_\nu (x_1, x_2)|^2 \, dx_1 dx_2 \le \gamma.$$

Apply Exercise 1.4.6 to get the result. $\quad \blacksquare$

Exercise 1.4.8. Since $u \in W^{1,p}(\Omega)$, we have that it is weakly differentiable and therefore

$$\int_\Omega [u_{x_i}\,\psi + u\,\psi_{x_i}] = 0, \quad \forall \psi \in C_0^\infty(\Omega).$$

Let $\varphi \in W_0^{1,p'}(\Omega)$ and $\epsilon > 0$ be arbitrary. We can then find $\psi \in C_0^\infty(\Omega)$ so that

$$\|\psi - \varphi\|_{L^{p'}} + \|\nabla\psi - \nabla\varphi\|_{L^{p'}} \le \epsilon.$$

We hence obtain, appealing to the two relations above, that

$$\left| \int_\Omega [u_{x_i}\,\varphi + u\,\varphi_{x_i}] \right| \le \int_\Omega [|u_{x_i}|\,|\varphi - \psi| + |u|\,|\varphi_{x_i} - \psi_{x_i}|]$$

$$\le \|u\|_{W^{1,p}} [\|\psi - \varphi\|_{L^{p'}} + \|\varphi_{x_i} - \psi_{x_i}\|_{L^{p'}}] \le \epsilon \|u\|_{W^{1,p}}.$$

Since ϵ is arbitrary, we have indeed obtained that

$$\int_\Omega u_{x_i}\,\varphi = -\int_\Omega u\,\varphi_{x_i}, \quad i = 1, \cdots, n. \quad \blacksquare$$

Exercise 1.4.9. The present exercise is very similar to Exercise 1.3.5 and we only prove that if $u \in W^{1,p}(\mathbb{R}^n)$, then

$$\nabla(\varphi_\nu * u) = \varphi_\nu * \nabla u.$$

Differentiating under the integral sign and then integrating by parts, we obtain

$$\frac{\partial(\varphi_\nu * u)}{\partial x_j}(x) = \int_{\mathbb{R}^n} \frac{\partial \varphi_\nu}{\partial x_j}(x - y)\,u(y)\,dy = -\int_{\mathbb{R}^n} \frac{\partial \varphi_\nu}{\partial y_j}(x - y)\,u(y)\,dy$$

$$= \int_{\mathbb{R}^n} \varphi_\nu(x - y)\,\frac{\partial u}{\partial y_j}(y)\,dy = \varphi_\nu * \frac{\partial u}{\partial x_j}(x)$$

which is what we had to prove. ∎

Exercise 1.4.10. *Step 1.* We start with the case $1 < p < \infty$. For every $\varphi \in C_0^\infty (\mathbb{R}^n)$, we have, from Exercise 1.4.8 and from the fact that $u \in W_0^{1,p} (\Omega)$,

$$\left| \int_{\mathbb{R}^n} \widetilde{u}\, \varphi_{x_i} \right| = \left| \int_\Omega u\, \varphi_{x_i} \right| = \left| \int_\Omega u_{x_i} \varphi \right|$$
$$\leq \| u_{x_i} \|_{L^p(\Omega)} \| \varphi \|_{L^{p'}(\Omega)} \leq \| u_{x_i} \|_{L^p(\Omega)} \| \varphi \|_{L^{p'}(\mathbb{R}^n)} .$$

Appealing to Theorem 1.37, we have the claim. The fact that, for every $i = 1, 2, \cdots, n$,

$$\widetilde{u}_{x_i} (x) = \begin{cases} u_{x_i} (x) & \text{if } x \in \Omega \\ 0 & \text{if } x \notin \Omega \end{cases}$$

is proved in a very analogous manner.

Step 2. We now turn to the case $p = \infty$. By definition $W_0^{1,\infty} (\Omega) = W^{1,\infty} (\Omega) \cap W_0^{1,1} (\Omega)$, or what amounts to the same $W_0^{1,\infty} (\Omega) = W^{1,\infty} (\Omega) \cap W_0^{1,p} (\Omega)$ for any $p > 1$. Applying Step 1, we have that $\widetilde{u} \in W^{1,p} (\mathbb{R}^n)$ where

$$\widetilde{u} = \begin{cases} u & \text{in } \Omega \\ 0 & \text{outside } \Omega \end{cases} \quad \text{and} \quad \widetilde{u}_{x_i} = \begin{cases} u_{x_i} & \text{in } \Omega \\ 0 & \text{outside } \Omega. \end{cases}$$

Since $\widetilde{u}, \widetilde{u}_{x_i} \in L^\infty (\mathbb{R}^n)$, we have indeed proved that $\widetilde{u} \in W^{1,\infty} (\mathbb{R}^n)$. ∎

Exercise 1.4.11. In view of Theorem 1.41, it is enough to prove that the function

$$\widetilde{u} (x) = \begin{cases} u (x) & \text{if } x \in \Omega \\ 0 & \text{if } x \notin \Omega \end{cases}$$

is such that $\widetilde{u} \in W^{1,p} (\mathbb{R}^n)$. From Exercise 1.4.10 we know that

$$\widetilde{u^\nu} (x) = \begin{cases} u^\nu (x) & \text{if } x \in \Omega \\ 0 & \text{if } x \notin \Omega \end{cases}$$

is such that $\widetilde{u^\nu} \in W^{1,p} (\mathbb{R}^n)$. Define

$$v_i (x) = \begin{cases} u_{x_i} (x) & \text{if } x \in \Omega \\ 0 & \text{if } x \notin \Omega \end{cases}$$

and observe that clearly $v_i \in L^p (\mathbb{R}^n)$. In order to finish the proof it remains to show that v_i is indeed the weak partial derivative, with respect to x_i, of $\widetilde{u}$. So let $\varphi \in C_0^\infty (\mathbb{R}^n)$ and note that

$$\int_{\mathbb{R}^n} v_i \, \varphi + \int_{\mathbb{R}^n} \widetilde{u} \, \varphi_{x_i} = \int_{\mathbb{R}^n} \left[v_i - \left(\widetilde{u^\nu} \right)_{x_i} \right] \varphi + \int_{\mathbb{R}^n} \left(\widetilde{u^\nu} \right)_{x_i} \varphi + \int_{\mathbb{R}^n} \widetilde{u} \, \varphi_{x_i}$$

$$= \int_{\mathbb{R}^n} \left[v_i - \left(\widetilde{u^\nu} \right)_{x_i} \right] \varphi + \int_{\mathbb{R}^n} \left[\widetilde{u} - \widetilde{u^\nu} \right] \varphi_{x_i}$$

$$= \int_{\Omega} \left[u_{x_i} - u^\nu_{x_i} \right] \varphi + \int_{\Omega} \left[u - u^\nu \right] \varphi_{x_i} .$$

Since $\varphi \in W^{1,p'} (\Omega)$ and $u^\nu \rightharpoonup u$ in $W^{1,p}$, the limit in the right-hand side is 0 and therefore $v_i = \left(\widetilde{u} \right)_{x_i}$ as wished. ∎

Exercise 1.4.12. (i) Let $\epsilon > 0$ and $u \in W_0^{1,p} (\Omega)$. By definition, we can find $u_\epsilon \in C_0^\infty (\Omega)$ such that

$$\|u\|_{L^p} \leq \|u_\epsilon\|_{L^p} + \epsilon \quad \text{and} \quad \|\nabla u_\epsilon\|_{L^p} \leq \|\nabla u\|_{L^p} + \epsilon.$$

Applying Poincaré inequality for C_0^∞ functions, we can find $\gamma = \gamma (\Omega, p) > 0$ such that

$$\|u\|_{L^p} \leq \|u_\epsilon\|_{L^p} + \epsilon \leq \gamma \|\nabla u_\epsilon\|_{L^p} + \epsilon \leq \gamma \|\nabla u\|_{L^p} + (\gamma + 1) \epsilon.$$

Since ϵ is arbitrary, we have the claim.

(ii) We have

$$u (x_1, x_2, \cdots , x_n) = u (-R, x_2, \cdots , x_n) + \int_{-R}^{x_1} \frac{d}{dt} u (t, x_2, \cdots , x_n) \, dt$$

and thus

$$|u (x)| \leq \int_{-R}^{x_1} |u_{x_1} (t, x_2, \cdots , x_n)| \, dt \leq \int_{-R}^{R} |u_{x_1} (t, x_2, \cdots , x_n)| \, dt.$$

Appealing to Jensen inequality (see Theorem 1.54), we find

$$|u (x)|^p \leq (2R)^p \left(\frac{1}{2R} \int_{-R}^{R} |u_{x_1} (t, x_2, \cdots , x_n)| \, dt \right)^p$$

$$\leq (2R)^{p-1} \int_{-R}^{R} |u_{x_1} (t, x_2, \cdots , x_n)|^p \, dt.$$

We hence get, integrating with respect to $x_2, \cdots , x_n$,

$$\int_{-R}^{R} \cdots \int_{-R}^{R} |u (x_1, x_2, \cdots , x_n)|^p \, dx_2 \cdots dx_n$$

$$\leq (2R)^{p-1} \int_Q |u_{x_1} (t, x_2, \cdots , x_n)|^p \, dt \, dx_2 \cdots dx_n .$$

Integrating once more, this time with respect to x_1, we have

$$\|u\|_{L^p(Q)}^p \le (2R)^p \|\nabla u\|_{L^p(Q)}^p .$$

(iii) We first choose $R > 0$ sufficiently large so that $\Omega \subset \overline{\Omega} \subset Q = (-R, R)^n$. Let $u \in C_0^\infty(\Omega)$ be extended by 0 outside Ω. Apply (ii) to find

$$\|u\|_{L^p(\Omega)} = \|u\|_{L^p(Q)} \le (2R) \|\nabla u\|_{L^p(Q)} = (2R) \|\nabla u\|_{L^p(\Omega)} .$$

Finally invoke (i) to get Poincaré inequality. ∎

Exercise 1.4.13. Let $u \in W_0^{1,\infty}(\Omega)$. Extend u to be 0 outside Ω (see Exercise 1.4.10) so that the extended function, not relabeled, is in $W^{1,\infty}(\mathbb{R}^n)$. Appealing to Exercise 1.4.9 we can find $u_\nu \in C^\infty(\mathbb{R}^n)$ so that

$$u_\nu \to u \text{ in } L^\infty(\mathbb{R}^n) \quad \text{and} \quad \|\nabla u_\nu\|_{L^\infty} \le \|\nabla u\|_{L^\infty} .$$

Let $x \in \Omega$ and let $\bar{t} > 0$ such that (e_1 denoting the first unit vector of the Euclidean basis)

$$x + te_1 \in \Omega, \ \forall t \in [0, \bar{t}) \quad \text{and} \quad x + \bar{t}e_1 \in \partial\Omega.$$

We then have, since $u(x + \bar{t}e_1) = 0$,

$$u(x) = u(x) - u(x + \bar{t}e_1)$$
$$= u(x) - u_\nu(x) + u_\nu(x + \bar{t}e_1) - u(x + \bar{t}e_1) + u_\nu(x) - u_\nu(x + \bar{t}e_1)$$

and therefore

$$u(x) = u(x) - u_\nu(x) + u_\nu(x + \bar{t}e_1) - u(x + \bar{t}e_1) - \int_0^{\bar{t}} \frac{d}{dt}(u_\nu(x + te_1)) \, dt.$$

We thus get, γ denoting a constant that depends only on Ω,

$$|u(x)| \le |u(x) - u_\nu(x)| + |u_\nu(x + \bar{t}e_1) - u(x + \bar{t}e_1)| + \bar{t} \|\nabla u_\nu\|_{L^\infty}$$
$$\le 2 \|u - u_\nu\|_{L^\infty} + \gamma \|\nabla u_\nu\|_{L^\infty} \le 2 \|u - u_\nu\|_{L^\infty} + \gamma \|\nabla u\|_{L^\infty}$$

and hence, letting $\nu \to \infty$,

$$\|u\|_{L^\infty} \le \gamma \|\nabla u\|_{L^\infty}$$

which is exactly Poincaré inequality. ∎

Exercise 1.4.14. In all three cases we can assume, without loss of generality, that $v = 0$.

(i) We use Poincaré inequality (1.14) to get

$$\|u\|_{L^p} \leq \gamma \|\nabla u\|_{L^p} = 0$$

and thus $u = 0$ a.e. in Ω.

(ii) We invoke the second form of Poincaré inequality (1.15) to find

$$\|u - u_\Omega\|_{L^p} \leq \gamma \|\nabla u\|_{L^p} = 0$$

where

$$u_\Omega = \frac{1}{\operatorname{meas} \Omega} \int_\Omega u.$$

We therefore obtain that $u = u_\Omega$ a.e., as wished.

(iii) It is enough to prove the result when $p = 1$. Let $\epsilon > 0$. According to Exercise 1.3.10, there exists $\delta > 0$ so that for any measurable set $E \subset \Omega$

$$\operatorname{meas} E \leq \delta \quad \Rightarrow \quad \int_E [|u| + |\nabla u|] \leq \epsilon.$$

Since A is measurable, we can find an open set $A \subset \overline{A} \subset O \subset \overline{O} \subset \Omega$ so that $\operatorname{meas}(O \backslash A) \leq \delta$ and thus

$$\int_{O \backslash A} [|u| + |\nabla u|] \leq \epsilon.$$

Let $\varphi \in C_0^\infty(O)$ and

$$\chi_A(x) = \begin{cases} 1 & \text{if } x \in A \\ 0 & \text{if } x \in \Omega \backslash A. \end{cases}$$

We therefore have, for every $i = 1, \cdots, n$,

$$\int_O u_{x_i} \chi_A \varphi = \int_A u_{x_i} \varphi = \int_O u_{x_i} \varphi - \int_{O \backslash A} u_{x_i} \varphi$$

$$= -\int_O u \varphi_{x_i} - \int_{O \backslash A} u_{x_i} \varphi$$

$$= -\int_A u \varphi_{x_i} - \int_{O \backslash A} [u \varphi_{x_i} + u_{x_i} \varphi] = -\int_{O \backslash A} [u \varphi_{x_i} + u_{x_i} \varphi].$$

We therefore get

$$\left| \int_O u_{x_i} \chi_A \varphi \right| \leq \|\varphi\|_{W^{1,\infty}} \int_{O \backslash A} [|u| + |u_{x_i}|] \leq \epsilon \|\varphi\|_{W^{1,\infty}}.$$

Since ϵ is arbitrary, we have indeed obtained that, for every $\varphi \in C_0^\infty(O)$,

$$\int_O u_{x_i} \chi_A \, \varphi = 0.$$

The fundamental lemma of the calculus of variations implies then that $u_{x_i} = 0$ a.e. in A, which is exactly what had to be proved. ∎

Exercise 1.4.15. We divide the proof into three steps.

Step 1. For $\epsilon > 0$, we can find, by definition of the space $W_0^{1,1}$, a function $v \in C_0^\infty(\Omega)$ (extended to be 0 outside Ω) such that

$$\|v - u\|_{L^1} \le \epsilon \quad \text{and} \quad \|v_{x_2}\|_{L^1} = \|v_{x_2} - u_{x_2}\|_{L^1} \le \|\nabla v - \nabla u\|_{L^1} \le \epsilon.$$

Step 2. Since Ω is bounded, we can find $R > 0$ such that

$$\Omega \subset \overline{\Omega} \subset (-R, R)^2.$$

Note that $v(x_1, -R) = 0$ for every $x_1 \in (-R, R)$. We therefore have, for any $(x_1, x_2) \in (-R, R)^2$,

$$v(x_1, x_2) = v(x_1, -R) + \int_{-R}^{x_2} \frac{d}{dt}[v(x_1, t)] \, dt = \int_{-R}^{x_2} v_{x_2}(x_1, t) \, dt.$$

Since $v = 0$ outside Ω, we deduce that

$$\int_{-R}^{R} |v(x_1, x_2)| \, dx_1 \le \int_{-R}^{R} \int_{-R}^{R} |v_{x_2}(x_1, t)| \, dt \, dx_1 = \iint_\Omega |v_{x_2}(x_1, t)| \, dt \, dx_1$$

$$= \|v_{x_2}\|_{L^1} \le \epsilon$$

and hence

$$\|v\|_{L^1} = \iint_\Omega |v(x_1, x_2)| \, dx_1 dx_2 = \int_{-R}^{R} \int_{-R}^{R} |v(x_1, x_2)| \, dx_1 dx_2 \le 2R\epsilon.$$

Step 3. Combining the two steps we have

$$\|u\|_{L^1} \le \|v\|_{L^1} + \|v - u\|_{L^1} \le (2R + 1)\epsilon.$$

Letting $\epsilon \to 0$ we have the claimed result. ∎

Exercise 1.4.16. Before starting our proof, we should emphasize that, in view of Example 1.29, the exercise cannot be extended to $n = 1$.

1) We start by observing that, by Hölder inequality, we have, for every $0 < \epsilon < 1/2$,

$$\int_{B_{2\epsilon}} |u| \le \left(\int_{B_{2\epsilon}} |u|^p \right)^{1/p} (\text{meas } B_{2\epsilon})^{1/p'} \le \|u\|_{L^p} (\text{meas } B_{2\epsilon})^{1/p'}$$

and thus there exists a constant $\gamma_1 = \gamma_1 \left(n, \|u\|_{L^p} \right) > 0$, such that

$$\int_{B_{2\epsilon}} |u| \leq \gamma_1 \epsilon^{n(p-1)/p}.$$

2) We next define, for $0 < \epsilon < 1/2$, a function $\rho_\epsilon \in C^\infty \left([-1,1] \right), 0 \leq \rho_\epsilon \leq 1$,

$$\rho_\epsilon (t) = \begin{cases} 1 & \text{if } |t| \leq \epsilon \\ 0 & \text{if } |t| \geq 2\epsilon \end{cases} \quad \text{and} \quad |\rho'_\epsilon (t)| \leq \frac{\gamma_2}{\epsilon}$$

for a certain $\gamma_2 > 0$. We then set

$$u_\epsilon (x) = \left(1 - \rho_\epsilon \left(|x| \right) \right) u \left(x \right).$$

Observe that $u_\epsilon \in C^1 \left(\overline{B_1} \right)$ and, since $u_{x_i} = 0$ in $\overline{B_1} \setminus \{0\}$,

$$\frac{\partial u_\epsilon}{\partial x_i} (x) = \begin{cases} -\rho'_\epsilon \left(|x| \right) \frac{x_i}{|x|} u \left(x \right) & \text{if } \epsilon \leq |x| \leq 2\epsilon \\ 0 & \text{if } |x| \geq 2\epsilon \text{ or } |x| \leq \epsilon. \end{cases}$$

We moreover obtain, for every $\varphi \in C_0^\infty \left(B_1 \right)$,

$$\left| \int_{B_1} u_\epsilon \, \varphi_{x_i} \right| = \left| \int_{B_1} \frac{\partial u_\epsilon}{\partial x_i} \varphi \right| \leq \frac{\gamma_2}{\epsilon} \|\varphi\|_{L^\infty} \int_{B_{2\epsilon}} |u|.$$

3) Combining together the two observations, we get

$$\left| \int_{B_1} u \, \varphi_{x_i} \right| = \left| \int_{B_1} u_\epsilon \, \varphi_{x_i} + \int_{B_1} \rho_\epsilon \, u \, \varphi_{x_i} \right| = \left| \int_{B_1} u_\epsilon \, \varphi_{x_i} + \int_{B_{2\epsilon}} \rho_\epsilon \, u \, \varphi_{x_i} \right|$$

$$\leq \frac{\gamma_2}{\epsilon} \|\varphi\|_{L^\infty} \int_{B_{2\epsilon}} |u| + \|\varphi\|_{W^{1,\infty}} \int_{B_{2\epsilon}} |u|$$

$$\leq \gamma_1 \epsilon^{n(p-1)/p} \|\varphi\|_{W^{1,\infty}} \left(\frac{\gamma_2}{\epsilon} + 1 \right).$$

The result follows by letting $\epsilon \to 0$, since $p > n/ (n-1)$ is equivalent to $n \left(p - 1 \right) / p > 1$. ∎

Exercise 1.4.17. The notion of functions of bounded variation is very old, specially when $n = 1$. The classical definition is not the one given here but it is equivalent. There are many books dealing with such functions and one can consult, for example, [5], [17], [38], [57] or [58].

(i) Let $u \in W^{1,1} \left(\Omega \right)$. According to the definition of the weak derivative we have, for every $\varphi \in C_0^\infty \left(\Omega; \mathbb{R}^n \right)$ or equivalently $\varphi \in C_0^1 \left(\Omega; \mathbb{R}^n \right)$, that (letting $\varphi = \left(\varphi^1, \cdots, \varphi^n \right)$)

$$\int_\Omega u \, \varphi_{x_i}^j = - \int_\Omega u_{x_i} \, \varphi^j, \ j = 1, \cdots, n \quad \Rightarrow \quad \int_\Omega u \, \text{div} \, \varphi = - \int_\Omega \langle \nabla u; \varphi \rangle.$$

We therefore deduce, if $\|\varphi\|_{L^\infty} \le 1$, that

$$\left| \int_\Omega u \operatorname{div} \varphi \right| \le \left| \int_\Omega \langle \nabla u; \varphi \rangle \right| \le \|\nabla u\|_{L^1} \|\varphi\|_{L^\infty} \le \|u\|_{W^{1,1}}$$

and thus

$$V(u, \Omega) \le \|u\|_{W^{1,1}}$$

showing that $W^{1,1}(\Omega) \subset BV(\Omega)$.

(ii) Let $\varphi \in C_0^1(-1, 1)$ with $\|\varphi\|_{L^\infty} \le 1$, then

$$\left| \int_{-1}^1 H \varphi' \right| = \left| \int_0^1 \varphi' \right| = |\varphi(0)| \le 1.$$

Since $H \in L^1(-1, 1)$, we find that

$$V(H, (-1, 1)) = \sup \left\{ \left| \int_{-1}^1 H \varphi' \right| : \varphi \in C_0^1(-1, 1) \text{ with } \|\varphi\|_{L^\infty} \le 1 \right\} = 1$$

and thus $H \in BV(-1, 1)$, although (as seen in Example 1.29) $H \notin W^{1,1}(-1, 1)$.

(iii) The fact that $u \in C([-1, 1])$ is elementary. Let us now prove that $u \notin BV(-1, 1)$ by showing that $V(u, (-1, 1)) = +\infty$. We proceed in two steps.

Step 1. Observe that it is enough to find a sequence $\varphi_N \in W_0^{1,\infty}(-1, 1)$ with $\|\varphi_N\|_{L^\infty} \le 1$ such that

$$\left| \int_{-1}^1 u \varphi_N' \right| \to +\infty.$$

Indeed, we can approximate φ_N by $\varphi_\epsilon \in C_0^\infty(-1, 1)$ such that

$$\|\varphi_\epsilon - \varphi_N\|_{W^{1,1}} \le \epsilon \quad \text{and} \quad \|\varphi_\epsilon\|_{L^\infty} \le \|\varphi_N\|_{L^\infty} \le 1$$

to find

$$V(u, (-1, 1)) \ge \left| \int_{-1}^1 u \varphi_\epsilon' \right| \ge \left| \int_{-1}^1 u \varphi_N' \right| - \left| \int_{-1}^1 u(\varphi_\epsilon' - \varphi_N') \right|$$

$$\ge \left| \int_{-1}^1 u \varphi_N' \right| - \|u\|_{L^\infty} \|\varphi_\epsilon - \varphi_N\|_{W^{1,1}} \to +\infty.$$

Step 2. Let N be an arbitrary integer and define $\varphi_N \in W_0^{1,\infty}(-1, 1)$ with $\|\varphi_N\|_{L^\infty} \le 1$ in the following way. In $\left[-1, \frac{1}{8N+9} \right]$ the function vanishes identically and on each interval of the form $\left[\frac{1}{8k+9}, \frac{1}{8k+1} \right]$, where $k = 0, 1, \cdots, N$, we let

$$\varphi_N(x) = \begin{cases} 0 & \text{if } x \in \left[\frac{1}{8k+9}, \frac{1}{8k+7}\right] \\ \frac{(8k+7)(8k+5)}{2}\left(\frac{1}{8k+7} - x\right) & \text{if } x \in \left[\frac{1}{8k+7}, \frac{1}{8k+5}\right] \\ -1 & \text{if } x \in \left[\frac{1}{8k+5}, \frac{1}{8k+3}\right] \\ \frac{(8k+3)(8k+1)}{2}\left(x - \frac{1}{8k+1}\right) & \text{if } x \in \left[\frac{1}{8k+3}, \frac{1}{8k+1}\right]. \end{cases}$$

Note that $-1 \le \varphi_N \le 0$ and that $\varphi_N \in W_0^{1,\infty}(-1,1) \cap W_0^{1,\infty}\left(\frac{1}{8k+9}, \frac{1}{8k+1}\right)$.
Observe also that, for $x \in \left[\frac{1}{8k+7}, \frac{1}{8k+5}\right]$,

$$\frac{\pi}{4x} \in \left[\frac{5\pi}{4}, \frac{7\pi}{4}\right] + 2k\pi \quad \Rightarrow \quad -1 \le \sin\left(\frac{\pi}{4x}\right) \le -\frac{\sqrt{2}}{2}$$

and hence

$$\int_{\frac{1}{8k+7}}^{\frac{1}{8k+5}} x \sin\left(\frac{\pi}{4x}\right) \varphi_N'(x)\, dx \ge \frac{1}{8k+7}\frac{\sqrt{2}}{2}.$$

Similarly we find, for $x \in \left[\frac{1}{8k+3}, \frac{1}{8k+1}\right]$,

$$\frac{\pi}{4x} \in \left[\frac{\pi}{4}, \frac{3\pi}{4}\right] + 2k\pi \quad \Rightarrow \quad \frac{\sqrt{2}}{2} \le \sin\left(\frac{\pi}{4x}\right) \le 1.$$

and thus

$$\int_{\frac{1}{8k+3}}^{\frac{1}{8k+1}} x \sin\left(\frac{\pi}{4x}\right) \varphi_N'(x)\, dx \ge \frac{1}{8k+3}\frac{\sqrt{2}}{2}.$$

We therefore have

$$\int_{-1}^{1} u\, \varphi_N' \ge \frac{\sqrt{2}}{2} \sum_{k=0}^{N} \left[\frac{1}{8k+7} + \frac{1}{8k+3}\right].$$

Letting $N \to \infty$, we have the claim. ∎

7.1.4 Convex analysis

Exercise 1.5.1. (i) $\Rightarrow$ (ii) Let us prove that if f is convex then, for every $x, y \in \mathbb{R}$,

$$f(x) \ge f(y) + \langle \nabla f(y); x - y \rangle.$$

Apply the inequality of convexity

$$f(x) - f(y) \ge \frac{1}{\lambda}[f(y + \lambda(x - y)) - f(y)]$$

and let $\lambda \to 0$ to get the result.

(ii) $\Rightarrow$ **(i)** Let $\lambda \in [0,1]$ and apply (ii) to find

$$f(x) \geq f(\lambda x + (1-\lambda)y) + (1-\lambda)\langle\nabla f(\lambda x + (1-\lambda)y); x - y\rangle$$

$$f(y) \geq f(\lambda x + (1-\lambda)y) - \lambda\langle\nabla f(\lambda x + (1-\lambda)y); x - y\rangle.$$

Multiplying the first inequality by λ, the second one by $(1-\lambda)$ and summing the two of them, we have indeed obtained the desired convexity inequality

$$\lambda f(x) + (1-\lambda)f(y) \geq f(\lambda x + (1-\lambda)y).$$

(ii) $\Rightarrow$ **(iii)** We have

$$f(x) \geq f(y) + \langle\nabla f(y); x - y\rangle$$
$$f(y) \geq f(x) - \langle\nabla f(x); x - y\rangle$$

and thus

$$\langle\nabla f(x) - \nabla f(y); x - y\rangle \geq 0.$$

(iii) $\Rightarrow$ **(ii)** Let $\lambda \in (0,1)$, $x, y \in \mathbb{R}^n$ and define

$$z = \frac{x - y}{\lambda} + y \quad \Leftrightarrow \quad x = y + \lambda(z - y)$$

$$\varphi(\lambda) = f(y + \lambda(z - y)).$$

Observe that

$$\varphi'(\lambda) - \varphi'(0) = \langle\nabla f(y + \lambda(z - y)) - \nabla f(y); z - y\rangle$$
$$= \frac{1}{\lambda}\langle\nabla f(y + \lambda(z - y)) - \nabla f(y); (y + \lambda(z - y)) - y\rangle \geq 0$$

since (iii) holds. Therefore, integrating the inequality, we find

$$\varphi(\lambda) \geq \varphi(0) + \lambda\varphi'(0).$$

Returning to the definition of φ and z, we find, as wished,

$$f(x) \geq f(y) + \langle\nabla f(y); x - y\rangle.$$

(iii) $\Rightarrow$ **(iv)** Choose $y = x + \epsilon v$ with $\epsilon \neq 0$ and apply (iii) to get

$$\left\langle\frac{\nabla f(x + \epsilon v) - \nabla f(x)}{\epsilon}; v\right\rangle \geq 0.$$

This implies, letting $\epsilon \to 0$,

$$\left\langle \nabla^2 f\left(x\right) v; v \right\rangle \geq 0.$$

(iv) $\Rightarrow$ **(iii)** Write

$$\left\langle \nabla f\left(x\right) - \nabla f\left(y\right); x - y \right\rangle = \int_0^1 \left\langle \frac{d}{dt} \nabla f\left(y + t\left(x - y\right)\right); x - y \right\rangle dt$$
$$= \int_0^1 \left\langle \nabla^2 f\left(y + t\left(x - y\right)\right)\left(x - y\right); x - y \right\rangle dt$$

and apply (iv) to have the claim. ■

Exercise 1.5.2. (i) Let f be strictly convex and assume, for the sake of contradiction, that there exist $x \neq y$ such that

$$f\left(x\right) = f\left(y\right) + \left\langle \nabla f\left(y\right); x - y \right\rangle.$$

Let $\lambda \in \left(0, 1\right)$ and observe that by strict convexity we have

$$f\left(y\right) + \lambda \left\langle \nabla f\left(y\right); x - y \right\rangle = \lambda f\left(x\right) + \left(1 - \lambda\right) f\left(y\right)$$
$$> f\left(\lambda x + \left(1 - \lambda\right) y\right) \geq f\left(y\right) + \lambda \left\langle \nabla f\left(y\right); x - y \right\rangle$$

which is the desired contradiction.

(ii) The reverse implication is proved exactly as in Exercise 1.5.1 ((ii) $\Rightarrow$(i)). ■

Exercise 1.5.3. (i) Let

$$u_\Omega = \left(u_\Omega^1, \cdots, u_\Omega^N\right) \quad \text{with} \quad u_\Omega^i = \frac{1}{\text{meas}\,\Omega} \int_\Omega u^i\left(x\right) dx.$$

Since f is convex, we have, for every $\alpha, \beta \in \mathbb{R}^N$,

$$f\left(\alpha\right) \geq f\left(\beta\right) + \left\langle \nabla f\left(\beta\right); \alpha - \beta \right\rangle$$

so, in particular,

$$f\left(u\left(x\right)\right) \geq f\left(u_\Omega\right) + \left\langle \nabla f\left(u_\Omega\right); u\left(x\right) - u_\Omega \right\rangle.$$

Integrate to get the result.

(ii) When equality occurs in Jensen inequality, in view of the above considerations, we must have that, for almost every $x \in \Omega$,

$$f\left(u\left(x\right)\right) = f\left(u_\Omega\right) + \left\langle \nabla f\left(u_\Omega\right); u\left(x\right) - u_\Omega \right\rangle.$$

Therefore, if f is strictly convex (see Exercise 1.5.2), we find that, up to a set of measure 0, $u = u_\Omega$, thus u is constant. ∎

Exercise 1.5.4. We first observe that f^* is even and therefore we consider only $x^* \geq 0$. We easily find that if $x^* > 1$, then $f^*(x^*) = +\infty$. When $x^* = 1$, we get, in a straightforward way, that $f^*(x^*) = 0$. We now discuss the case $0 \leq x^* < 1$. It is clear, in this case, that the supremum in the definition of f^* is attained at the point

$$x^* = \frac{x}{\sqrt{1+x^2}} \quad \Leftrightarrow \quad x = \frac{x^*}{\sqrt{1-(x^*)^2}} .$$

We thus get

$$f^*(x^*) = xx^* - \sqrt{1+x^2} = -\sqrt{1-(x^*)^2}.$$

We have therefore found that

$$f^*(x^*) = \begin{cases} -\sqrt{1-(x^*)^2} & \text{if } |x^*| \leq 1 \\ +\infty & \text{otherwise.} \end{cases}$$

Note, in passing, that $f(x) = \sqrt{1+x^2}$ is strictly convex over $\mathbb{R}$. ∎

Exercise 1.5.5. (i) We have that

$$f^*(x^*) = \sup_{x \in \mathbb{R}} \left\{ xx^* - \frac{|x|^p}{p} \right\} .$$

The supremum is, in fact, attained at a point y where

$$x^* = |y|^{p-2} y \quad \Leftrightarrow \quad y = |x^*|^{p'-2} x^*.$$

Replacing this value in the definition of f^* we have obtained that

$$f^*(x^*) = \frac{|x^*|^{p'}}{p'} .$$

(ii) We do not compute f^*, but instead use Theorem 1.59. We let

$$g(x) = \begin{cases} (x^2 - 1)^2 & \text{if } |x| \geq 1 \\ 0 & \text{if } |x| < 1 \end{cases}$$

and we wish to show that $f^{**} = g$. We start by observing that g is convex, $0 \leq g \leq f$, and therefore according to Theorem 1.58 (ii) we must have

$$g \leq f^{**} \leq f.$$

First consider the case where $|x| \geq 1$; the functions g and f coincide there and hence $f^{**}(x) = g(x)$, for such x. We next consider the case $|x| < 1$. Choose in Theorem 1.59

$$x_1 = 1, \quad x_2 = -1, \quad \lambda_1 = \frac{1+x}{2}, \quad \lambda_2 = \frac{1-x}{2}$$

to find immediately that $f^{**}(x) = g(x) = 0$. We have therefore proved the claim.

(iv) This is straightforward since clearly

$$f^*(X^*) = \sup_{X \in \mathbb{R}^{2 \times 2}} \{\langle X; X^* \rangle - \det X\} \equiv +\infty$$

and therefore

$$f^{**}(X) = \sup_{X^* \in \mathbb{R}^{2 \times 2}} \{\langle X; X^* \rangle - f^*(X^*)\} \equiv -\infty. \quad \blacksquare$$

Exercise 1.5.6. It is easy to see that

$$f^*(X^*) = \sup_{X \in \mathbb{R}^{2 \times 2}} \left\{\langle X; X^* \rangle - (\det X)^2\right\} = \begin{cases} 0 & \text{if } X^* = 0 \\ +\infty & \text{if } X^* \neq 0 \end{cases}$$

and therefore

$$f^{**}(X) = \sup_{X^* \in \mathbb{R}^{2 \times 2}} \{\langle X; X^* \rangle - f^*(X^*)\} \equiv 0. \quad \blacksquare$$

Exercise 1.5.7. (i) Let $x^*, y^* \in \mathbb{R}^n$ and $\lambda \in [0, 1]$. It follows from the definition that

$$f^*(\lambda x^* + (1 - \lambda) y^*) = \sup_{x \in \mathbb{R}^n} \{\langle x; \lambda x^* + (1 - \lambda) y^* \rangle - f(x)\}$$

$$= \sup_x \{\lambda (\langle x; x^* \rangle - f(x)) + (1 - \lambda)(\langle x; y^* \rangle - f(x))\}$$

$$\leq \lambda \sup_x \{\langle x; x^* \rangle - f(x)\} + (1 - \lambda) \sup_x \{\langle x; y^* \rangle - f(x)\}$$

$$\leq \lambda f^*(x^*) + (1 - \lambda) f^*(y^*).$$

(ii) For this part we can refer to Theorem I.10 in Brézis [14], Theorem 2.43 in [32, 2nd edition] or Theorem 12.2 (coupled with Corollaries 10.1.1 and 12.1.1) in Rockafellar [98].

(iii) Since $f^{**} \leq f$, we find that $f^{***} \geq f^*$. Furthermore, by definition of f^{**}, we find, for every $x \in \mathbb{R}^n$, $x^* \in \mathbb{R}^n$,

$$\langle x; x^* \rangle - f^{**}(x) \leq f^*(x^*).$$

Taking the supremum over all x in the left-hand side of the inequality, we get $f^{***} \leq f^*$, and hence the claim.

(iv) By definition of f^*, we have

$$f^* (\nabla f (x)) = \sup_y \{ \langle y; \nabla f (x) \rangle - f (y) \} \geq \langle x; \nabla f (x) \rangle - f (x)$$

and hence

$$f (x) + f^* (\nabla f (x)) \geq \langle x; \nabla f (x) \rangle .$$

We next show the reverse inequality. Since f is convex, we have

$$f (y) \geq f (x) + \langle y - x; \nabla f (x) \rangle$$

which means that

$$\langle x; \nabla f (x) \rangle - f (x) \geq \langle y; \nabla f (x) \rangle - f (y) .$$

Taking the supremum over all y, we have indeed obtained the reverse inequality and thus the proof is complete.

(v) We refer to the bibliography, in particular to Mawhin-Willem [83] page 35, Rockafellar [98] (Theorems 23.5, 26.3 and 26.5 as well as Corollary 25.5.1) and for the second part to Theorems 2.48 and 2.50 in [32, 2nd edition] or Theorem 23.5 in Rockafellar [98]; see also the exercise below. ∎

Exercise 1.5.8. We divide the proof into three steps. We recall that $f'' > 0$ and

$$\lim_{|x| \to \infty} \frac{f (x)}{|x|} = +\infty. \qquad (7.11)$$

Step 1. We claim that f' is a bijection from $\mathbb{R}$ onto $\mathbb{R}$. Indeed, since $f'' > 0$, we have that f' is strictly increasing and therefore one to one. It is moreover onto. To prove this last fact it is enough to show that $\lim_{x \to \pm\infty} f' (x) = \pm\infty$. In fact, f being convex and satisfying (7.11), we have

$$f (0) \geq f (x) - x f' (x) \quad \Rightarrow \quad \lim_{x \to \pm\infty} f' (x) = \pm\infty.$$

Observe also that, since $f'' > 0$, $(f')^{-1}$ is $C^1 (\mathbb{R})$.

Step 2. We know that

$$f^* (x^*) = \sup_x \{ x x^* - f (x) \} .$$

With our hypotheses, we may infer that there exists x such that

$$f^* (x^*) = x x^* - f (x) \quad \text{and} \quad x^* = f' (x) .$$

In other words we find, calling $g = (f')^{-1}$, that $x = g(x^*)$ and thus

$$f^*(x^*) = x^* g(x^*) - f(g(x^*)) \quad \text{and} \quad x^* = f'(g(x^*)). \tag{7.12}$$

The right-hand side of the first equation being C^1, we deduce that f^* is $C^1(\mathbb{R})$.

Step 3. We now conclude that

$$(f^*)' = g = (f')^{-1}. \tag{7.13}$$

Indeed, we have, differentiating (7.12), that

$$(f^*)'(x^*) = g(x^*) + x^* g'(x^*) - f'(g(x^*)) g'(x^*)$$
$$= g(x^*) + g'(x^*)[x^* - f'(g(x^*))] = g(x^*).$$

We therefore have that f^* verifies (7.13) and is as regular as f (so, in particular, f^* is C^2). ∎

Exercise 1.5.9. (i) Let $h > 0$ and let $\{e_1, \cdots, e_n\}$ be the Euclidean basis. Use the convexity of f (in the form of (ii) of Theorem 1.52) and (1.17) to write

$$h \frac{\partial f}{\partial x_i}(x) \le f(x + he_i) - f(x) \le \alpha_1 (1 + |x + he_i|^p) + \alpha_1 (1 + |x|^p)$$

$$-h \frac{\partial f}{\partial x_i}(x) \le f(x - he_i) - f(x) \le \alpha_1 (1 + |x - he_i|^p) + \alpha_1 (1 + |x|^p).$$

We can therefore find $\tilde{\alpha}_1 > 0$, so that

$$\left| \frac{\partial f}{\partial x_i}(x) \right| \le \tilde{\alpha}_1 \frac{(1 + |x|^p + |h|^p)}{h}.$$

Choosing $h = 1 + |x|$, we can surely find $\alpha_2 > 0$ so that (1.18) is satisfied, i.e.

$$\left| \frac{\partial f}{\partial x_i}(x) \right| \le \alpha_2 \left(1 + |x|^{p-1}\right), \quad \forall x \in \mathbb{R}.$$

The inequality (1.19) is then a consequence of (1.18), since

$$f(x) - f(y) = \int_0^1 \frac{d}{ds} f(y + s(x - y)) \, ds = \int_0^1 \langle \nabla f(y + s(x - y)); x - y \rangle \, ds.$$

(ii) Note that the convexity of f is essential in the above argument. Indeed, taking, for example, $n = 1$ and $f(x) = x + \sin x^2$, we find that f satisfies (1.17) with $p = 1$, but it does not verify (1.18). More sophisticated examples show that the result does not carry either to the second derivative of a convex function f.

(iii) Of course if $\partial f / \partial x_i$ satisfies (1.18), we have by straight integration

$$f(x) = f(0) + \int_0^1 \frac{d}{ds} f(sx)\, ds = f(0) + \int_0^1 \langle \nabla f(sx); x \rangle\, ds$$

that f verifies (1.17), even if f is not convex. ∎

Exercise 1.5.10. With an easy argument, see [32] for details, the proof below, in fact, gives that f is locally Lipschitz continuous. We divide the proof into two steps.

Step 1. Let $x_0 \in \mathbb{R}$, $\{e_1, \cdots, e_n\}$ be the Euclidean basis and

$$a > \max_{i=1,\cdots,n} \{|f(x_0 + e_i) - f(x_0)|, |f(x_0 - e_i) - f(x_0)|\}.$$

It is then easy to see that if we denote by

$$|x|_1 = \sum_{i=1}^{n} |x_i|$$

then

$$|x|_1 \leq 1 \quad \Rightarrow \quad f(x_0 + x) - f(x_0) < a.$$

This is surely true if $x = 0$, so let $x \neq 0$ and let

$$I = \{i \in \{1, \cdots, n\} : x_i \neq 0\} \quad \Rightarrow \quad |x|_1 = \sum_{i \in I} |x_i|.$$

We have, from the convexity of f,

$$f(x_0 + x) = f\left(x_0 + \sum_{i \in I} |x_i| \frac{x_i}{|x_i|} e_i + (1 - |x|_1)\, 0 \right)$$

$$\leq \sum_{i \in I} |x_i|\, f\left(x_0 + \frac{x_i}{|x_i|} e_i \right) + (1 - |x|_1)\, f(x_0)$$

and hence, since $|x|_1 \leq 1$,

$$f(x_0 + x) - f(x_0) \leq \sum_{i \in I} |x_i| \left[f\left(x_0 + \frac{x_i}{|x_i|} e_i \right) - f(x_0) \right]$$

$$\leq \max_{i=1,\cdots,n} \{|f(x_0 \pm e_i) - f(x_0)|\} < a.$$

Step 2. We now show that for a as in Step 1 and for every $0 < \epsilon < a$, the following holds true

$$|x|_1 \leq \frac{\epsilon}{a} \quad \Rightarrow \quad |f(x_0 + x) - f(x_0)| \leq \epsilon.$$

1) We have, from the convexity of f and from Step 1,

$$f\left(x_0 + x\right) - f\left(x_0\right) = f\left(x_0 + \frac{\epsilon}{a}\frac{ax}{\epsilon} + \left(1 - \frac{\epsilon}{a}\right)0\right) - f\left(x_0\right)$$

$$\leq \frac{\epsilon}{a}\left[f\left(x_0 + \frac{ax}{\epsilon}\right) - f\left(x_0\right)\right] \leq \frac{\epsilon}{a}a = \epsilon.$$

2) In a similar way, we obtain

$$0 = f\left(x_0 + \frac{a}{a+\epsilon}x + \frac{\epsilon}{a+\epsilon}\left(-\frac{ax}{\epsilon}\right)\right) - f\left(x_0\right)$$

$$\leq \frac{a}{a+\epsilon}\left[f\left(x_0 + x\right) - f\left(x_0\right)\right] + \frac{\epsilon}{a+\epsilon}\left[f\left(x_0 - \frac{ax}{\epsilon}\right) - f\left(x_0\right)\right]$$

$$\leq \frac{a}{a+\epsilon}\left[f\left(x_0 + x\right) - f\left(x_0\right)\right] + \frac{\epsilon a}{a+\epsilon}$$

and thus

$$f\left(x_0 + x\right) - f\left(x_0\right) \geq -\epsilon.$$

Combining the two inequalities, we have the desired continuity. ∎

7.2 Chapter 2. Classical methods

7.2.1 Euler-Lagrange equation

Exercise 2.2.1. The proof is almost identical to that of the theorem. The Euler-Lagrange equation then becomes a system of ordinary differential equations, namely, if $u = \left(u^1, \cdots, u^N\right)$ and $\xi = \left(\xi^1, \cdots, \xi^N\right)$, we have

$$\frac{d}{dx}\left[f_{\xi^i}\left(x, \overline{u}, \overline{u}'\right)\right] = f_{u^i}\left(x, \overline{u}, \overline{u}'\right), \quad i = 1, \cdots, N. \quad ∎$$

Exercise 2.2.2. We proceed as in the theorem. We let

$$X = \left\{u \in C^n\left([a,b]\right) : u^{(j)}\left(a\right) = \alpha_j, \ u^{(j)}\left(b\right) = \beta_j, \ 0 \leq j \leq n-1\right\}.$$

If $\overline{u} \in X \cap C^{2n}\left([a,b]\right)$ is a minimizer of (P) we have $I\left(\overline{u} + \epsilon v\right) \geq I\left(\overline{u}\right)$, $\forall \epsilon \in \mathbb{R}$ and $\forall v \in C_0^\infty\left(a,b\right)$. Letting $f = f\left(x, u, \xi_1, \cdots, \xi_n\right)$ and using the fact that

$$\frac{d}{d\epsilon}I\left(\overline{u} + \epsilon v\right)\big|_{\epsilon=0} = 0$$

we find

$$\int_a^b \left\{f_u\left(x, \overline{u}, \cdots, \overline{u}^{(n)}\right)v + \sum_{i=1}^n f_{\xi_i}\left(x, \overline{u}, \cdots, \overline{u}^{(n)}\right)v^{(i)}\right\} = 0, \quad \forall v \in C_0^\infty\left(a,b\right).$$

Integrating by parts and appealing to the fundamental lemma of the calculus of variations (Theorem 1.24) we find

$$\sum_{i=1}^{n} (-1)^{i+1} \frac{d^i}{dx^i} \left[f_{\xi_i} \left(x, \overline{u}, \cdots, \overline{u}^{(n)} \right) \right] = f_u \left(x, \overline{u}, \cdots, \overline{u}^{(n)} \right). \quad \blacksquare$$

Exercise 2.2.3. (i) Let

$$X_0 = \left\{ v \in C^1 \left([a, b] \right) : v \left(a \right) = 0 \right\}.$$

Let $\overline{u} \in X \cap C^2 \left([a, b] \right)$ be a minimizer for (P), since $I \left(\overline{u} + \epsilon v \right) \geq I \left(\overline{u} \right), \forall v \in X_0$ and $\forall \epsilon \in \mathbb{R}$, we deduce as in the theorem that

$$\int_a^b \left\{ f_u \left(x, \overline{u}, \overline{u}' \right) v + f_\xi \left(x, \overline{u}, \overline{u}' \right) v' \right\} = 0, \quad \forall v \in X_0.$$

Integrating by parts (bearing in mind that $v \left(a \right) = 0$) we find

$$\int_a^b \left\{ \left[f_u - \frac{d}{dx} f_\xi \right] v \right\} + f_\xi \left(b, \overline{u} \left(b \right), \overline{u}' \left(b \right) \right) v \left(b \right) = 0, \quad \forall v \in X_0.$$

Using the fundamental lemma of the calculus of variations and the fact that $v \left(b \right)$ is arbitrary we find

$$\begin{cases} \dfrac{d}{dx} \left[f_\xi \left(x, \overline{u}, \overline{u}' \right) \right] = f_u \left(x, \overline{u}, \overline{u}' \right), \quad \forall x \in [a, b] \\ f_\xi \left(b, \overline{u} \left(b \right), \overline{u}' \left(b \right) \right) = 0. \end{cases}$$

We sometimes say that $f_\xi \left(b, \overline{u} \left(b \right), \overline{u}' \left(b \right) \right) = 0$ is a *natural* boundary condition.

(ii) The proof is completely analogous to the preceding one and we find, in addition to the above conditions, that

$$f_\xi \left(a, \overline{u} \left(a \right), \overline{u}' \left(a \right) \right) = 0.$$

(iii) As above we assume that there exists a minimizer $\overline{u} \in C^2 \left([a, b] \right)$ of (P) and we find, for every $v \in C^1 \left([a, b] \right)$, that

$$0 = \int_a^b \left\{ f_u \left(x, \overline{u}, \overline{u}' \right) v + f_\xi \left(x, \overline{u}, \overline{u}' \right) v' \right\}$$
$$+ g_y \left(\overline{u} \left(a \right), \overline{u} \left(b \right) \right) v \left(a \right) + g_z \left(\overline{u} \left(a \right), \overline{u} \left(b \right) \right) v \left(b \right).$$

Integrating by parts, we get

$$0 = \int_a^b \left\{ \left[f_u - \frac{d}{dx} f_\xi \right] v \right\} + \left[f_\xi \left(b, \overline{u} \left(b \right), \overline{u}' \left(b \right) \right) + g_z \left(\overline{u} \left(a \right), \overline{u} \left(b \right) \right) \right] v \left(b \right)$$
$$+ \left[-f_\xi \left(a, \overline{u} \left(a \right), \overline{u}' \left(a \right) \right) + g_y \left(\overline{u} \left(a \right), \overline{u} \left(b \right) \right) \right] v \left(a \right).$$

The fundamental lemma of the calculus of variations and the fact that $v(a)$ and $v(b)$ are arbitrary then imply

$$
\begin{cases}
\dfrac{d}{dx}\left[f_\xi\left(x,\overline{u},\overline{u}'\right)\right] = f_u\left(x,\overline{u},\overline{u}'\right), & \forall\, x \in [a,b] \\[2mm]
f_\xi\left(b,\overline{u}\left(b\right),\overline{u}'\left(b\right)\right) + g_z\left(\overline{u}\left(a\right),\overline{u}\left(b\right)\right) = 0 \\[2mm]
f_\xi\left(a,\overline{u}\left(a\right),\overline{u}'\left(a\right)\right) - g_y\left(\overline{u}\left(a\right),\overline{u}\left(b\right)\right) = 0.
\end{cases}
$$

∎

Exercise 2.2.4. Let $\overline{u} \in X \cap C^2\left([a,b]\right)$ be a minimizer of (P). Recall that

$$
X = \left\{ u \in C^1\left([a,b]\right) : u\left(a\right) = \alpha,\ u\left(b\right) = \beta,\ \int_a^b g\left(x, u\left(x\right), u'\left(x\right)\right) dx = 0 \right\}.
$$

We assume that there exists $w \in C_0^\infty\left(a,b\right)$ such that

$$
\int_a^b \left[g_\xi\left(x,\overline{u}\left(x\right),\overline{u}'\left(x\right)\right) w'\left(x\right) + g_u\left(x,\overline{u}\left(x\right),\overline{u}'\left(x\right)\right) w\left(x\right)\right] dx \neq 0
$$

or equivalently there exists $x \in (a,b)$ such that

$$
\frac{d}{dx}\left[g_\xi\left(x,\overline{u},\overline{u}'\right)\right] \neq g_u\left(x,\overline{u},\overline{u}'\right).
$$

By homogeneity we choose one such w so that

$$
\int_a^b \left[g_\xi\left(x,\overline{u}\left(x\right),\overline{u}'\left(x\right)\right) w'\left(x\right) + g_u\left(x,\overline{u}\left(x\right),\overline{u}'\left(x\right)\right) w\left(x\right)\right] dx = 1.
$$

Let $v \in C_0^\infty\left(a,b\right)$ be arbitrary, w as above and define for $\epsilon, h \in \mathbb{R}$

$$
F\left(\epsilon,h\right) = I\left(\overline{u} + \epsilon v + hw\right) = \int_a^b f\left(x,\overline{u} + \epsilon v + hw,\overline{u}' + \epsilon v' + hw'\right)
$$

$$
G\left(\epsilon,h\right) = \int_a^b g\left(x,\overline{u} + \epsilon v + hw,\overline{u}' + \epsilon v' + hw'\right).
$$

Observe that $G\left(0,0\right) = 0$ and that by hypothesis

$$
G_h\left(0,0\right) = \int_a^b \left[g_\xi\left(x,\overline{u}\left(x\right),\overline{u}'\left(x\right)\right) w'\left(x\right) + g_u\left(x,\overline{u}\left(x\right),\overline{u}'\left(x\right)\right) w\left(x\right)\right] dx = 1.
$$

Applying the implicit function theorem we can find $\epsilon_0 > 0$ and a function $t_v \in C^1\left([-\epsilon_0,\epsilon_0]\right)$ with $t_v\left(0\right) = 0$ such that

$$
G\left(\epsilon, t_v\left(\epsilon\right)\right) = 0, \quad \forall\, \epsilon \in [-\epsilon_0,\epsilon_0]
$$

which implies that $\overline{u} + \epsilon v + t_v(\epsilon) w \in X$. Note also that

$$G_\epsilon(\epsilon, t_v(\epsilon)) + G_h(\epsilon, t_v(\epsilon)) t'_v(\epsilon) = 0, \quad \forall \epsilon \in [-\epsilon_0, \epsilon_0]$$

and hence

$$t'_v(0) = -G_\epsilon(0, 0).$$

Since we know that

$$F(0, 0) \leq F(\epsilon, t_v(\epsilon)), \quad \forall \epsilon \in [-\epsilon_0, \epsilon_0]$$

we deduce that

$$F_\epsilon(0, 0) + F_h(0, 0) t'_v(0) = 0.$$

Note that $F_\epsilon(0, 0)$ depends on v while $F_h(0, 0)$ does not. We therefore let $\lambda = F_h(0, 0)$ be the Lagrange multiplier and we find

$$F_\epsilon(0, 0) - \lambda G_\epsilon(0, 0) = 0$$

or in other words

$$\int_a^b \{ [f_\xi(x, \overline{u}, \overline{u}') v' + f_u(x, \overline{u}, \overline{u}') v] - \lambda [g_\xi(x, \overline{u}, \overline{u}') v' + g_u(x, \overline{u}, \overline{u}') v] \} dx = 0.$$

Appealing once more to the fundamental lemma of the calculus of variations and to the fact that $v \in C_0^\infty(a, b)$ is arbitrary we get

$$\frac{d}{dx} [f_\xi(x, \overline{u}, \overline{u}')] - f_u(x, \overline{u}, \overline{u}') = \lambda \left\{ \frac{d}{dx} [g_\xi(x, \overline{u}, \overline{u}')] - g_u(x, \overline{u}, \overline{u}') \right\}. \quad \blacksquare$$

Exercise 2.2.5. Let $v \in C_0^1(a, b)$, $\epsilon \in \mathbb{R}$ and set $\varphi(\epsilon) = I(\overline{u} + \epsilon v)$. Since $\overline{u}$ is a minimizer of (P) we have $\varphi(\epsilon) \geq \varphi(0)$, $\forall \epsilon \in \mathbb{R}$, and hence we have that $\varphi'(0) = 0$ (which leads to the Euler-Lagrange equation) and $\varphi''(0) \geq 0$. Computing the last expression we find

$$\int_a^b \left\{ f_{uu} v^2 + 2 f_{u\xi} v v' + f_{\xi\xi} (v')^2 \right\} \geq 0, \quad \forall v \in C_0^1(a, b).$$

Noting that $2vv' = (v^2)'$ and recalling that $v(a) = v(b) = 0$, we find

$$\int_a^b \left\{ f_{\xi\xi} (v')^2 + \left(f_{uu} - \frac{d}{dx} f_{u\xi} \right) v^2 \right\} \geq 0, \quad \forall v \in C_0^1(a, b). \quad \blacksquare$$

Exercise 2.2.6. (i) Setting

$$u_1(x) = \begin{cases} x & \text{if } x \in [0, 1/2] \\ 1 - x & \text{if } x \in (1/2, 1] \end{cases}$$

we find that $I(u_1) = 0$. Observe, however, that $u_1 \notin X$ where

$$X = \left\{ u \in C^1([0,1]) : u(0) = u(1) = 0 \right\}.$$

Let $\epsilon > 0$. Since $u_1 \in W_0^{1,\infty}(0,1)$, we can find $v \in C_0^\infty(0,1)$ (hence in particular $v \in X$) such that

$$\|u_1 - v\|_{W^{1,4}} \le \epsilon.$$

Note also that since $f(\xi) = (\xi^2 - 1)^2$, we can find $\gamma_1 > 0$ such that

$$|f(\xi) - f(n)| \le \gamma_1 \left(1 + |\xi|^3 + |\eta|^3\right) |\xi - \eta|.$$

Combining the above inequalities with Hölder inequality we get (γ_2 denoting a constant independent of ϵ)

$$0 \le m \le I(v) = I(v) - I(u_1) \le \int_0^1 |f(v') - f(u_1')|$$

$$\le \gamma_1 \int_0^1 \left\{ \left(1 + |v'|^3 + |u_1'|^3\right) |v' - u_1'| \right\}$$

$$\le \gamma_1 \left(\int_0^1 \left(1 + |v'|^3 + |u_1'|^3\right)^{4/3} \right)^{3/4} \left(\int_0^1 |v' - u_1'|^4 \right)^{1/4}$$

$$\le \gamma_2 \|u_1 - v\|_{W^{1,4}} \le \gamma_2 \epsilon.$$

Since ϵ is arbitrary we deduce the result, i.e. $m = 0$.

(ii) The argument is analogous to the preceding one and we skip the details. We first let

$$X = \left\{ v \in C^1([0,1]) : v(0) = 1, \ v(1) = 0 \right\}$$
$$X_{\text{piec}} = \left\{ v \in C_{\text{piec}}^1([0,1]) : v(0) = 1, \ v(1) = 0 \right\}$$

where C_{piec}^1 stands for the set of piecewise C^1 functions. We have already proved that

$$(P_{\text{piec}}) \quad \inf_{u \in X_{\text{piec}}} \left\{ I(u) = \int_0^1 x \left(u'(x)\right)^2 dx \right\} = 0.$$

We need to establish that $m = 0$ where

$$(P) \quad \inf_{u \in X} \left\{ I(u) = \int_0^1 x \left(u'(x)\right)^2 dx \right\} = m.$$

We start by observing that for any $\epsilon > 0$ and $u \in X_{\text{piec}}$, we can find $v \in X$ such that

$$\|u - v\|_{W^{1,2}} \le \epsilon.$$

It is an easy matter (exactly as above) to show that if

$$I(u) = \int_0^1 x \left(u'(x) \right)^2 dx$$

then we can find a constant γ so that

$$0 \leq I(v) \leq I(u) + \gamma \epsilon.$$

Taking the infimum over all elements $v \in X$ and $u \in X_{\text{piec}}$ we get that

$$0 \leq m \leq \gamma \epsilon$$

which is the desired result, since ϵ is arbitrary. ∎

Exercise 2.2.7. Let $u \in C^1([0,1])$ with $u(0) = u(1) = 0$. Invoking Poincaré inequality (Theorem 1.49), we can find a constant $\gamma > 0$ such that

$$\int_0^1 u^2 \leq \gamma \int_0^1 \left(u' \right)^2.$$

We hence obtain that $m_\lambda = 0$ if λ is small (more precisely $\lambda^2 \leq 1/\gamma$). Observe that $u_0 \equiv 0$ satisfies $I_\lambda(u_0) = m_\lambda = 0$. Furthermore, it is the unique solution of (P_λ) since, by inspection, it is the only solution (if $\lambda^2 < \pi^2$) of the Euler-Lagrange equation

$$\begin{cases} u'' + \lambda^2 u = 0 \\ u(0) = u(1) = 0. \end{cases}$$

The claim then follows. ∎

Exercise 2.2.8. Let

$$X_{\text{piec}} = \left\{ u \in C^1_{\text{piec}}([-1,1]) : u(-1) = 0,\ u(1) = 1 \right\}$$

and

$$\overline{u}(x) = \begin{cases} 0 & \text{if } x \in [-1,0] \\ x & \text{if } x \in (0,1]. \end{cases}$$

It is then obvious to see that

$$\inf_{u \in X_{\text{piec}}} \left\{ I(u) = \int_{-1}^1 f(u(x), u'(x))\, dx \right\} = I(\overline{u}) = 0.$$

Note also that the only solution in X_{piec} is $\overline{u}$.

Since any element in X_{piec} can be approximated arbitrarily closely by an element of X (see the argument of Exercise 2.2.6) we deduce that $m = 0$ in

$$(P) \quad \inf_{u \in X} \left\{ I(u) = \int_{-1}^{1} f(u(x), u'(x)) \, dx \right\} = m.$$

To conclude, it is sufficient to observe that if for a certain $u \in C^1([-1, 1])$ we have $I(u) = 0$, then either $u \equiv 0$ or $u' \equiv 1$, and both possibilities are incompatible with the boundary data.

Another possibility of showing that $m = 0$ is to consider the sequence

$$u_\nu(x) = \begin{cases} 0 & \text{if } x \in [-1, 0] \\ -\nu^2 x^3 + 2\nu x^2 & \text{if } x \in \left(0, \frac{1}{\nu}\right] \\ x & \text{if } x \in \left(\frac{1}{\nu}, 1\right] \end{cases}$$

and observe that $u_\nu \in X$ and that

$$I(u_\nu) = \int_0^{1/\nu} f(u_\nu(x), u'_\nu(x)) \, dx \to 0.$$

This proves that $m = 0$, as wished. ∎

Exercise 2.2.9. Note first that, by Jensen inequality, $m \geq 1$, where

$$(P) \quad \inf_{u \in X} \left\{ I(u) = \int_0^1 |u'(x)| \, dx \right\} = m$$

and $X = \left\{ u \in C^1([0, 1]) : u(0) = 0, \, u(1) = 1 \right\}$. Let $n \geq 1$ be an integer and observe that u_n defined by $u_n(x) = x^n$ belongs to X and satisfies $I(u_n) = 1$. Therefore, u_n is a solution of (P) for every n. In fact, any $u \in X$ with $u' \geq 0$ in $[0, 1]$ is a minimizer of (P). ∎

Exercise 2.2.10. Set $v(x) = A(u(x))$. We then have, using Jensen inequality,

$$I(u) = \int_a^b a(u(x)) |u'(x)|^p \, dx$$

$$= \int_a^b \left| (a(u(x)))^{1/p} u'(x) \right|^p \, dx = \int_a^b |v'(x)|^p \, dx$$

$$\geq (b - a) \left| \frac{1}{b - a} \int_a^b v'(x) \, dx \right|^p = (b - a) \left| \frac{v(b) - v(a)}{b - a} \right|^p$$

and hence

$$I(u) \geq (b - a) \left| \frac{A(\beta) - A(\alpha)}{b - a} \right|^p.$$

Setting

$$\overline{v}(x) = \frac{A(\beta) - A(\alpha)}{b - a}(x - a) + A(\alpha) \quad \text{and} \quad \overline{u}(x) = A^{-1}(\overline{v}(x))$$

we have from the preceding inequality

$$I(u) \geq (b - a) \left| \frac{A(\beta) - A(\alpha)}{b - a} \right|^{p} = \int_{a}^{b} |\overline{v}'(x)|^{p} \, dx = I(\overline{u})$$

as claimed. ∎

7.2.2 Second form of the Euler-Lagrange equation

Exercise 2.3.1. Write $f_{\xi} = (f_{\xi^1}, \cdots, f_{\xi^N})$ and start by the simple observation that for any $u \in C^2([a, b]; \mathbb{R}^N)$

$$\frac{d}{dx}\left[f(x, u, u') - \langle u'; f_{\xi}(x, u, u')\rangle\right]$$

$$= f_x(x, u, u') + \sum_{i=1}^{N}(u^i)'\left[f_{u^i}(x, u, u') - \frac{d}{dx}\left[f_{\xi^i}(x, u, u')\right]\right].$$

Since the Euler-Lagrange system (see Exercise 2.2.1) is given by

$$\frac{d}{dx}\left[f_{\xi^i}(x, \overline{u}, \overline{u}')\right] = f_{u^i}(x, \overline{u}, \overline{u}'), \quad i = 1, \cdots, N$$

we obtain

$$\frac{d}{dx}\left[f(x, \overline{u}, \overline{u}') - \langle \overline{u}'; f_{\xi}(x, \overline{u}, \overline{u}')\rangle\right] = f_x(x, \overline{u}, \overline{u}'). \quad \blacksquare$$

Exercise 2.3.2. The second form of the Euler-Lagrange equation is

$$0 = \frac{d}{dx}\left[f(u(x), u'(x)) - u'(x) f_{\xi}(u(x), u'(x))\right] = \frac{d}{dx}\left[-u(x) - \frac{1}{2}(u'(x))^2\right]$$

$$= -u'(x) - u''(x) u'(x) = -u'(x)[u''(x) + 1]$$

and it is satisfied by $u \equiv 1$. However $u \equiv 1$ does not verify the Euler-Lagrange equation, which is in the present case

$$u''(x) = -1. \quad \blacksquare$$

7.2.3 Hamiltonian formulation

Exercise 2.4.1. The proof is a mere repetition of that of Theorem 2.11 and we skip the details. We just state the result. Let $u = \left(u^1, \cdots, u^N\right)$ and $\xi = \left(\xi^1, \cdots, \xi^N\right)$. We assume that $f \in C^2\left([a,b] \times \mathbb{R}^N \times \mathbb{R}^N\right)$, $f = f(x, u, \xi)$, and that it verifies

$$D^2 f(x, u, \xi) = \left(f_{\xi^i \xi^j}\right)_{1 \leq i, j \leq N} > 0, \quad \text{for every } (x, u, \xi) \in [a, b] \times \mathbb{R}^N \times \mathbb{R}^N$$

$$f(x, u, \xi) \geq \omega\left(|\xi|\right) + g(x, u), \quad \text{for every } (x, u, \xi) \in [a, b] \times \mathbb{R}^N \times \mathbb{R}^N$$

where $g : [a, b] \times \mathbb{R}^N \to \mathbb{R}$ is continuous and ω is a non-negative continuous and increasing function with $\lim_{t \to \infty} \omega(t)/t = \infty$. If we let

$$H(x, u, v) = \sup_{\xi \in \mathbb{R}^N} \left\{\langle v; \xi \rangle - f(x, u, \xi)\right\}$$

then $H \in C^2\left([a, b] \times \mathbb{R}^N \times \mathbb{R}^N\right)$ and, denoting by

$$H_v(x, u, v) = \left(H_{v^1}(x, u, v), \cdots, H_{v^N}(x, u, v)\right)$$

and similarly for $H_u(x, u, v)$, we also have

$$H_x(x, u, v) = -f_x(x, u, H_v(x, u, v))$$

$$H_u(x, u, v) = -f_u(x, u, H_v(x, u, v))$$

$$H(x, u, v) = \langle v; H_v(x, u, v) \rangle - f(x, u, H_v(x, u, v))$$

$$v = f_\xi(x, u, \xi) \quad \Leftrightarrow \quad \xi = H_v(x, u, v).$$

The Euler-Lagrange system is

$$\frac{d}{dx}\left[f_{\xi^i}(x, u, u')\right] = f_{u^i}(x, u, u'), \quad i = 1, \cdots, N$$

and the associated Hamiltonian system is given, for every $i = 1, \cdots, N$, by

$$\begin{cases} \left(u^i\right)' = H_{v^i}(x, u, v) \\ \left(v^i\right)' = -H_{u^i}(x, u, v). \end{cases} \quad \blacksquare$$

Exercise 2.4.2. (i) The Euler-Lagrange equations are, for $i = 1, \cdots, N$,

$$\begin{cases} m_i x_i'' = -U_{x_i}(t, u) \\ m_i y_i'' = -U_{y_i}(t, u) \\ m_i z_i'' = -U_{z_i}(t, u). \end{cases}$$

In terms of the Hamiltonian, if we let $u_i = (x_i, y_i, z_i)$, $\xi_i = (\xi_i^x, \xi_i^y, \xi_i^z)$ and $v_i = (v_i^x, v_i^y, v_i^z)$, for $i = 1, \cdots, N$, we find

$$H(t, u, v) = \sup_{\xi \in \mathbb{R}^{3N}} \left\{ \sum_{i=1}^{N} \left[\langle v_i; \xi_i \rangle - \frac{1}{2} m_i |\xi_i|^2 \right] + U(t, u) \right\}$$

$$= \sum_{i=1}^{N} \frac{|v_i|^2}{2m_i} + U(t, u).$$

(ii) Note that along the trajectories we have $v_i = m_i u_i'$, i.e.

$$v_i^x = m_i x_i', \quad v_i^y = m_i y_i', \quad v_i^z = m_i z_i'$$

and hence

$$H(t, u, v) = \frac{1}{2} \sum_{i=1}^{N} m_i |u_i'|^2 + U(t, u). \quad \blacksquare$$

Exercise 2.4.3. Although the hypotheses of Theorem 2.11 are not satisfied in the present context, the procedure is exactly the same and leads to the following analysis. The Hamiltonian is

$$H(x, u, v) = \begin{cases} -\sqrt{g(x, u) - v^2} & \text{if } v^2 \le g(x, u) \\ +\infty & \text{otherwise.} \end{cases}$$

We therefore have, provided $v^2 < g(x, u)$, that

$$\begin{cases} u' = H_v = \dfrac{v}{\sqrt{g(x, u) - v^2}} \\ v' = -H_u = \dfrac{1}{2} \dfrac{g_u}{\sqrt{g(x, u) - v^2}}. \end{cases}$$

We hence obtain that $2vv' = g_u u'$ and thus

$$\left[v^2(x) - g(x, u(x)) \right]' + g_x(x, u(x)) = 0.$$

If $g(x, u) = g(u)$, we get (c being a constant)

$$v^2(x) = c + g(u(x)). \quad \blacksquare$$

7.2.4 Hamilton-Jacobi equation

Exercise 2.5.1. We state the results without proofs (they are similar to the case $N = 1$ and we refer to Gelfand-Fomin [54] page 88, if necessary). Let $H \in$

$C^1\left([a,b]\times\mathbb{R}^N\times\mathbb{R}^N\right)$, $H=H\left(x,u,v\right)$ and $u=\left(u^1,\cdots,u^N\right)$. The Hamilton-Jacobi equation is

$$S_x+H\left(x,u,S_u\right)=0,\quad\forall\left(x,u,\alpha\right)\in[a,b]\times\mathbb{R}^N\times\mathbb{R}^N,$$

where $S=S\left(x,u,\alpha\right)$ and $S_u=\left(S_{u^1},\cdots,S_{u^N}\right)$. Jacobi Theorem then reads as follows. Let $S\in C^2\left([a,b]\times\mathbb{R}^N\times\mathbb{R}^N\right)$ be a solution of the Hamilton-Jacobi equation and

$$\det\left(S_{u\alpha}\left(x,u,\alpha\right)\right)\neq0,\quad\forall\left(x,u,\alpha\right)\in[a,b]\times\mathbb{R}^N\times\mathbb{R}^N,$$

where $S_{u\alpha}=\left(\partial^2S/\partial\alpha^i\partial u^j\right)_{1\leq i,j\leq N}$. If $u\in C^1\left([a,b];\mathbb{R}^N\right)$ satisfies

$$\frac{d}{dx}\left[S_{\alpha^i}\left(x,u\left(x\right),\alpha\right)\right]=0,\quad\forall\left(x,\alpha\right)\in[a,b]\times\mathbb{R}^N,\ i=1,\cdots,N$$

and if $v\left(x\right)=\left(v^1,\cdots,v^N\right)=S_u\left(x,u\left(x\right),\alpha\right)=\left(S_{u^1},\cdots,S_{u^N}\right)$, then

$$\begin{cases}u'\left(x\right)=H_v\left(x,u\left(x\right),v\left(x\right)\right)\\v'\left(x\right)=-H_u\left(x,u\left(x\right),v\left(x\right)\right).\end{cases}\quad\blacksquare$$

Exercise 2.5.2. The procedure is formal because the hypotheses of Theorem 2.20 are not satisfied. We have seen in Exercise 2.4.3 that

$$H\left(u,v\right)=\begin{cases}-\sqrt{g\left(u\right)-v^2}&\text{if }v^2\leq g\left(u\right)\\+\infty&\text{otherwise.}\end{cases}$$

The Hamilton-Jacobi equation (in this context, it is called the *eikonal equation*) is then

$$S_x-\sqrt{g\left(u\right)-S_u^2}=0\quad\Leftrightarrow\quad S_x^2+S_u^2=g\left(u\right).$$

Its reduced form is then, for $\alpha>0$, $g\left(u\right)-\left(S_u^*\right)^2=\alpha^2$ and this leads to

$$S^*\left(u,\alpha\right)=\int_{u_0}^u\sqrt{g\left(s\right)-\alpha^2}\,ds.$$

We therefore get

$$S\left(x,u,\alpha\right)=\alpha x+\int_{u_0}^u\sqrt{g\left(s\right)-\alpha^2}\,ds.$$

The equation $S_\alpha=\beta$ (where β is a constant) reads as

$$x-\alpha\int_{u_0}^{u(x)}\frac{ds}{\sqrt{g\left(s\right)-\alpha^2}}=\beta$$

which implies

$$1 - \frac{\alpha u'(x)}{\sqrt{g(u(x)) - \alpha^2}} = 0.$$

Note that, indeed, any such $u = u(x)$ and

$$v = v(x) = S_u(x, u(x), \alpha) = \sqrt{g(u(x)) - \alpha^2}$$

satisfy

$$\begin{cases} u'(x) = H_v(x, u(x), v(x)) = \frac{\sqrt{g(u(x)) - \alpha^2}}{\alpha} \\ v'(x) = -H_u(x, u(x), v(x)) = \frac{g'(u(x))u'(x)}{2\sqrt{g(u(x)) - \alpha^2}} = \frac{g'(u(x))}{2\alpha} \end{cases}.$$

Exercise 2.5.3. The Hamiltonian is easily seen to be

$$H(u, v) = \frac{v^2}{2a(u)}.$$

The Hamilton-Jacobi equation and its reduced form are given by

$$S_x + \frac{(S_u)^2}{2a(u)} = 0 \quad \text{and} \quad \frac{(S_u^*)^2}{2a(u)} = \frac{\alpha^2}{2}.$$

Therefore, defining A by $A'(u) = \sqrt{a(u)}$, we find

$$S^*(u, \alpha) = \alpha A(u) \quad \text{and} \quad S(x, u, \alpha) = -\frac{\alpha^2}{2}x + \alpha A(u).$$

Hence, according to Theorem 2.20 (note that $S_{u\alpha} = \sqrt{a(u)} > 0$) the solution is given implicitly by

$$S_\alpha(x, u(x), \alpha) = -\alpha x + A(u(x)) \equiv \beta = \text{constant}.$$

Since A is invertible we find (compare with Exercise 2.2.10)

$$u(x) = A^{-1}(\alpha x + \beta). \quad \blacksquare$$

Exercise 2.5.4. (i) Set

$$S(x, u, \alpha) = \alpha u - \int_0^x H(t, \alpha) \, dt.$$

We clearly have $S_{u\alpha} \neq 0$, $S_u = \alpha$ and

$$S_x = -H(x, \alpha) = -H(x, S_u).$$

(ii) Note that

$$\frac{d}{dx}\left[S_\alpha\left(x,u\left(x\right),\alpha\right)\right]=0 \;\Leftrightarrow\; \frac{d}{dx}\left[u-\int_0^x H_v\left(t,\alpha\right)\right]=0 \;\Leftrightarrow\; u'\left(x\right)=H_v\left(x,\alpha\right).$$

If $v\left(x\right)=S_u\left(x,u\left(x\right),\alpha\right)=\alpha$, then $\left(u,v\right)$ is indeed a solution of

$$\begin{cases} u'\left(x\right)=H_v\left(x,v\left(x\right)\right) \\ v'\left(x\right)=-H_u\left(x,v\left(x\right)\right)=0. \end{cases} \quad\blacksquare$$

Exercise 2.5.5. Observe that S is Lipschitz, belongs to $C^1\left(\overline{\Omega}\setminus\{0\}\right)$ and that clearly $S\left(u\right)=0$ on $\partial\Omega$. For every $u\neq 0$, we clearly have

$$S\left(u\right)=\text{dist}\left(u,\partial\Omega\right)=\left|u-\frac{u}{|u|}\right|=1-|u|$$

and thus

$$\nabla S\left(u\right)=-\frac{u}{|u|}$$

which leads to the claim, namely $H\left(\nabla S\left(u\right)\right)=0$. $\blacksquare$

7.2.5   Fields theories

Exercise 2.6.1. Let $f\in C^2\left(\left[a,b\right]\times\mathbb{R}^N\times\mathbb{R}^N\right),\alpha,\beta\in\mathbb{R}^N$. Assume that there exists $\Phi\in C^3\left(\left[a,b\right]\times\mathbb{R}^N\right)$ satisfying $\Phi\left(a,\alpha\right)=\Phi\left(b,\beta\right)$. Suppose also that

$$\widetilde{f}\left(x,u,\xi\right)=f\left(x,u,\xi\right)+\langle\Phi_u\left(x,u\right);\xi\rangle+\Phi_x\left(x,u\right)$$

is such that $\left(u,\xi\right)\to\widetilde{f}\left(x,u,\xi\right)$ is convex. The claim is then that any solution $\overline{u}$ of the Euler-Lagrange system

$$\frac{d}{dx}\left[f_{\xi^i}\left(x,\overline{u},\overline{u}'\right)\right]=f_{u^i}\left(x,\overline{u},\overline{u}'\right), \quad i=1,\cdots,N$$

is a minimizer of

$$(P)\quad \inf_{u\in X}\left\{I\left(u\right)=\int_a^b f\left(x,u\left(x\right),u'\left(x\right)\right)dx\right\}=m$$

where $X=\left\{u\in C^1\left(\left[a,b\right];\mathbb{R}^N\right):u\left(a\right)=\alpha,\ u\left(b\right)=\beta\right\}$.

The proof is exactly as that of the one dimensional case and we skip the details. $\blacksquare$

Exercise 2.6.2. The procedure is very similar to that of Theorem 2.28. An exact field $\Phi = \Phi(x, u)$ covering a connected open set $D \subset \mathbb{R}^{N+1}$ is a map $\Phi : D \to \mathbb{R}^N$ so that there exists $S \in C^1\left(D; \mathbb{R}^N\right)$ satisfying

$$S_{u^i}(x, u) = f_{\xi^i}(x, u, \Phi(x, u)), \quad i = 1, \cdots, N$$
$$S_x(x, u) = f(x, u, \Phi(x, u)) - \langle S_u(x, u); \Phi(x, u) \rangle.$$

The Weierstrass function is defined, for $u, \eta, \xi \in \mathbb{R}^N$, as

$$E(x, u, \eta, \xi) = f(x, u, \xi) - f(x, u, \eta) - \langle f_\xi(x, u, \eta); (\xi - \eta) \rangle.$$

The proof is then identical to the one dimensional case. ∎

Exercise 2.6.3. (i) We have by definition

$$\begin{cases} S_u(x, u) = f_\xi(x, u, \Phi(x, u)) \\ S_x(x, u) = -\left[S_u(x, u)\, \Phi(x, u) - f(x, u, \Phi(x, u))\right]. \end{cases}$$

From the first equation and (2.10) of Lemma 2.9, we get

$$\Phi(x, u) = H_v(x, u, S_u(x, u)).$$

The second equation, combined with (2.9) of Lemma 2.9, becomes

$$\begin{aligned} S_x(x, u) &= -\left[S_u H_v(x, u, S_u) - f(x, u, H_v(x, u, S_u))\right] \\ &= -H(x, u, S_u(x, u)). \end{aligned}$$

(ii) Using again Lemma 2.9 we obtain

$$\begin{cases} H(x, u, S_u) = S_u(x, u)\, \Phi(x, u) - f(x, u, \Phi(x, u)) \\ \qquad\quad S_u(x, u) = f_\xi(x, u, \Phi(x, u)). \end{cases}$$

Since S is a solution of Hamilton-Jacobi equation, we get $S_x = -H(x, u, S_u)$ as wished. ∎

Exercise 2.6.4. The result can be found in Proposition 2.5 of [32, 1st edition]. We first observe that $\overline{u} \equiv 0 \in X$ verifies

$$(E) \quad \frac{d}{dx}\left[f_\xi(\overline{u}, \overline{u}')\right] = \frac{d}{dx}\left[h'(0)\right] = -g'(0) = f_u(\overline{u}, \overline{u}'), \quad x \in (0, 1).$$

We consider three cases.

Case 1: $g_0 \le 0$. The function f is then convex in both variables and we may apply Part 2 of Theorem 2.1 to get the result, namely

$$\int_0^1 f(u(x), u'(x))\, dx \ge \int_0^1 f(\overline{u}(x), \overline{u}'(x))\, dx = f(0, 0), \quad \forall u \in X$$

i.e.

$$\int_0^1 \left[h\left(u'\left(x\right)\right) - h\left(0\right)\right] dx \geq \int_0^1 \left[g\left(u\left(x\right)\right) - g\left(0\right)\right] dx, \quad \forall u \in X.$$

Case 2: $g_0 > 0$ *and* $\pi^2 h_0 - g_0 > 0$. The hypotheses imply then that $h_0 > 0$. Define, for every $(x, u, \xi) \in [0, 1] \times \mathbb{R} \times \mathbb{R}$,

$$\Phi\left(x, u\right) = \frac{\sqrt{g_0 h_0}}{2} \tan\left[\sqrt{\frac{g_0}{h_0}} \left(x - \frac{1}{2}\right)\right] u^2$$

$$\widetilde{f}\left(x, u, \xi\right) = f\left(u, \xi\right) + \Phi_u\left(x, u\right)\xi + \Phi_x\left(x, u\right)$$

and observe that Φ and $\widetilde{f}$ satisfy all the hypotheses of Theorem 2.22. This theorem therefore leads, for every $u \in X$, to

$$\int_0^1 f\left(u\left(x\right), u'\left(x\right)\right) dx = \int_0^1 \widetilde{f}\left(x, u\left(x\right), u'\left(x\right)\right) dx$$

$$\geq \int_0^1 \widetilde{f}\left(x, \overline{u}\left(x\right), \overline{u}'\left(x\right)\right) dx = f\left(0, 0\right)$$

which is our claim

$$\int_0^1 \left[h\left(u'\left(x\right)\right) - h\left(0\right)\right] dx \geq \int_0^1 \left[g\left(u\left(x\right)\right) - g\left(0\right)\right] dx, \quad \forall u \in X.$$

Case 3: $g_0 > 0$ *and* $\pi^2 h_0 - g_0 = 0$. We let, for $\epsilon > 0$ small,

$$g^{\epsilon}\left(u\right) = g\left(u\right) - \frac{\epsilon}{2} u^2.$$

We then apply Case 2 to g^{ϵ} and h. A straightforward passage to the limit leads to the claim. ∎

7.3   Chapter 3. Direct methods: existence

7.3.1   The model case: Dirichlet integral

Exercise 3.2.1. The proof is almost completely identical to that of Theorem 3.1; only the first step is slightly different. So let $\{u_\nu\}$ be a minimizing sequence

$$I\left(u_\nu\right) \to m = \inf\left\{I\left(u\right) : u \in W_0^{1,2}\left(\Omega\right)\right\}.$$

Since $I(0) = 0$, we have that $m \leq 0$. Consequently, we have from Hölder inequality that

$$m + 1 \geq I(u_\nu) = \frac{1}{2} \int_\Omega |\nabla u_\nu|^2 - \int_\Omega h\,u_\nu$$

$$\geq \int_\Omega \frac{1}{2} |\nabla u_\nu|^2 - \|h\|_{L^2} \|u_\nu\|_{L^2} = \frac{1}{2} \|\nabla u_\nu\|_{L^2}^2 - \|h\|_{L^2} \|u_\nu\|_{L^2}.$$

Using Poincaré inequality (Theorem 1.49) we can find constants (independent of ν) $\gamma_k > 0$, $k = 1, \cdots, 5$, so that

$$m + 1 \geq \gamma_1 \|u_\nu\|_{W^{1,2}}^2 - \gamma_2 \|u_\nu\|_{W^{1,2}} \geq \gamma_3 \|u_\nu\|_{W^{1,2}}^2 - \gamma_4$$

and hence, as wished,

$$\|u_\nu\|_{W^{1,2}} \leq \gamma_5. \quad \blacksquare$$

Exercise 3.2.2. (i) The only difference from the proof of Theorem 3.1 and Exercise 3.2.1 is in the compactness argument and the uniqueness.

Step 1 (Compactness). Let $u_\nu \in X$ be a minimizing sequence of (N), this means that

$$I(u_\nu) \to \inf \{I(u)\} = m \leq I(0) = 0, \quad \text{as } \nu \to \infty.$$

In particular we can find ν sufficiently large so that $I(u_\nu) \leq 1$. Appealing to Poincaré inequality under the form (1.15), we can find constants $\gamma_1, \gamma_2, \gamma_3, \gamma_4 > 0$ (independent of ν) so that

$$1 \geq I(u_\nu) \geq \frac{1}{2} \|\nabla u_\nu\|_{L^2}^2 - \|h\|_{L^2} \|u_\nu\|_{L^2}$$

$$\geq \frac{1}{4} \|\nabla u_\nu\|_{L^2}^2 + \gamma_1 \|u_\nu\|_{L^2}^2 - \gamma_2 \|u_\nu\|_{L^2} \geq \gamma_3 \|u_\nu\|_{W^{1,2}}^2 - \gamma_4$$

and thus there exists $\gamma_5 > 0$ so that

$$\|u_\nu\|_{W^{1,2}} \leq \gamma_5.$$

Applying Exercise 1.4.5 (ii), we deduce that there exists $\overline{u} \in X$ and a subsequence (still denoted u_ν) so that our claim holds, namely

$$u_\nu \rightharpoonup \overline{u} \text{ in } W^{1,2}, \quad \text{as } \nu \to \infty.$$

Step 2 (Uniqueness). Assume that there exist $\overline{u}, \overline{v} \in X$ so that

$$I(\overline{u}) = I(\overline{v}) = m.$$

As in Part 2 of Theorem 3.1, we deduce that $\nabla \overline{u} = \nabla \overline{v}$ a.e. in Ω. It follows from Exercise 1.4.14 (ii) that $\overline{u} = \overline{v}$ a.e. in Ω as wished.

(ii) Recall that we denote the average of a function u by

$$u_\Omega = \frac{1}{\operatorname{meas}\Omega} \int_\Omega u.$$

Let $\epsilon \in \mathbb{R}$ and $\psi \in W^{1,2}(\Omega)$. Note that $\varphi = \psi - \psi_\Omega \in X$ and therefore $\overline{u} + \epsilon\varphi \in X$, which combined with the fact that $\overline{u}$ is the minimizer of (N), leads to

$$I(\overline{u}) \le I(\overline{u} + \epsilon\varphi) = \int_\Omega \left[\frac{1}{2} |\nabla \overline{u} + \epsilon \nabla\varphi|^2 - h(\overline{u} + \epsilon\varphi) \right]$$

$$= I(\overline{u}) + \epsilon \int_\Omega [\langle \nabla\overline{u}; \nabla\varphi\rangle - h\,\varphi] + \epsilon^2 \int_\Omega \frac{1}{2} |\nabla\varphi|^2$$

and thus to

$$0 = \frac{d}{d\epsilon} I(\overline{u} + \epsilon\varphi) \Big|_{\epsilon=0} = \int_\Omega [\langle \nabla\overline{u}; \nabla\varphi\rangle - h\,\varphi] = 0.$$

Observing that

$$\int_\Omega h\,\varphi = \int_\Omega (h - h_\Omega)\,\psi,$$

we find that the weak form of the Euler-Lagrange equation is given by

$$\int_\Omega [\langle \nabla\overline{u}; \nabla\psi\rangle - (h - h_\Omega)\,\psi] = 0, \quad \forall \psi \in W^{1,2}(\Omega). \tag{7.14}$$

(iii) When $\overline{u} \in C^2(\overline{\Omega})$ and $h \in C(\overline{\Omega})$, we can integrate and get, ν denoting the outward unit normal,

$$\int_{\partial\Omega} \frac{\partial \overline{u}}{\partial \nu} \psi - \int_\Omega (\Delta\overline{u} + h - h_\Omega)\,\psi = 0, \quad \forall \psi \in C^\infty(\overline{\Omega}).$$

Choosing first $\psi \in C_0^\infty(\overline{\Omega})$, we obtain

$$\Delta\overline{u} = h_\Omega - h \quad \text{in } \Omega.$$

Returning then to the integrated equation we find that

$$\frac{\partial \overline{u}}{\partial \nu} = \langle \nabla\overline{u}; \nu\rangle = 0 \quad \text{on } \partial\Omega. \quad \blacksquare$$

7.3.2   A general existence theorem

Exercise 3.3.1. As in Exercise 3.2.1, it is the compactness proof in Theorem 3.3 that has to be modified, the remaining part of the proof is essentially unchanged. Let therefore $\{u_\nu\}$ be a minimizing sequence, i.e. $I(u_\nu) \to m$. We have from (H_2) that for ν sufficiently large

$$m + 1 \geq I(u_\nu) \geq \alpha_1 \|\nabla u_\nu\|_{L^p}^p - |\alpha_2| \|u_\nu\|_{L^q}^q - |\alpha_3| \operatorname{meas} \Omega.$$

From now on we denote by $\gamma_k > 0$ constants that are independent of ν. Since by Hölder inequality we have

$$\|u_\nu\|_{L^q}^q = \int_\Omega |u_\nu|^q \leq \left(\int_\Omega |u_\nu|^p \right)^{q/p} \left(\int_\Omega dx \right)^{(p-q)/p} = (\operatorname{meas} \Omega)^{(p-q)/p} \|u_\nu\|_{L^p}^q$$

we deduce that we can find constants γ_1 and γ_2 such that

$$m + 1 \geq \alpha_1 \|\nabla u_\nu\|_{L^p}^p - \gamma_1 \|u_\nu\|_{L^p}^q - \gamma_2$$
$$\geq \alpha_1 \|\nabla u_\nu\|_{L^p}^p - \gamma_1 \|u_\nu\|_{W^{1,p}}^q - \gamma_2.$$

Invoking Poincaré inequality (Theorem 1.49) we can find $\gamma_3, \gamma_4, \gamma_5$, so that

$$m + 1 \geq \gamma_3 \|u_\nu\|_{W^{1,p}}^p - \gamma_4 \|u_0\|_{W^{1,p}}^p - \gamma_1 \|u_\nu\|_{W^{1,p}}^q - \gamma_5$$

and hence, γ_6 being a constant,

$$m + 1 \geq \gamma_3 \|u_\nu\|_{W^{1,p}}^p - \gamma_1 \|u_\nu\|_{W^{1,p}}^q - \gamma_6.$$

Since $1 \leq q < p$, we can find γ_7, γ_8 so that

$$m + 1 \geq \gamma_7 \|u_\nu\|_{W^{1,p}}^p - \gamma_8$$

which, combined with the fact that $m < +\infty$, leads to the claim, namely

$$\|u_\nu\|_{W^{1,p}} \leq \gamma_9. \quad \blacksquare$$

Exercise 3.3.2. This time it is the lower semicontinuity step in Theorem 3.3 that has to be changed. Let $u_\nu \rightharpoonup \overline{u}$ in $W^{1,p}$ and thus $u_\nu \to \overline{u}$ in L^p. Let

$$I(u) = I_1(u) + I_2(u)$$

where

$$I_1(u) = \int_\Omega h(x, \nabla u(x)) \, dx \quad \text{and} \quad I_2(u) = \int_\Omega g(x, u(x)) \, dx.$$

We have to show that

$$\liminf_{\nu\to\infty} \left[I_1\left(u_\nu\right) + I_2\left(u_\nu\right) \right] \geq \liminf_{\nu\to\infty} I_1\left(u_\nu\right) + \liminf_{\nu\to\infty} I_2\left(u_\nu\right) \geq I_1\left(\overline{u}\right) + I_2\left(\overline{u}\right).$$

It is clear that, by the proof of the theorem,

$$\liminf_{\nu\to\infty} I_1\left(u_\nu\right) \geq I_1\left(\overline{u}\right).$$

Our claim will follow if we can prove that

$$L = \liminf_{\nu\to\infty} I_2\left(u_\nu\right) \geq I_2\left(\overline{u}\right).$$

First choose a subsequence in order to have

$$L = \lim_{\nu_i\to\infty} I_2\left(u_{\nu_i}\right).$$

Up to extracting a further subsequence, we have, since $u_\nu \to \overline{u}$ in L^p, that $u_{\nu_{i_j}} \to \overline{u}$ a.e. and thus

$$g\left(x, u_{\nu_{i_j}}\left(x\right)\right) \to g\left(x, \overline{u}\left(x\right)\right) \quad \text{a.e. in } \Omega.$$

We then invoke Fatou lemma to get

$$L = \lim_{\nu_i\to\infty} I_2\left(u_{\nu_i}\right) = \lim_{\nu_{i_j}\to\infty} I_2\left(u_{\nu_{i_j}}\right) = \lim_{\nu_{i_j}\to\infty} \int_\Omega g\left(x, u_{\nu_{i_j}}\left(x\right)\right) dx$$

$$\geq \int_\Omega \lim_{\nu_{i_j}\to\infty} g\left(x, u_{\nu_{i_j}}\left(x\right)\right) dx = \int_\Omega g\left(x, \overline{u}\left(x\right)\right) dx = I_2\left(\overline{u}\right)$$

which is exactly our claim. ∎

Exercise 3.3.3. We have here weakened the hypothesis (H_3) in the proof of the theorem. We used this hypothesis only in the lower semicontinuity part of the proof, so let us establish this property under the new condition. Let $u_\nu \rightharpoonup \overline{u}$ in $W^{1,p}\left((a,b); \mathbb{R}^N\right)$. Using the convexity of $(u, \xi) \to f\left(x, u, \xi\right)$ we find

$$\int_a^b f\left(x, u_\nu, u_\nu'\right) dx \geq \int_a^b f\left(x, \overline{u}, \overline{u}'\right) dx$$

$$+ \int_a^b \left[\langle f_u\left(x, \overline{u}, \overline{u}'\right) ; u_\nu - \overline{u} \rangle + \langle f_\xi\left(x, \overline{u}, \overline{u}'\right) ; u_\nu' - \overline{u}' \rangle \right] dx.$$

Since, by Rellich theorem, $u_\nu \to \overline{u}$ in L^∞, to pass to the limit in the second term of the right-hand side of the inequality we need only have $f_u\left(x, \overline{u}, \overline{u}'\right) \in L^1$. This is ascertained by the hypothesis

$$\left| f_u\left(x, u, \xi\right) \right| \leq \beta\left(1 + \left|\xi\right|^p\right).$$

Similarly, to pass to the limit in the last term we need to have $f_\xi\left(x,\overline{u},\overline{u}'\right) \in L^{p'}$, $p' = p/\left(p-1\right)$; and this is precisely true because of the hypothesis

$$\left|f_\xi\left(x,u,\xi\right)\right| \leq \beta\left(1 + |\xi|^{p-1}\right). \quad \blacksquare$$

Exercise 3.3.4. We proceed as in Theorem 3.3. Assume that there exist $\overline{u},\overline{v} \in u_0 + W_0^{1,p}\left(\Omega\right)$ so that

$$I\left(\overline{u}\right) = I\left(\overline{v}\right) = m$$

and we prove that this implies $\overline{u} = \overline{v}$. We obtain, as in the theorem, from the convexity of $\left(u,\xi\right) \to f\left(x,u,\xi\right)$, that

$$\frac{1}{2}f\left(x,\overline{u},\nabla\overline{u}\right) + \frac{1}{2}f\left(x,\overline{v},\nabla\overline{v}\right) - f\left(x, \frac{\overline{u}+\overline{v}}{2}, \frac{\nabla\overline{u}+\nabla\overline{v}}{2}\right) = 0 \quad \text{a.e. in } \Omega.$$

We now use the special structure of f to find, since $u \to g\left(x,u\right)$ and $\xi \to h\left(x,\xi\right)$ are convex, that

$$\frac{1}{2}g\left(x,\overline{u}\right) + \frac{1}{2}g\left(x,\overline{v}\right) - g\left(x, \frac{\overline{u}+\overline{v}}{2}\right) = 0 \quad \text{a.e. in } \Omega$$

$$\frac{1}{2}h\left(x,\nabla\overline{u}\right) + \frac{1}{2}h\left(x,\nabla\overline{v}\right) - h\left(x, \frac{\nabla\overline{u}+\nabla\overline{v}}{2}\right) = 0 \quad \text{a.e. in } \Omega.$$

1) If $u \to g\left(x,u\right)$ is strictly convex, we deduce that $\overline{u} = \overline{v}$ a.e. in Ω, as wished.

2) If $\xi \to h\left(x,\xi\right)$ is strictly convex, we conclude that $\nabla\overline{u} = \nabla\overline{v}$ a.e. in Ω, and thus (see Exercise 1.4.14 (i)) also $\overline{u} = \overline{v}$ a.e. in Ω, since $\overline{u} - \overline{v} \in W_0^{1,p}\left(\Omega\right)$. $\quad \blacksquare$

7.3.3 Euler-Lagrange equation

Exercise 3.4.1. We have to prove that for $\overline{u} \in W^{1,p}$, the following expression is meaningful

$$\int_\Omega \left\{f_u\left(x,\overline{u},\nabla\overline{u}\right)\varphi + \langle f_\xi\left(x,\overline{u},\nabla\overline{u}\right);\nabla\varphi\rangle\right\}dx = 0, \quad \forall\varphi \in W_0^{1,p}.$$

Case 1: $p > n$. We have from Sobolev imbedding theorem (Theorem 1.44) that $\overline{u},\varphi \in C\left(\overline{\Omega}\right)$. We therefore only need to have

$$f_u\left(x,\overline{u},\nabla\overline{u}\right) \in L^1 \quad \text{and} \quad f_\xi\left(x,\overline{u},\nabla\overline{u}\right) \in L^{p'}$$

where $p' = p/(p-1)$. This is true if we assume that for every $R > 0$, there exists $\beta = \beta(R)$ so that for every (x, u, ξ) with $|u| \leq R$ the following inequalities hold

$$|f_u(x, u, \xi)| \leq \beta(1 + |\xi|^p) \quad \text{and} \quad |f_\xi(x, u, \xi)| \leq \beta\left(1 + |\xi|^{p-1}\right).$$

Case 2: $p = n$. This time we have $\overline{u}, \varphi \in L^q, \forall q \in [1, \infty)$. We therefore have to ascertain that, for a certain $r > 1$,

$$f_u(x, \overline{u}, \nabla\overline{u}) \in L^r \quad \text{and} \quad f_\xi(x, \overline{u}, \nabla\overline{u}) \in L^{p'}.$$

To guarantee this claim we impose that there exist $\beta > 0$, $p > s_2 \geq 1$, $s_1 \geq 1$ such that

$$|f_u(x, u, \xi)| \leq \beta(1 + |u|^{s_1} + |\xi|^{s_2}) \quad \text{and} \quad |f_\xi(x, u, \xi)| \leq \beta\left(1 + |u|^{s_1} + |\xi|^{p-1}\right).$$

Case 3: $p < n$. We now only have $\overline{u}, \varphi \in L^q, \forall q \in [1, np/(n-p)]$. We therefore should have, for $q' = q/(q-1)$,

$$f_u(x, \overline{u}, \nabla\overline{u}) \in L^{q'} \quad \text{and} \quad f_\xi(x, \overline{u}, \nabla\overline{u}) \in L^{p'}.$$

This happens if there exist $\beta > 0$,

$$1 \leq s_1 \leq \frac{np - n + p}{n - p}, \quad 1 \leq s_2 \leq \frac{np - n + p}{n}, \quad 1 \leq s_3 \leq \frac{n(p-1)}{n-p}$$

so that

$$|f_u(x, u, \xi)| \leq \beta(1 + |u|^{s_1} + |\xi|^{s_2}) \quad \text{and} \quad |f_\xi(x, u, \xi)| \leq \beta\left(1 + |u|^{s_3} + |\xi|^{p-1}\right). \quad\blacksquare$$

Exercise 3.4.2. Use Exercise 3.4.1 to deduce the following growth conditions on $g \in C^1\left(\overline{\Omega} \times \mathbb{R}\right)$.

Case 1: $p > n$. No growth condition is imposed on g.

Case 2: $p = n$. There exist $\beta > 0$ and $s_1 \geq 1$ such that

$$|g_u(x, u)| \leq \beta(1 + |u|^{s_1}), \quad \forall(x, u) \in \overline{\Omega} \times \mathbb{R}.$$

Case 3: $p < n$. There exist $\beta > 0$ and $1 \leq s_1 \leq (np - n + p)/(n - p)$, so that

$$|g_u(x, u)| \leq \beta(1 + |u|^{s_1}), \quad \forall(x, u) \in \overline{\Omega} \times \mathbb{R}. \quad\blacksquare$$

Exercise 3.4.3. Let $\varphi \in C^\infty\left(\overline{\Omega}\right)$ and $\epsilon > 0$. Since $\overline{u}$ is a minimizer we have that, ν denoting the outward unit normal and $\partial\overline{u}/\partial\nu$ the normal component of the gradient,

$$0 = \frac{d}{d\epsilon} I \left(\overline{u} + \epsilon\varphi \right) \Big|_{\epsilon=0} = \int_{\Omega} \left[\langle \nabla\overline{u}; \nabla\varphi \rangle - h\,\varphi \right] + \lambda \int_{\partial\Omega} \overline{u}\,\varphi$$

$$= \int_{\Omega} \left[-h - \Delta\overline{u} \right] \varphi + \int_{\partial\Omega} \left(\frac{\partial\overline{u}}{\partial\nu} + \lambda\overline{u} \right) \varphi.$$

Choosing first $\varphi \in C_0^{\infty}\left(\overline{\Omega}\right)$, we obtain

$$-\Delta\overline{u} = h \quad \text{in } \Omega.$$

Returning then to the integrated equation we find that

$$\frac{\partial\overline{u}}{\partial\nu} + \lambda\overline{u} = 0 \quad \text{on } \partial\Omega. \quad \blacksquare$$

Exercise 3.4.4. (i) Let ν be an integer and

$$u_{\nu}\left(x, t\right) = \sin\left(\nu x\right)\sin t.$$

We obviously have $u_{\nu} \in W_0^{1,2}\left(\Omega\right)$ (in fact $u_{\nu} \in C^{\infty}\left(\overline{\Omega}\right)$ and $u_{\nu} = 0$ on $\partial\Omega$). An elementary computation shows that $\lim_{\nu\to\infty} I\left(u_{\nu}\right) = -\infty$.

(ii) The second part is elementary.

It is also clear that for the wave equation it is not reasonable to impose an initial condition (at $t = 0$) and a final condition (at $t = \pi$). $\quad \blacksquare$

7.3.4 The vectorial case

Exercise 3.5.1. Let

$$\xi_1 = \left(\begin{array}{cc} 1 & 0 \\ 0 & 0 \end{array} \right) \quad \text{and} \quad \xi_2 = \left(\begin{array}{cc} 0 & 0 \\ 0 & 1 \end{array} \right).$$

We have that both functions are not convex, since

$$\frac{1}{2} f_1\left(\xi_1\right) + \frac{1}{2} f_1\left(\xi_2\right) = \frac{1}{2}\left(\det\xi_1\right)^2 + \frac{1}{2}\left(\det\xi_2\right)^2 = 0 < f_1\left(\frac{1}{2}\xi_1 + \frac{1}{2}\xi_2\right) = \frac{1}{16}$$

$$\frac{1}{2} f_2\left(\xi_1\right) + \frac{1}{2} f_2\left(\xi_2\right) = 1 < f_2\left(\frac{1}{2}\xi_1 + \frac{1}{2}\xi_2\right) = \frac{5}{4}. \quad \blacksquare$$

Exercise 3.5.2. (i) *Step 1.* We first prove the result assuming that $u, v \in C^2\left(\overline{\Omega}; \mathbb{R}^2\right)$ with $u = v$ on $\partial\Omega$. Write

$$u = u\left(x_1, x_2\right) = \left(u^1\left(x_1, x_2\right), u^2\left(x_1, x_2\right)\right)$$
$$v = v\left(x_1, x_2\right) = \left(v^1\left(x_1, x_2\right), v^2\left(x_1, x_2\right)\right).$$

Use the fact that

$$\det \nabla u = u^1_{x_1} u^2_{x_2} - u^1_{x_2} u^2_{x_1} = \left(u^1 u^2_{x_2}\right)_{x_1} - \left(u^1 u^2_{x_1}\right)_{x_2}$$

and the divergence theorem to get

$$\iint_\Omega \det \nabla u \, dx_1 dx_2 = \int_{\partial \Omega} \left(u^1 u^2_{x_2} \nu_1 - u^1 u^2_{x_1} \nu_2\right) d\sigma$$

where $\nu = (\nu_1, \nu_2)$ is the outward unit normal to $\partial \Omega$. Since $u^1 = v^1$ on $\partial \Omega$, we have, applying twice the divergence theorem,

$$\iint_\Omega \det \nabla u = \iint_\Omega \left[\left(v^1 u^2_{x_2}\right)_{x_1} - \left(v^1 u^2_{x_1}\right)_{x_2}\right] = \iint_\Omega \left[\left(v^1_{x_1} u^2\right)_{x_2} - \left(v^1_{x_2} u^2\right)_{x_1}\right]$$

$$= \int_{\partial \Omega} \left(v^1_{x_1} u^2 \nu_2 - v^1_{x_2} u^2 \nu_1\right).$$

Since $u^2 = v^2$ on $\partial \Omega$, we obtain, again using the divergence theorem, that

$$\iint_\Omega \det \nabla u = \iint_\Omega \left[\left(v^1_{x_1} v^2\right)_{x_2} - \left(v^1_{x_2} v^2\right)_{x_1}\right] = \iint_\Omega \det \nabla v.$$

Step 2. We now show the result by approximation. We first regularize v, meaning that for every $\epsilon > 0$ we find $v^\epsilon \in C^2\left(\overline{\Omega}; \mathbb{R}^2\right)$ so that

$$\|v - v^\epsilon\|_{W^{1,p}} \le \epsilon.$$

Since $u - v \in W^{1,p}_0\left(\Omega; \mathbb{R}^2\right)$, we can find $w^\epsilon \in C^\infty_0\left(\Omega; \mathbb{R}^2\right)$ so that

$$\|(u - v) - w^\epsilon\|_{W^{1,p}} \le \epsilon.$$

Define $u^\epsilon = v^\epsilon + w^\epsilon$ and observe that $u^\epsilon, v^\epsilon \in C^2\left(\overline{\Omega}; \mathbb{R}^2\right)$, with $u^\epsilon = v^\epsilon$ on $\partial \Omega$, and

$$\|u - u^\epsilon\|_{W^{1,p}} = \|(u - v) - w^\epsilon + (v - v^\epsilon)\|_{W^{1,p}} \le 2\epsilon.$$

Using Exercise 3.5.4 below, we deduce that there exists γ_1 (independent of ϵ) so that

$$\|\det \nabla u - \det \nabla u^\epsilon\|_{L^{p/2}} + \|\det \nabla v - \det \nabla v^\epsilon\|_{L^{p/2}} \le \gamma_1 \epsilon.$$

Combining Step 1 with the above estimates we obtain that there exists a constant γ_2 (independent of ϵ) such that

$$\left|\iint_\Omega (\det \nabla u - \det \nabla v)\right| \le \left|\iint_\Omega (\det \nabla u^\epsilon - \det \nabla v^\epsilon)\right|$$

$$+ \left|\iint_\Omega (\det \nabla u - \det \nabla u^\epsilon)\right| + \left|\iint_\Omega (\det \nabla v^\epsilon - \det \nabla v)\right| \le \gamma_2 \epsilon.$$

Since ϵ is arbitrary, we have indeed the result.

(ii) *Step 1.* As before we start by considering the case where $u, v \in C^2\left(\overline{\Omega}; \mathbb{R}^2\right)$ with $u = v$ on $\partial\Omega$. We also assume, for the moment, that $g \in C^1\left(\mathbb{R}^2\right)$. Let

$$G\left(y_1, y_2\right) = \int_0^{y_1} g\left(t, y_2\right) dt$$

and observe that

$$\frac{\partial}{\partial x_1}\left[G\left(u^1\left(x\right), u^2\left(x\right)\right)\right] = \left[G\left(u\right)\right]_{x_1} = g\left(u\right) u^1_{x_1} + G_{y_2}\left(u\right) u^2_{x_1}$$

$$\frac{\partial}{\partial x_2}\left[G\left(u^1\left(x\right), u^2\left(x\right)\right)\right] = \left[G\left(u\right)\right]_{x_2} = g\left(u\right) u^1_{x_2} + G_{y_2}\left(u\right) u^2_{x_2}.$$

We therefore find, after computation,

$$g\left(u\right) \det \nabla u = g\left(u\right)\left[u^1_{x_1} u^2_{x_2} - u^1_{x_2} u^2_{x_1}\right]$$
$$= \left[G\left(u\right)\right]_{x_1} u^2_{x_2} - \left[G\left(u\right)\right]_{x_2} u^2_{x_1} = \det \nabla \widetilde{u}$$

where $\widetilde{u} = \left(G\left(u\right), u^2\left(x\right)\right)$ and similarly we let $\widetilde{v} = \left(G\left(v\right), v^2\left(x\right)\right)$. We then apply (i) of the present exercise to $\widetilde{u}$ and $\widetilde{v}$ to get the claim, namely

$$\iint_\Omega g\left(u\right) \det \nabla u = \iint_\Omega \det \nabla \widetilde{u} = \iint_\Omega \det \nabla \widetilde{v} = \iint_\Omega g\left(v\right) \det \nabla v.$$

Note that by a straightforward passage to the limit we have that the above identity is valid if $g \in C^0$ and not only when $g \in C^1$.

Step 2. To pass to the general case we find, as in Step 2 (i) above, $u^\epsilon, v^\epsilon \in C^2\left(\overline{\Omega}; \mathbb{R}^2\right)$, with $u^\epsilon = v^\epsilon$ on $\partial\Omega$, such that (recall that here $p > 2$)

$$\|u - u^\epsilon\|_{W^{1,p}} + \|v - v^\epsilon\|_{W^{1,p}} + \|u - u^\epsilon\|_{L^\infty} + \|v - v^\epsilon\|_{L^\infty} \leq \epsilon$$

and thus there exists a constant γ so that

$$\|g\left(u\right) \det \nabla u - g\left(u^\epsilon\right) \det \nabla u^\epsilon\|_{L^{p/2}} + \|g\left(v\right) \det \nabla v - g\left(v^\epsilon\right) \det \nabla v^\epsilon\|_{L^{p/2}} \leq \gamma \epsilon.$$

The claim then follows at once from Step 1 applied to u^ϵ, v^ϵ and the above estimate. ∎

Exercise 3.5.3. Let $u \in W^{1,p}\left(\Omega; \mathbb{R}^2\right)$,

$$u\left(x_1, x_2\right) = \left(\varphi\left(x_1, x_2\right), \psi\left(x_1, x_2\right)\right),$$

be a minimizer of (P) and let $v \in C_0^\infty\left(\Omega; \mathbb{R}^2\right)$

$$v\left(x_1, x_2\right) = \left(\alpha\left(x_1, x_2\right), \beta\left(x_1, x_2\right)\right)$$

be arbitrary. Since $I(u + \epsilon v) \geq I(u)$ for every ϵ, we must have

$$\frac{d}{d\epsilon} I(u + \epsilon v) \bigg|_{\epsilon = 0} = 0.$$

Since

$$I(u + \epsilon v) = \iint_\Omega [(\varphi_{x_1} + \epsilon\alpha_{x_1})(\psi_{x_2} + \epsilon\beta_{x_2}) - (\varphi_{x_2} + \epsilon\alpha_{x_2})(\psi_{x_1} + \epsilon\beta_{x_1})]$$

we therefore deduce that

$$\iint_\Omega [(\psi_{x_2}\alpha_{x_1} - \psi_{x_1}\alpha_{x_2}) + (\varphi_{x_1}\beta_{x_2} - \varphi_{x_2}\beta_{x_1})] = 0.$$

Integrating by parts, we find that the left-hand side vanishes identically. The result is not surprising in view of Exercise 3.5.2, which shows that $I(u)$ is in fact constant. ∎

Exercise 3.5.4. The proof is divided into two steps.

Step 1. It is easily proved that the following algebraic inequality holds

$$|\det A - \det B| \leq \gamma(|A| + |B|)|A - B|, \quad \forall A, B \in \mathbb{R}^{2\times 2}$$

where γ is a constant.

Step 2. We therefore deduce that

$$|\det \nabla u - \det \nabla v|^{p/2} \leq \gamma^{p/2}(|\nabla u| + |\nabla v|)^{p/2}|\nabla u - \nabla v|^{p/2}.$$

Hölder inequality then implies

$$\iint_\Omega |\det \nabla u - \det \nabla v|^{p/2} \leq \gamma^{p/2} \left(\iint_\Omega (|\nabla u| + |\nabla v|)^p \right)^{1/2} \left(\iint_\Omega |\nabla u - \nabla v|^p \right)^{1/2}.$$

We therefore obtain that

$$\|\det \nabla u - \det \nabla v\|_{L^{p/2}} = \left(\iint_\Omega |\det \nabla u - \det \nabla v|^{p/2} \right)^{2/p}$$

$$\leq \gamma \left(\iint_\Omega (|\nabla u| + |\nabla v|)^p \right)^{1/p} \left(\iint_\Omega |\nabla u - \nabla v|^p \right)^{1/p}$$

and hence the claim. ∎

Exercise 3.5.5. (i) For more details concerning this question see [32] page 158 or Example 8.6 in [32, 2nd edition]. The example given here is due to Tartar.

We have seen (Exercise 1.4.7) that the sequence $u^\nu \rightharpoonup 0$ in $W^{1,2}$. An elementary computation gives

$$\det \nabla u^\nu = -\nu \left(1 - x_2\right)^{2\nu - 1}.$$

Let us show that $\det \nabla u^\nu \rightharpoonup 0$ in L^1 does not hold. Indeed, let $\varphi \equiv 1 \in L^\infty\left(\Omega\right)$. It is not difficult to see that

$$\lim_{\nu \to \infty} \iint_\Omega \det \nabla u^\nu \left(x_1, x_2\right) \varphi \left(x_1, x_2\right) dx_1 dx_2 = -\frac{1}{2} \neq 0,$$

and thus the result.

(ii) Note first that by Rellich theorem (Theorem 1.45) we have that if $u^\nu \rightharpoonup u$ in $W^{1,p}$ then $u^\nu \to u$ in L^q, $\forall q \in [1, 2p/\left(2 - p\right))$ provided $p < 2$ and $\forall q \in [1, \infty)$ if $p = 2$ (the case $p > 2$ has already been considered in Lemma 3.23). Consequently, if $p > 4/3$, we have $u^\nu \to u$ in L^4. Therefore let

$$u^\nu = u^\nu \left(x_1, x_2\right) = \left(\varphi^\nu \left(x_1, x_2\right), \psi^\nu \left(x_1, x_2\right)\right)$$

and $v \in C_0^\infty\left(\Omega\right)$. Since

$$\det \nabla u^\nu = \varphi_{x_1}^\nu \psi_{x_2}^\nu - \varphi_{x_2}^\nu \psi_{x_1}^\nu = \left(\varphi^\nu \psi_{x_2}^\nu\right)_{x_1} - \left(\varphi^\nu \psi_{x_1}^\nu\right)_{x_2}$$

(this is allowed since $u^\nu \in C^2$) we have, after integrating by parts,

$$\iint_\Omega \det \nabla u^\nu \, v = \iint_\Omega \left(\varphi^\nu \psi_{x_1}^\nu v_{x_2} - \varphi^\nu \psi_{x_2}^\nu v_{x_1}\right).$$

However, we know that $\psi^\nu \rightharpoonup \psi$ in $W^{1,4/3}$ (since $u^\nu \rightharpoonup u$ in $W^{1,p}$ and $p > 4/3$) and $\varphi^\nu \to \varphi$ in L^4, we therefore deduce (see Exercise 1.3.3) that

$$\left(\varphi^\nu \psi_{x_1}^\nu, \varphi^\nu \psi_{x_2}^\nu\right) \rightharpoonup \left(\varphi \psi_{x_1}, \varphi \psi_{x_2}\right) \quad \text{in } L^1.$$

Passing to the limit and integrating by parts once more we get the claim, namely

$$\lim_{\nu \to \infty} \iint_\Omega \det \nabla u^\nu \, v = \iint_\Omega \left(\varphi \psi_{x_1} v_{x_2} - \varphi \psi_{x_2} v_{x_1}\right) = \iint_\Omega \det \nabla u \, v. \quad \blacksquare$$

Exercise 3.5.6. (i) Let $x = \left(x_1, x_2\right)$, we then find

$$\nabla u = \begin{pmatrix} \dfrac{x_2^2}{|x|^3} & -\dfrac{x_1 x_2}{|x|^3} \\ -\dfrac{x_1 x_2}{|x|^3} & \dfrac{x_1^2}{|x|^3} \end{pmatrix} \quad \Rightarrow \quad |\nabla u|^2 = \frac{1}{|x|^2}.$$

We therefore deduce (see Exercise 1.4.1) that $u \in L^\infty$ and $u \in W^{1,p}$ provided $p \in [1, 2)$, but $u \notin W^{1,2}$ and $u \notin C^0$.

(ii) Since

$$\iint_\Omega |u^\nu(x) - u(x)|^q \, dx = 2\pi \int_0^1 \frac{r}{(\nu r + 1)^q} dr = \frac{2\pi}{\nu^2} \int_1^{\nu+1} \frac{s-1}{s^q} ds$$

we deduce that $u^\nu \to u$ in L^q, for every $q \geq 1$; however, the convergence $u^\nu \to u$ in L^∞ does not hold. We next show that $u^\nu \rightharpoonup u$ in $W^{1,p}$ if $p \in [1,2)$. We readily have

$$\nabla u^\nu = \frac{1}{|x|(|x|+1/\nu)^2} \begin{pmatrix} x_2^2 + \dfrac{|x|}{\nu} & -x_1 x_2 \\ -x_1 x_2 & x_1^2 + \dfrac{|x|}{\nu} \end{pmatrix}$$

and thus

$$|\nabla u^\nu| = \frac{\left(|x|^2 + \dfrac{2|x|}{\nu} + \dfrac{2}{\nu^2}\right)^{1/2}}{(|x|+1/\nu)^2}.$$

We therefore find, if $1 \leq p < 2$, that, γ denoting a constant independent of ν,

$$\iint_\Omega |\nabla u^\nu|^p \, dx_1 dx_2 = 2\pi \int_0^1 \frac{\left((r+1/\nu)^2 + 1/\nu^2\right)^{p/2}}{(r+1/\nu)^{2p}} r \, dr$$

$$\leq 2\pi \int_0^1 \frac{2^{p/2}(r+1/\nu)^p}{(r+1/\nu)^{2p}} r \, dr = 2^{(2+p)/2} \pi \nu^p \int_0^1 \frac{r \, dr}{(\nu r + 1)^p}$$

$$\leq 2^{(2+p)/2} \pi \nu^{p-2} \int_1^{\nu+1} \frac{(s-1)\, ds}{s^p} \leq \gamma.$$

This implies, according to Exercise 1.4.6, that $u^\nu \rightharpoonup u$ in $W^{1,p}$, as claimed.

(iii) A direct computation gives

$$\det \nabla u^\nu = |\det \nabla u^\nu| = \frac{1}{\nu(|x|+1/\nu)^3}$$

and hence

$$\iint_\Omega |\det \nabla u^\nu| \, dx_1 dx_2 = 2\pi\nu^2 \int_0^1 \frac{r dr}{(\nu r + 1)^3} = 2\pi \int_1^{\nu+1} \frac{(s-1)\, ds}{s^3}$$

$$= 2\pi \left[\frac{1}{2s^2} - \frac{1}{s}\right]_1^{\nu+1}.$$

We therefore have

$$\lim_{\nu \to \infty} \iint_\Omega |\det \nabla u^\nu| = \pi. \tag{7.15}$$

Let $\epsilon > 0$ be arbitrary and small. Observe that if

$$\Omega_\epsilon = \left\{ x \in \mathbb{R}^2 : |x| < \epsilon \right\},$$

then

$$\det \nabla u^\nu = |\det \nabla u^\nu| \to 0 \quad \text{in } L^\infty \left(\Omega \setminus \Omega_\epsilon \right). \tag{7.16}$$

Let $\varphi \in C_0^\infty \left(\Omega \right)$. We then have, for every $x \in \Omega_\epsilon$,

$$|\varphi(x) - \varphi(0)| \le \|\nabla \varphi\|_{L^\infty} |x| \le \|\nabla \varphi\|_{L^\infty} \epsilon. \tag{7.17}$$

We then combine (7.15), (7.16) and (7.17) to get the result. Indeed, let $\varphi \in C_0^\infty \left(\Omega \right)$ and obtain

$$\iint_\Omega \det \nabla u^\nu \varphi \, dx = \varphi(0) \iint_\Omega \det \nabla u^\nu dx + \iint_{\Omega_\epsilon} \det \nabla u^\nu \left(\varphi(x) - \varphi(0) \right) dx$$
$$+ \iint_{\Omega \setminus \Omega_\epsilon} \det \nabla u^\nu \left(\varphi(x) - \varphi(0) \right) dx.$$

This leads to the following estimate

$$\left| \iint_\Omega \left(\det \nabla u^\nu \varphi \right) - \varphi(0) \iint_\Omega \det \nabla u^\nu \right|$$
$$\le \|\nabla \varphi\|_{L^\infty} \epsilon \iint_\Omega |\det \nabla u^\nu| + 2 \|\varphi\|_{L^\infty} \iint_{\Omega \setminus \Omega_\epsilon} |\det \nabla u^\nu|.$$

Keeping ϵ fixed, we let $\nu \to \infty$ and obtain

$$\limsup_{\nu \to \infty} \left| \iint_\Omega \left(\det \nabla u^\nu \varphi \right) - \pi \varphi(0) \right| \le \pi \|\nabla \varphi\|_{L^\infty} \epsilon,$$

ϵ being arbitrary, we have indeed obtained the result. ∎

Exercise 3.5.7. (i) Start by observing that, in view of Exercise 3.5.2, we have

$$\int_\Omega \det \left(\xi + \nabla \varphi(x) \right) dx = \det \left(\xi \right) \operatorname{meas} \Omega$$

for every $\xi \in \mathbb{R}^{2 \times 2}$ and every $\varphi \in W_0^{1,\infty} \left(\Omega; \mathbb{R}^2 \right)$. We also trivially have

$$\int_\Omega \left(\xi + \nabla \varphi(x) \right) dx = \xi \operatorname{meas} \Omega.$$

We thus deduce from Jensen inequality that, for every $\xi \in \mathbb{R}^{2 \times 2}$ and every $\varphi \in W_0^{1,\infty} \left(\Omega; \mathbb{R}^2 \right)$,

$$\int_\Omega f \left(\xi + \nabla \varphi(x) \right) dx = \int_\Omega F \left(\xi + \nabla \varphi(x), \det \left(\xi + \nabla \varphi(x) \right) \right) dx$$
$$\ge F \left(\xi, \det \xi \right) \operatorname{meas} \Omega = f \left(\xi \right) \operatorname{meas} \Omega$$

as claimed.

(ii) *Step 1.* We start with a preliminary computation. Let $t \in \mathbb{R}$, $\xi \in \mathbb{R}^{2\times2}$ and $\lambda \in \mathbb{R}^{2\times2}$ with $\det \lambda = 0$. Observe first that

$$\det(\xi + t\lambda) = \det(\xi) + t\left\langle \tilde{\lambda}; \xi \right\rangle$$

where we have denoted

$$\left\langle \tilde{\lambda}; \xi \right\rangle = \lambda_1^1 \xi_2^2 + \lambda_2^2 \xi_1^1 - \lambda_2^1 \xi_1^2 - \lambda_1^2 \xi_2^1 .$$

The above observation leads immediately to the following claim. For every $t, s \in \mathbb{R}$, $\xi, \lambda \in \mathbb{R}^{2\times2}$ with $\det \lambda = 0$ and $\theta \in [0, 1]$

$$\xi + (\theta t + (1 - \theta) s) \lambda = \theta (\xi + t\lambda) + (1 - \theta)(\xi + s\lambda)$$

$$\det[\xi + (\theta t + (1 - \theta) s) \lambda] = \theta \det(\xi + t\lambda) + (1 - \theta)\det(\xi + s\lambda)$$

or in other words

$$(\xi + (\theta t + (1 - \theta) s) \lambda, \det[\xi + (\theta t + (1 - \theta) s) \lambda])$$
$$= \theta(\xi + t\lambda, \det(\xi + t\lambda)) + (1 - \theta)(\xi + s\lambda, \det(\xi + s\lambda)).$$

Step 2. We have to show that, for every $t, s \in \mathbb{R}$ and $\theta \in [0, 1]$,

$$\psi(\theta t + (1 - \theta) s) \le \theta \psi(t) + (1 - \theta) \psi(s).$$

This is indeed the case in view of Step 1 and the convexity of $F : \mathbb{R}^{2\times2} \times \mathbb{R} \to \mathbb{R}$, since

$$\psi(t) = F(\xi + t\lambda, \det[\xi + t\lambda])$$
$$\psi(s) = F(\xi + s\lambda, \det[\xi + s\lambda])$$

$$\psi(\theta t + (1 - \theta) s) = F(\xi + (\theta t + (1 - \theta) s) \lambda, \det[\xi + (\theta t + (1 - \theta) s) \lambda])$$
$$= F(\theta(\xi + t\lambda, \det(\xi + t\lambda)) + (1 - \theta)(\xi + s\lambda, \det(\xi + s\lambda))).$$

(iii) Since $f \in C^2(\mathbb{R}^{2\times2})$, we have that $\psi \in C^2(\mathbb{R})$ for every $\xi, \lambda \in \mathbb{R}^{2\times2}$ with $\det \lambda = 0$. We have just proved that ψ is convex and thus

$$\psi''(t) = \sum_{i,j,\alpha,\beta=1}^{2} \frac{\partial^2 f(\xi + t\lambda)}{\partial \xi_i^\alpha \partial \xi_j^\beta} \lambda_i^\alpha \lambda_j^\beta \ge 0.$$

Choosing $t = 0$ and

$$\lambda = \begin{pmatrix} a_1 b^1 & a_2 b^1 \\ a_1 b^2 & a_2 b^2 \end{pmatrix}$$

we have the claim. ∎

7.3.5 Relaxation theory

Exercise 3.6.1. (i) Let

$$\overline{u}\left(x\right) = \begin{cases} ax + \alpha & \text{if } x \in [0, \lambda] \\ b\left(x - 1\right) + \beta & \text{if } x \in [\lambda, 1] . \end{cases}$$

Note that $\overline{u}\left(0\right) = \alpha$, $\overline{u}\left(1\right) = \beta$ and $\overline{u}$ is continuous at $x = \lambda$ since $\beta - \alpha = \lambda a + \left(1 - \lambda\right) b$, hence $\overline{u} \in X$. Since $f^{**} \leq f$ and f^{**} is convex, we have, appealing to Jensen inequality, that, for any $u \in X$,

$$I\left(u\right) = \int_0^1 f\left(u'\left(x\right)\right) dx \geq \int_0^1 f^{**}\left(u'\left(x\right)\right) dx$$

$$\geq f^{**}\left(\int_0^1 u'\left(x\right) dx\right) = f^{**}\left(\beta - \alpha\right) = \lambda f\left(a\right) + \left(1 - \lambda\right) f\left(b\right) = I\left(\overline{u}\right).$$

Hence $\overline{u}$ is a minimizer of (P).

(ii) The preceding result does not apply to $f\left(\xi\right) = e^{-\xi^2}$ and $\alpha = \beta = 0$. Indeed, we have $f^{**}\left(\xi\right) \equiv 0$ and we therefore cannot find $\lambda \in [0, 1]$, $a, b \in \mathbb{R}$ so that

$$\begin{cases} \lambda a + \left(1 - \lambda\right) b = 0 \\ \lambda e^{-a^2} + \left(1 - \lambda\right) e^{-b^2} = 0. \end{cases}$$

In fact, we should need $a = -\infty$ and $b = +\infty$. Recall that in Section 2.2 we already saw that (P) has no minimizer.

(iii) If $f\left(\xi\right) = \left(\xi^2 - 1\right)^2$, we then find

$$f^{**}\left(\xi\right) = \begin{cases} \left(\xi^2 - 1\right)^2 & \text{if } |\xi| \geq 1 \\ 0 & \text{if } |\xi| < 1. \end{cases}$$

Therefore, if $|\beta - \alpha| \geq 1$, choose in (i)

$$\lambda = 1/2 \quad \text{and} \quad a = b = \beta - \alpha.$$

However, if $|\beta - \alpha| < 1$, choose

$$a = 1, \quad b = -1 \quad \text{and} \quad \lambda = \left(1 + \beta - \alpha\right)/2.$$

In conclusion, in both cases, we find that $\overline{u}$ (defined in (i)) is a minimizer of (P). ∎

Exercise 3.6.2. Note that if

$$(\overline{P}) \quad \inf\left\{\overline{I}\left(u\right) = \int_\Omega f^{**}\left(\nabla u\left(x\right)\right) dx : u \in u_0 + W_0^{1,p}\left(\Omega\right)\right\} = \overline{m},$$

then, by Jensen inequality, we have for every $u \in u_0 + W_0^{1,p}(\Omega)$

$$\int_\Omega f^{**}(\nabla u(x)) \, dx \geq \operatorname{meas} \Omega \, f^{**}\left(\frac{1}{\operatorname{meas} \Omega} \int_\Omega \nabla u(x) \, dx\right) = \operatorname{meas} \Omega \, f^{**}(\xi_0).$$

The above computation, coupled with the relaxation theorem (Theorem 3.28), leads to

$$m = \overline{m} = \overline{I}(u_0) = f^{**}(\xi_0) \operatorname{meas} \Omega. \quad \blacksquare$$

Exercise 3.6.3. Let

$$h(\xi) = \alpha_2 |\xi|^p + \alpha_3.$$

Since $\alpha_2 \geq 0$ (in fact $\alpha_2 > 0$), we deduce that h is convex. By hypothesis $h \leq f$ and thus, by definition of f^{**}, we have

$$f(x, u, \xi) \geq f^{**}(x, u, \xi) \geq h(\xi), \quad \forall (x, u, \xi) \in \overline{\Omega} \times \mathbb{R} \times \mathbb{R}^n. \tag{7.18}$$

We can therefore apply Theorem 3.3 to f^{**} to show that $(\overline{P})$ has at least one minimizer $\overline{u} \in u_0 + W_0^{1,p}(\Omega)$.

By Theorem 3.28 (i) we have that there exists a sequence $u_\nu \in u_0 + W_0^{1,p}(\Omega)$ so that

$$u_\nu \to \overline{u} \text{ in } L^p \quad \text{and} \quad I(u_\nu) \to \overline{I}(\overline{u}), \text{ as } \nu \to \infty.$$

From (7.18) and Poincaré inequality we deduce that $\|u_\nu\|_{W^{1,p}}$ is uniformly bounded. Since $p > 1$, we obtain, according to Exercise 1.4.6,

$$u_\nu \rightharpoonup \overline{u} \text{ in } W^{1,p} \quad \text{and} \quad I(u_\nu) \to \overline{I}(\overline{u}), \text{ as } \nu \to \infty. \quad \blacksquare$$

Exercise 3.6.4. If we set $\xi = (\xi_1, \xi_2)$, we easily have

$$f^{**}(\xi) = \begin{cases} f(\xi) & \text{if } |\xi_1| \geq 1 \\ (\xi_2)^4 & \text{if } |\xi_1| < 1 \end{cases}$$

and thus $f^{**}(0) = 0$. Applying Exercise 3.6.2, we get that $m = 0$. However, according to Exercise 1.4.15, no function $\overline{u} \in W_0^{1,4}(\Omega)$ can satisfy $I(\overline{u}) = 0$, which is equivalent to

$$|\overline{u}_{x_1}| = 1 \quad \text{and} \quad |\overline{u}_{x_2}| = 0 \text{ a.e. in } \Omega;$$

hence (P) has no minimizer. $\quad \blacksquare$

Exercise 3.6.5. It is easy to see that

$$f^{**}(\xi) = \begin{cases} f(\xi) & \text{if } |\xi_1|, |\xi_2| \geq 1 \\ \left((\xi_1)^2 - 1\right)^2 & \text{if } |\xi_2| < 1 \leq |\xi_1| \\ \left((\xi_2)^2 - 1\right)^2 & \text{if } |\xi_1| < 1 \leq |\xi_2| \\ 0 & \text{if } |\xi_1|, |\xi_2| < 1 \end{cases}$$

and thus $f^{**}(0) = 0$. We therefore deduce, from Exercise 3.6.2, that $m = 0$. By construction $\overline{u} \in W_0^{1,\infty} \subset W_0^{1,4}$ and satisfies, for almost every $(x_1, x_2) \in \Omega$,

$$|\overline{u}_{x_1}| = |\overline{u}_{x_2}| = 1.$$

We therefore have, for almost every $(x_1, x_2) \in \Omega$,

$$f(\nabla\overline{u}) = f^{**}(\nabla\overline{u}) = 0$$

and thus $\overline{u}$ is a minimizer of (P). $\blacksquare$

Exercise 3.6.6. Note first that $f^{**}(\xi) \equiv 0$ and thus, according to Exercise 3.6.2, $m = 0$. Observe then that any sequence $\{u_\nu\}_{\nu=1}^\infty \subset W_0^{1,1}(0,1)$ such that

$$\int_0^1 f(u_\nu'(x)) \, dx \to 0, \quad \text{as } \nu \to \infty$$

also satisfies by Jensen inequality

$$0 \le e^{-\int_0^1 |u_\nu'|} \le \int_0^1 e^{-|u_\nu'|} = \int_0^1 f(u_\nu') \to 0$$

and therefore must satisfy

$$\int_0^1 |u_\nu'| \to \infty.$$

Thus it cannot converge weakly to any u in $W^{1,1}$. Note that the sequence defined on each interval $[k/\nu, (k+1)/\nu]$, $0 \le k \le \nu - 1$, by

$$u_\nu(x) = \begin{cases} \sqrt{\nu}\left(x - \frac{k}{\nu}\right) & \text{if } x \in \left[\frac{2k}{2\nu}, \frac{2k+1}{2\nu}\right] \\ \sqrt{\nu}\left(\frac{k+1}{\nu} - x\right) & \text{if } x \in \left(\frac{2k+1}{2\nu}, \frac{2k+2}{2\nu}\right] \end{cases}$$

satisfies $u_\nu \in W_0^{1,1}(0,1)$ with $|u_\nu'| = \sqrt{\nu}$ a.e., $|u_\nu| \le 1/2\sqrt{\nu}$ and thus, as $\nu \to \infty$,

$$u_\nu \to u = 0 \text{ in } L^\infty(0,1) \quad \text{and} \quad \int_0^1 f(u_\nu') = e^{-\sqrt{\nu}} \to 0. \quad \blacksquare$$

Exercise 3.6.7. We refer for more details to Marcellini-Sbordone [82], from where example (ii) below is taken; see also Section 9.2.2 in [32, 2nd edition].

(i) Let $(x_\nu, u_\nu) \to (x, u)$. From Theorem 1.59, we deduce that

$$f^{**}(x_\nu, u_\nu, \xi) \le \sum_{i=1}^{n+1} \lambda_i f(x_\nu, u_\nu, \xi_i) \quad \text{for every } (\lambda_i, \xi_i) \in \Lambda_\xi$$

where

$$\Lambda_\xi = \left\{ (\lambda_1, \xi_1), \cdots, (\lambda_{n+1}, \xi_{n+1}) \in [0,1] \times \mathbb{R}^n : \sum_{i=1}^{n+1} \lambda_i (1, \xi_i) = (1, \xi) \right\}.$$

Since f is continuous, we get

$$\limsup_{\nu \to \infty} f^{**} (x_\nu, u_\nu, \xi) \leq \sum_{i=1}^{n+1} \lambda_i f (x, u, \xi_i) \quad \text{for every } (\lambda_i, \xi_i) \in \Lambda_\xi.$$

Taking the infimum on all possible choices of $(\lambda_i, \xi_i) \in \Lambda_\xi$, we obtain, appealing once more to Theorem 1.59, that

$$\limsup_{\nu \to \infty} f^{**} (x_\nu, u_\nu, \xi) \leq f^{**} (x, u, \xi)$$

as wished.

(ii) Observe that when $|u| \geq 1$ the function

$$\xi \to f (u, \xi) = (|\xi| + 1)^{|u|}$$

is convex. If $|u| < 1$, we have, by Theorem 1.59, for any $\nu \geq 1$,

$$1 \leq f^{**} (u, \xi) = f^{**} (u, |\xi|) \leq \frac{|\xi|}{\nu + |\xi|} f (u, \nu + |\xi|) + \frac{\nu}{\nu + |\xi|} f (u, 0).$$

Letting $\nu \to \infty$, we have that the right-hand side, since $|u| < 1$, tends to 1. We therefore find

$$f^{**} (u, \xi) = \begin{cases} f (u, \xi) & \text{if } |u| \geq 1 \\ 1 & \text{if } |u| < 1 \end{cases}$$

and that, for any $\xi \neq 0$, $u \to f^{**} (u, \xi)$ is not continuous at $|u| = 1$, but only upper semicontinuous. ∎

7.4 Chapter 4. Direct methods: regularity

7.4.1 The one dimensional case

Exercise 4.2.1. **(i)** We first show that $\overline{u} \in W^{2,\infty} (a, b)$, by proving (iii) of Theorem 1.37. Observe that from (H_1') and the fact that $\overline{u} \in W^{1,\infty} (a, b)$, we can find a constant $\gamma_1 > 0$ such that, for every $z \in \mathbb{R}$ with $|z| \leq \|\overline{u}'\|_{L^\infty}$,

$$f_{\xi\xi} (x, \overline{u} (x), z) \geq \gamma_1 > 0, \quad \forall x \in [a, b]. \tag{7.19}$$

We have to prove that we can find a constant $\gamma > 0$ so that

$$|\overline{u}'(x+h) - \overline{u}'(x)| \le \gamma |h|, \quad \text{a.e. } x \in \omega$$

for every open set $\omega \subset \overline{\omega} \subset (a, b)$ and every $h \in \mathbb{R}$ satisfying $|h| < \text{dist}(\omega, (a, b)^c)$. Using (7.19) we have

$$\gamma_1 |\overline{u}'(x+h) - \overline{u}'(x)| \le \left| \int_{\overline{u}'(x)}^{\overline{u}'(x+h)} f_{\xi\xi}(x, \overline{u}(x), z)\, dz \right|$$

$$\le |f_\xi(x, \overline{u}(x), \overline{u}'(x+h)) - f_\xi(x, \overline{u}(x), \overline{u}'(x))|$$

and thus

$$\gamma_1 |\overline{u}'(x+h) - \overline{u}'(x)|$$
$$\le |f_\xi(x+h, \overline{u}(x+h), \overline{u}'(x+h)) - f_\xi(x, \overline{u}(x), \overline{u}'(x))|$$
$$+ |f_\xi(x, \overline{u}(x), \overline{u}'(x+h)) - f_\xi(x+h, \overline{u}(x+h), \overline{u}'(x+h))|.$$

Now let us evaluate both terms on the right-hand side of the inequality. Since we know from Lemma 4.3 that $x \to f_u(x, \overline{u}(x), \overline{u}'(x))$ is in $L^\infty(a, b)$ and the equation

$$\frac{d}{dx}[f_\xi(x, \overline{u}, \overline{u}')] = f_u(x, \overline{u}, \overline{u}'), \quad \text{a.e. } x \in (a, b)$$

holds, we deduce that $x \to \varphi(x) = f_\xi(x, \overline{u}(x), \overline{u}'(x))$ is in $W^{1,\infty}(a, b)$. Therefore, applying Theorem 1.37 to φ, we can find a constant $\gamma_2 > 0$, such that

$$|\varphi(x+h) - \varphi(x)| = |f_\xi(x+h, \overline{u}(x+h), \overline{u}'(x+h)) - f_\xi(x, \overline{u}(x), \overline{u}'(x))|$$
$$\le \gamma_2 |h|.$$

Similarly, since $\overline{u} \in W^{1,\infty}$ and $f \in C^\infty$, we can find constants $\gamma_3, \gamma_4 > 0$, such that

$$|f_\xi(x, \overline{u}(x), \overline{u}'(x+h)) - f_\xi(x+h, \overline{u}(x+h), \overline{u}'(x+h))|$$
$$\le \gamma_3(|h| + |\overline{u}(x+h) - \overline{u}(x)|) \le \gamma_4 |h|.$$

Combining these two inequalities we find

$$|\overline{u}'(x+h) - \overline{u}'(x)| \le \frac{\gamma_2 + \gamma_4}{\gamma_1} |h|$$

as wished; thus $\overline{u} \in W^{2,\infty}(a, b)$.

   (ii) Since $\overline{u} \in W^{2,\infty}(a, b)$, we find that, for almost every $x \in (a, b)$,

$$\frac{d}{dx}[f_\xi(x, \overline{u}, \overline{u}')] = f_{\xi\xi}(x, \overline{u}, \overline{u}')\overline{u}'' + f_{u\xi}(x, \overline{u}, \overline{u}')\overline{u}' + f_{x\xi}(x, \overline{u}, \overline{u}')$$
$$= f_u(x, \overline{u}, \overline{u}').$$

Since (H_1') holds and $\overline{u} \in C^1([a,b])$, we deduce that there exists $\gamma_5 > 0$ such that

$$f_{\xi\xi}(x, \overline{u}(x), \overline{u}'(x)) \geq \gamma_5 > 0, \quad \forall x \in [a,b].$$

The equation can then be rewritten as

$$\overline{u}'' = \frac{f_u(x, \overline{u}, \overline{u}') - f_{x\xi}(x, \overline{u}, \overline{u}') - f_{u\xi}(x, \overline{u}, \overline{u}')\overline{u}'}{f_{\xi\xi}(x, \overline{u}, \overline{u}')}$$

and hence $\overline{u} \in C^2([a,b])$. Returning to the equation we find that the right-hand side is then C^1, and hence $\overline{u} \in C^3$. Iterating the process we conclude that $\overline{u} \in C^\infty([a,b])$, as claimed. ∎

Exercise 4.2.2. Note that

$$\overline{u}'(x) = |x|^{-\frac{2}{7}} x \quad \text{and} \quad (\overline{u}')^7 = x^5.$$

We therefore have $\overline{u} \in C^1([-1,1])$, more precisely $\overline{u} \in C^{1,5/7}$ and $(\overline{u}')^7 \in C^\infty$, but $\overline{u} \notin C^2$. Moreover, $\overline{u}$ satisfies the Euler-Lagrange equation associated with (P_1) and (P_2), namely

$$\left((\overline{u}')^7\right)' = 5x^4.$$

From Theorem 3.3 we deduce that (P_1) and (P_2) have at least one minimizer and from Exercise 3.3.4 that it is unique. From Theorem 3.11 we get that this minimizer is $\overline{u}$. ∎

Exercise 4.2.3. **(i)** We have

$$\overline{u}' = |x|^{\frac{p}{p-q}-2} x \quad \text{and} \quad \overline{u}'' = \left(\frac{p}{p-q} - 1\right)|x|^{\frac{p}{p-q}-2} = \frac{q}{p-q}|x|^{\frac{2q-p}{p-q}}$$

which implies, since $p > 2q > 2$, that $\overline{u} \in C^1([-1,1])$ but $\overline{u} \notin C^2([-1,1])$.

(ii) We find that

$$|\overline{u}'|^{p-2}\overline{u}' = |x|^{\frac{q(p-2)}{p-q}}|x|^{\frac{2q-p}{p-q}} x = |x|^{\frac{p(q-1)}{p-q}} x$$

$$|\overline{u}|^{q-2}\overline{u} = \left(\frac{p-q}{p}\right)^{q-1}|x|^{\frac{p(q-1)}{p-q}}.$$

If we choose, for instance, $\frac{p(q-1)}{p-q} = 4$ (which is realized, for example, if $p = 16$ and $q = 4$), then

$$|\overline{u}'|^{p-2}\overline{u}', \ |\overline{u}|^{q-2}\overline{u} \in C^\infty([-1,1])$$

although $\overline{u} \notin C^2([-1,1])$.

(iii) Since the function $(u, \xi) \to f(u, \xi)$ is strictly convex and satisfies all the hypotheses of Theorem 3.3 and Theorem 3.11, we have that (P) has a unique minimizer and that it should be the solution of the Euler-Lagrange equation

$$\left(|u'|^{p-2} u'\right)' = \lambda |u|^{q-2} u.$$

A direct computation shows that, indeed, $\bar{u}$ is a solution of this equation and therefore it is the unique minimizer of (P). ∎

Exercise 4.2.4. 1) We start by proving that $\bar{u} \in W_0^{1,\infty}(-1, 1)$. Since $\bar{u}(-1) = \bar{u}(1) = 0$, we only need (according to Remark 1.38) to prove that $\bar{u} \in C^{0,1}([-1, 1])$, i.e. $\bar{u}$ is Lipschitz on $[-1, 1]$. Let $-1 \le x < y \le 1$ and let us examine all different possibilities.

Case 1: $0 = x < y \le 1$ (and similarly for $-1 \le x < y = 0$). We have

$$|\bar{u}(x) - \bar{u}(y)| = |\bar{u}(y)| = |y^2 \sin(\pi/y)| \le |y| = |y - x|.$$

Case 2: $-1 \le x < 0 < y \le 1$. We get

$$|\bar{u}(x) - \bar{u}(y)| \le |x^2 \sin(\pi/x)| + |y^2 \sin(\pi/y)| \le |x| + |y| = y - x = |y - x|.$$

Case 3: $0 < x < y \le 1$ (and similarly for $-1 \le x < y < 0$). Observe then that $\bar{u} \in C^1([x, y])$ and thus, for any $z \in [x, y] \subset (0, 1]$,

$$|\bar{u}'(z)| = |2z \sin(\pi/z) - \pi \cos(\pi/z)| \le 2 + \pi.$$

We have therefore obtained the claim, namely

$$|\bar{u}(y) - \bar{u}(x)| = \left| \int_0^1 \frac{d}{dt} \bar{u}(x + t(y - x)) \, dt \right| \le (2 + \pi) |y - x|.$$

2) It is clear that $\bar{u} \notin C^1([-1, 1])$ and not even $C^1_{\text{piec}}([-1, 1])$, in the sense of Definition 1.6, since $\lim_{x \to 0\pm} \bar{u}'(x)$ does not exist.

3) For $x \ne 0$, we have

$$f(x, \bar{u}'(x)) \equiv 0.$$

Since $\bar{u} \in W_0^{1,\infty}(-1, 1)$ and $f \ge 0$, we therefore get that

$$\inf(P) = \int_0^1 f(x, \bar{u}'(x)) \, dx = 0.$$

To prove the uniqueness, we just observe that any minimizer should then necessarily satisfy

$$f(x, u'(x)) = 0, \quad \text{a.e. in } (-1, 1)$$

which implies that

$$u'(x) = 2x \sin(\pi/x) - \pi \cos(\pi/x) = \bar{u}'(x), \quad \text{a.e. in } (-1, 1).$$

Thus, uniqueness is established. ∎

7.4.2 The difference quotient method: interior regularity

Exercise 4.3.1. (i) We require that $G : \mathbb{R}_+ \to \mathbb{R}_+$ is C^2, convex and there exist constants $g_1, g_2, g_3 > 0$ such that, for every $t \geq 0$,

$$0 < g_1 \leq G'(t) \leq g_2 \quad \text{and} \quad 0 \leq G''(t) t \leq g_3.$$

Then the function $g(\xi) = G\left(|\xi|^2\right)$ satisfies all the hypotheses of Theorem 4.9, since

$$g_{\xi_i} = 2G'\left(|\xi|^2\right) \xi_i \quad \text{and} \quad g_{\xi_i \xi_j} = 2G'\left(|\xi|^2\right) \delta_{ij} + 4G''\left(|\xi|^2\right) \xi_i \xi_j.$$

In particular, the function $G(t) = t + \epsilon(1+t)^{-1}$ satisfies all the hypotheses, provided $0 \leq \epsilon < 1$, so that we have

$$g(\xi) = G\left(|\xi|^2\right) = |\xi|^2 + \frac{\epsilon}{1 + |\xi|^2}.$$

(ii) Let $g_{ij} = g_{ji} \in C^1\left(\overline{\Omega}\right)$ and there exists $\alpha > 0$ such that

$$\sum_{i,j=1}^{n} g_{ij}(x) \xi_i \xi_j \geq \alpha |\xi|^2, \quad \forall (x, \xi) \in \overline{\Omega} \times \mathbb{R}^n$$

then all the hypotheses of Theorem 4.9 are verified.

(iii) We now consider the case

$$g(\xi) = \frac{1}{2} |\xi|^2 + G(\langle a; \xi \rangle).$$

We have

$$g_{\xi_i} = \xi_i + G'(\langle a; \xi \rangle) a_i \quad \text{and} \quad g_{\xi_i \xi_j} = \delta_{ij} + G''(\langle a; \xi \rangle) a_i a_j$$

and therefore the hypotheses of the theorem are satisfied if we assume that there exists a constant $\alpha > 0$ such that, for every $t \in \mathbb{R}$,

$$0 \leq G''(t) \leq \alpha. \quad \blacksquare$$

Exercise 4.3.2. Let $\rho \in C_0^\infty(\Omega)$ be such that

$$0 \leq \rho \leq 1 \quad \text{and} \quad \rho \equiv 1 \text{ in } O$$

and choose $\varphi = \rho^2 \overline{u} \in W_0^{1,2}(\Omega)$ in the equation so that

$$\int_\Omega h \, \rho^2 \overline{u} = \int_\Omega \left\langle \nabla_\xi g(x, \nabla \overline{u}) ; \nabla(\rho^2 \overline{u}) \right\rangle$$

$$= \int_\Omega \left\langle \nabla_\xi g(x, \nabla \overline{u}) ; \rho^2 \, \nabla \overline{u} \right\rangle + 2 \int_\Omega \left\langle \nabla_\xi g(x, \nabla \overline{u}) ; \rho \, \overline{u} \nabla \rho \right\rangle.$$

In the following, all L^2 norms are understood in $L^2(\Omega)$ unless stated otherwise. Use the estimates $\langle \nabla_\xi g(x,\xi) ; \xi \rangle \geq \alpha |\xi|^2$ and $|\nabla_\xi g(x,\xi)| \leq \beta |\xi|$, recalling that $0 \leq \rho \leq 1$, to get

$$\alpha \|\rho \nabla \overline{u}\|_{L^2}^2 = \int_\Omega \alpha \rho^2 |\nabla \overline{u}|^2 \leq \int_\Omega h \rho^2 \overline{u} - 2 \int_\Omega \langle \nabla_\xi g(x, \nabla \overline{u}) ; \rho \overline{u} \nabla \rho \rangle$$

$$\leq \|\overline{u}\|_{L^2} \|h\|_{L^2} + 2 \int_\Omega |\rho \nabla_\xi g(x, \nabla \overline{u})| \, |\overline{u} \nabla \rho|$$

$$\leq \|\overline{u}\|_{L^2} \|h\|_{L^2} + 2\beta \|\rho \nabla \overline{u}\|_{L^2} \|\overline{u} \nabla \rho\|_{L^2} .$$

Fix $\epsilon > 0$. We can therefore find $\gamma_1 = \gamma_1(O, \Omega, \alpha, \beta) > 0$ so that

$$\|\rho \nabla \overline{u}\|_{L^2}^2 \leq \gamma_1 \left(\|\overline{u}\|_{L^2}^2 + \|h\|_{L^2}^2 + \epsilon \|\rho \nabla \overline{u}\|_{L^2}^2 + \frac{1}{\epsilon} \|\overline{u}\|_{L^2}^2 \right).$$

Choosing ϵ sufficiently small we have indeed obtained

$$\|\nabla \overline{u}\|_{L^2(O)} \leq \|\rho \nabla \overline{u}\|_{L^2(\Omega)} \leq \gamma \left(\|\overline{u}\|_{L^2(\Omega)} + \|h\|_{L^2(\Omega)} \right). \quad \blacksquare$$

Exercise 4.3.3. Recall that the equation is

$$\int_\Omega \langle \nabla_\xi g(x, \nabla \overline{u}) ; \nabla \varphi \rangle \, dx = \int_\Omega h \varphi \, dx, \quad \forall \varphi \in W_0^{1,2}(\Omega)$$

and we already have the estimate (4.17), namely

$$\|\overline{u}\|_{W^{2,2}(O)} \leq \gamma \left(\|\overline{u}\|_{W^{1,2}(\Omega)} + \|h\|_{L^2(\Omega)} \right). \tag{7.20}$$

Recall also (4.20), namely (with our hypothesis $\nabla_\xi g(x,0) = 0$)

$$\langle \nabla_\xi g(x,\xi) ; \xi \rangle = \langle \nabla_\xi g(x,\xi) - \nabla_\xi g(x,0) ; \xi \rangle \geq \alpha_5 |\xi|^2, \quad \forall (x,\xi) \in \overline{\Omega} \times \mathbb{R}^n.$$

Note also that (G_2) and the fact that $\nabla_\xi g(x,0) = 0$ imply that there exists $\beta > 0$ such that

$$|\nabla_\xi g(x,\xi)| \leq \beta |\xi|, \quad \forall (x,\xi) \in \overline{\Omega} \times \mathbb{R}^n.$$

(i) Let $O \subset \overline{O} \subset \Lambda \subset \overline{\Lambda} \subset \Omega$ be open sets. The very same argument of Theorem 4.9 (or of Theorem 4.8) that leads to (4.17) gives

$$\|\overline{u}\|_{W^{2,2}(O)} \leq \gamma_1 \left(\|\overline{u}\|_{W^{1,2}(\Lambda)} + \|h\|_{L^2(\Lambda)} \right).$$

Apply Exercise 4.3.2 to get

$$\|\nabla \overline{u}\|_{L^2(\Lambda)} \leq \gamma_2 \left(\|\overline{u}\|_{L^2(\Omega)} + \|h\|_{L^2(\Omega)} \right).$$

Finally, combine the above two inequalities to find the result.

(ii) Choosing $\varphi = \overline{u} \in W_0^{1,2}(\Omega)$, we find for $\epsilon > 0$ fixed

$$\alpha_5 \|\nabla \overline{u}\|_{L^2}^2 \leq \int_\Omega \langle \nabla_\xi g(x, \nabla \overline{u}); \nabla \overline{u} \rangle \, dx = \int_\Omega h\, \overline{u} \leq \epsilon \|\overline{u}\|_{W^{1,2}}^2 + \frac{1}{\epsilon} \|h\|_{L^2}^2 \, ,$$

where, unless stated otherwise, all L^2 norms are understood in $L^2(\Omega)$. Invoking Poincaré inequality, we have that there exists $\theta > 0$ such that

$$\theta \|\overline{u}\|_{W^{1,2}}^2 \leq \|\nabla \overline{u}\|_{L^2}^2 \, .$$

Combining the two inequalities we find, choosing ϵ small enough, that

$$\|\overline{u}\|_{W^{1,2}}^2 \leq \frac{1}{\epsilon(\alpha_5 \theta - \epsilon)} \|h\|_{L^2}^2 \, . \tag{7.21}$$

This estimate, together with (7.20), leads to the existence of $\gamma > 0$ such that

$$\|\overline{u}\|_{W^{2,2}(O)} \leq \gamma \|h\|_{L^2(\Omega)} \, . \quad \blacksquare$$

7.4.3 The difference quotient method: boundary regularity

Exercise 4.4.1. We change variables and set

$$x = H(y), \quad u(x) = v\left(H^{-1}(x)\right), \quad \varphi(x) = \psi\left(H^{-1}(x)\right)$$

$$y = H^{-1}(x), \quad v(y) = u(H(y)), \quad \psi(y) = \varphi(H(y)) \, .$$

We therefore immediately have

$$u_{x_i}(x) = \sum_{k=1}^{n} v_{y_k}\left(H^{-1}(x)\right) \frac{\partial H_k^{-1}}{\partial x_i}(x)$$

$$\varphi_{x_j}(x) = \sum_{l=1}^{n} \psi_{y_l}\left(H^{-1}(x)\right) \frac{\partial H_l^{-1}}{\partial x_j}(x)$$

Using the above and the change of variables $x = H(y)$, we obtain

$$\sum_{i,j=1}^{n} \int_U a_{ij}(x)\, u_{x_i}(x)\, \varphi_{x_j}(x)\, dx = \sum_{k,l=1}^{n} \int_Q b_{kl}(y)\, v_{y_k}(y)\, \psi_{y_l}(y)\, dy$$

where

$$b_{kl}(y) = \sum_{i,j=1}^{n} a_{ij}(H(y)) \frac{\partial H_k^{-1}}{\partial x_i}(H(y)) \frac{\partial H_l^{-1}}{\partial x_j}(H(y)) \det \nabla H(y) \, .$$

We also get, since

$$\sum_{i,j=1}^{n} a_{ij}\left(x\right)\lambda_i\lambda_j \geq \alpha\left|\lambda\right|^2$$

that (denoting the transpose of a matrix X by X^t)

$$\sum_{k,l=1}^{n} b_{kl}\left(y\right)\lambda_k\lambda_l = \det\nabla H \sum_{i,j,k,l=1}^{n} a_{ij}\frac{\partial H_k^{-1}}{\partial x_i}\frac{\partial H_l^{-1}}{\partial x_j}\lambda_k\lambda_l$$

$$= \det\nabla H \sum_{i,j=1}^{n} a_{ij}\left[\left(\nabla H^{-1}\right)^t\lambda\right]_i\left[\left(\nabla H^{-1}\right)^t\lambda\right]_j$$

$$\geq \alpha\det\nabla H\left(y\right)\left|\left(\nabla H^{-1}\left(H\left(y\right)\right)\right)^t\lambda\right|^2.$$

The result

$$\sum_{k,l=1}^{n} b_{kl}\left(y\right)\lambda_k\lambda_l \geq \beta\left|\lambda\right|^2$$

follows, since H is a regular change of variables. ∎

7.4.4 Higher regularity for the Dirichlet integral

Exercise 4.5.1. Write, in Theorem 4.12, $v = u - u_0 \in W_0^{1,2}\left(\Omega\right)$ and observe that

$$\int_\Omega \left|\nabla u\right|^2 = \int_\Omega \left[\left|\nabla v\right|^2 + 2\left\langle\nabla v;\nabla u_0\right\rangle + \left|\nabla u_0\right|^2\right] = \int_\Omega \left[\left|\nabla v\right|^2 - 2v\Delta u_0 + \left|\nabla u_0\right|^2\right].$$

We therefore have

$$\int_\Omega \left[\frac{1}{2}\left|\nabla u\right|^2 - h\,u\right] = \int_\Omega \left[\frac{1}{2}\left|\nabla v\right|^2 - \left(h + \Delta u_0\right)v\right] + \int_\Omega \left[\frac{1}{2}\left|\nabla u_0\right|^2 - h\,u_0\right].$$

Since the second integral is constant in the minimization problem, it is irrelevant and we are back to Theorem 4.12 with h replaced by $h + \Delta u_0$. The estimate is then

$$\left\|u - u_0\right\|_{W^{k+2,2}\left(\Omega\right)} \leq \gamma_1\left\|h + \Delta u_0\right\|_{W^{k,2}\left(\Omega\right)}$$

or equivalently

$$\left\|u\right\|_{W^{k+2,2}\left(\Omega\right)} \leq \gamma_2\left[\left\|h\right\|_{W^{k,2}\left(\Omega\right)} + \left\|u_0\right\|_{W^{k+2,2}\left(\Omega\right)}\right]. \quad\blacksquare$$

Exercise 4.5.2. We proceed as in Theorem 4.12 and we discuss only the case $k = 1$, the more general case being handled by induction. Let $O \subset \overline{O} \subset \Lambda \subset$

$\overline{\Lambda} \subset \Omega$ be open sets. We know from Theorem 4.9 that $u \in W_0^{1,2}(\Omega) \cap W^{2,2}(\Lambda)$. Observe that for every $\psi \in C_0^\infty(\Lambda)$, extended to be 0 outside Λ,

$$\int_\Lambda \sum_{i,j=1}^n g_{ij} (u_{x_l})_{x_i} \psi_{x_j} = - \int_\Lambda \sum_{i,j=1}^n \left[g_{ij} u_{x_i} \psi_{x_l x_j} + (g_{ij})_{x_l} u_{x_i} \psi_{x_j} \right].$$

We next appeal to the Euler-Lagrange equation

$$2 \int_\Omega \sum_{i,j=1}^n g_{ij} u_{x_i} \varphi_{x_j} = \int_\Omega h \varphi, \quad \forall \varphi \in W_0^{1,2}(\Omega),$$

replacing φ by ψ_{x_l}, to get

$$2 \int_\Lambda \sum_{i,j=1}^n g_{ij} (u_{x_l})_{x_i} \psi_{x_j} = - \int_\Lambda h \, \psi_{x_l} - 2 \int_\Lambda \sum_{i,j=1}^n (g_{ij})_{x_l} u_{x_i} \psi_{x_j}$$
$$= \int_\Lambda \left[h_{x_l} + 2 \sum_{i,j=1}^n \left((g_{ij})_{x_l} u_{x_i} \right)_{x_j} \right] \psi.$$

Since the term in brackets on the right-hand side is in $L^2(\Lambda)$ and since $C_0^\infty(\Lambda)$ is dense in $W_0^{1,2}(\Lambda)$, we obtain, as in the proof of Theorem 4.9, that $u_{x_l} \in W^{2,2}(O)$ and thus, since this holds for every $l = 1, \cdots, n$ and every $O \subset \overline{O} \subset \Omega$, we have that $u \in W_{loc}^{3,2}$. ∎

Exercise 4.5.3. Let $V(r) = |\log r|^\alpha$ and

$$u(x_1, x_2) = x_1 x_2 V(|x|).$$

A direct computation shows that

$$u_{x_1} = x_2 V(|x|) + \frac{x_1^2 x_2}{|x|} V'(|x|) \quad \text{and} \quad u_{x_2} = x_1 V(|x|) + \frac{x_1 x_2^2}{|x|} V'(|x|)$$

while

$$u_{x_1 x_1} = \frac{x_1^3 x_2}{|x|^2} V''(|x|) + \frac{x_1 x_2}{|x|^3} \left(2x_1^2 + 3x_2^2 \right) V'(|x|)$$

$$u_{x_2 x_2} = \frac{x_1 x_2^3}{|x|^2} V''(|x|) + \frac{x_1 x_2}{|x|^3} \left(2x_2^2 + 3x_1^2 \right) V'(|x|)$$

$$u_{x_1 x_2} = \frac{x_1^2 x_2^2}{|x|^2} V''(|x|) + \frac{x_1^4 + x_1^2 x_2^2 + x_2^4}{|x|^3} V'(|x|) + V(|x|).$$

We therefore get that

$$u_{x_1 x_1}, u_{x_2 x_2} \in C^0(\overline{\Omega}), \quad u_{x_1 x_2} \notin L^\infty(\Omega). \quad ∎$$

Exercise 4.5.4. Let $V(r) = \log|\log r|$. A direct computation shows that

$$u_{x_1} = \frac{x_1}{|x|} V'(|x|) \quad \text{and} \quad u_{x_2} = \frac{x_2}{|x|} V'(|x|)$$

and therefore

$$u_{x_1 x_1} = \frac{x_1^2}{|x|^2} V''(|x|) + \frac{x_2^2}{|x|^3} V'(|x|)$$

$$u_{x_2 x_2} = \frac{x_2^2}{|x|^2} V''(|x|) + \frac{x_1^2}{|x|^3} V'(|x|)$$

$$u_{x_1 x_2} = \frac{x_1 x_2}{|x|^2} V''(|x|) - \frac{x_1 x_2}{|x|^3} V'(|x|).$$

This leads to

$$\Delta u = V''(|x|) + \frac{V'(|x|)}{|x|} = \frac{-1}{|x|^2 |\log|x||^2} \in L^1(\Omega)$$

while $u_{x_1 x_1}, u_{x_1 x_2}, u_{x_2 x_2} \notin L^1(\Omega)$. Summarizing the results we indeed have that $u \notin W^{2,1}(\Omega)$ while $\Delta u \in L^1(\Omega)$. We also observe (compare with Example 1.33 (ii)) that, trivially, $u \notin L^\infty(\Omega)$ while $u \in W^{1,2}(\Omega)$, since $u \in L^2(\Omega)$ and

$$\iint_\Omega |\nabla u|^2\, dx = 2\pi \int_0^{1/2} \frac{dr}{r |\log r|^2} = \frac{2\pi}{\log 2}.$$

A much more involved example due to Ornstein [90] produces a u such that

$$u_{x_1 x_1}, u_{x_2 x_2} \in L^1(\Omega), \quad u_{x_1 x_2} \notin L^1(\Omega). \quad \blacksquare$$

Exercise 4.5.5. *Step 1.* We know that $\overline{u} \in X$ satisfies

$$\int_\Omega [\langle \nabla \overline{u}; \nabla \psi \rangle - (h - h_\Omega)\psi] = 0, \quad \forall \psi \in W^{1,2}(\Omega) \tag{7.22}$$

where

$$h_\Omega = \frac{1}{\text{meas}\,\Omega} \int_\Omega h.$$

The fact that $\overline{u} \in W_{\text{loc}}^{k+2,2}(\Omega)$ and that

$$-\Delta u = h - h_\Omega, \quad \text{a.e. in } \Omega$$

follows at once from Theorem 4.12, since (7.22) holds. We moreover have the existence of a constant $\gamma_1 = \gamma_1(O, \Omega) > 0$ such that

$$\|\overline{u}\|_{W^{k+2,2}(O)} \le \gamma_1 \left(\|\overline{u}\|_{W^{1,2}(\Omega)} + \|h - h_\Omega\|_{W^{k,2}(\Omega)} \right). \tag{7.23}$$

Step 2. We now prove the more precise estimate

$$\|\bar{u}\|_{W^{k+2,2}(O)} \leq \gamma \|h - h_\Omega\|_{L^2(\Omega)} . \tag{7.24}$$

Since $\bar{u}$ is a minimizer of (N) and $\int \bar{u} = 0$, we have

$$I(\bar{u}) = \int_\Omega \left[\frac{1}{2} |\nabla \bar{u}|^2 - h\bar{u} \right] = \int_\Omega \left[\frac{1}{2} |\nabla \bar{u}|^2 - (h - h_\Omega)\bar{u} \right] \leq I(0) = 0.$$

This implies, using Hölder inequality, that

$$\|\nabla \bar{u}\|^2_{L^2(\Omega)} \leq 2 \|h - h_\Omega\|_{L^2(\Omega)} \|\bar{u}\|_{L^2(\Omega)} \leq 2 \|h - h_\Omega\|_{L^2(\Omega)} \|\bar{u}\|_{W^{1,2}(\Omega)}$$

which, invoking Poincaré inequality under the form (1.15), leads to the existence of $\gamma_2 = \gamma_2(\Omega) > 0$

$$\frac{1}{\gamma_2} \|\bar{u}\|^2_{W^{1,2}(\Omega)} \leq \|\nabla \bar{u}\|^2_{L^2(\Omega)} \leq 2 \|h - h_\Omega\|_{L^2(\Omega)} \|\bar{u}\|_{W^{1,2}(\Omega)}$$

and thus

$$\|\bar{u}\|_{W^{1,2}(\Omega)} \leq 2\gamma_2 \|h - h_\Omega\|_{L^2(\Omega)} .$$

Combining the above inequality with (7.23) we have the inequality (7.24). ∎

7.4.5 Weyl lemma

Exercise 4.6.1. This is a classical result. In the following, we let

$$\sigma_{n-1} = \text{meas}\,(\partial B_1(0)) \quad \text{and} \quad \omega_n = \text{meas}\,(B_1(0))$$

recalling that $\sigma_{n-1} = n\,\omega_n$. We divide the proof into three steps.

Step 1. We first prove that if $u \in C^0(\Omega)$ satisfies the mean value formula on balls, namely

$$u(x) = \frac{n}{\sigma_{n-1}r^n} \int_{B_r(x)} u(y)\,dy = \frac{n}{\sigma_{n-1}r^n} \int_0^r \int_{|z|=1} u(x + \rho z)\,d\sigma_z\,\rho^{n-1}d\rho \tag{7.25}$$

it also satisfies the mean value formula on spheres (the converse is also true), which is

$$u(x) = \frac{1}{\sigma_{n-1}r^{n-1}} \int_{\partial B_r(x)} u\,d\sigma = \frac{1}{\sigma_{n-1}} \int_{|z|=1} u(x + rz)\,d\sigma_z \tag{7.26}$$

for every $x \in \Omega$ and for every $r > 0$ sufficiently small so that

$$B_r(x) = \{y \in \mathbb{R}^n : |y - x| < r\} \subset \overline{B_r(x)} \subset \Omega.$$

It suffices to differentiate (7.25) with respect to r to get

$$0 = \frac{n}{\sigma_{n-1}} \left(-nr^{-n-1} \right) \int_{B_r(x)} u(y) \, dy + \frac{n}{\sigma_{n-1} r^n} r^{n-1} \int_{|z|=1} u(x+rz) \, d\sigma_z .$$

We thus have from the above identity and (7.25) that

$$\frac{1}{\sigma_{n-1} r} \int_{|z|=1} u(x+rz) \, d\sigma_z = \frac{1}{r} \frac{n}{\sigma_{n-1} r^n} \int_{B_r(x)} u(y) \, dy = \frac{1}{r} u(x)$$

and hence the claim.

Step 2. We next choose $\psi \in C_0^\infty (0,1)$, $\psi \geq 0$, so that

$$\int_0^1 r^{n-1} \psi(r) \, dr = \frac{1}{\sigma_{n-1}} . \tag{7.27}$$

We then let $\psi \equiv 0$ outside of $(0,1)$ and we define for every $\epsilon > 0$

$$\varphi_\epsilon(x) = \frac{1}{\epsilon^n} \psi \left(\frac{|x|}{\epsilon} \right) .$$

Step 3. Let $\Omega_\epsilon = \left\{ x \in \mathbb{R}^n : \overline{B_\epsilon(x)} \subset \Omega \right\}$. Let $x \in \Omega_\epsilon$, the function $y \to \varphi_\epsilon(x-y)$ then has its support in Ω since $\operatorname{supp} \varphi_\epsilon \subset \overline{B_\epsilon(0)}$. We therefore have

$$\int_{\mathbb{R}^n} u(y) \varphi_\epsilon(x-y) \, dy = \int_{\mathbb{R}^n} u(x-y) \varphi_\epsilon(y) \, dy$$

$$= \frac{1}{\epsilon^n} \int_{|y|<\epsilon} u(x-y) \psi \left(\frac{|y|}{\epsilon} \right) dy$$

$$= \int_{|z|<1} u(x-\epsilon z) \psi(|z|) \, dz$$

and thus

$$\int_{\mathbb{R}^n} u(y) \varphi_\epsilon(x-y) \, dy = \int_0^1 \int_{|y|=1} u(x-\epsilon ry) \psi(r) r^{n-1} \, dr d\sigma_y .$$

Using (7.26), (7.27) and the above identity, we find

$$\int_{\mathbb{R}^n} u(y) \varphi_\epsilon(x-y) \, dy = u(x) .$$

Since $\varphi_\epsilon \in C_0^\infty (\mathbb{R}^n)$, we immediately get that $u \in C^\infty (\Omega_\epsilon)$. Since ϵ is arbitrary, we find that $u \in C^\infty (\Omega)$, as claimed. ∎

7.4.6   Some general results

Exercise 4.7.1. All the hypotheses of Theorems 3.3, 3.11 and 4.9 are satisfied. Therefore, there exists a unique minimizer $\overline{u} \in u_0 + W_0^{1,2}(\Omega)$ of

$$(P) \quad \inf\left\{ I(u) = \int_\Omega f(\nabla u(x))\, dx : u \in u_0 + W_0^{1,2}(\Omega) \right\}$$

satisfying $\overline{u} \in W_{\text{loc}}^{2,2}(\Omega)$ and

$$\sum_{i=1}^n \int_\Omega f_{\xi_i}(\nabla \overline{u})\, \varphi_{x_i} = 0, \quad \forall \varphi \in W_0^{1,2}(\Omega).$$

Let $O \subset \overline{O} \subset \omega \subset \overline{\omega} \subset \Omega$. Fix $k \in \{1, \cdots, n\}$, choose any $\psi \in C_0^\infty(\omega)$ and set in the above equation $\varphi = \psi_{x_k}$. Since $\overline{u} \in W^{2,2}(\omega)$, we can integrate by parts the equation and get, for every $\psi \in C_0^\infty(\omega)$ (and hence by density in $W_0^{1,2}(\omega)$),

$$0 = \sum_{i=1}^n \int_\omega \frac{\partial}{\partial x_k}\left[f_{\xi_i}(\nabla \overline{u}) \right] \psi_{x_i} = \sum_{i,j=1}^n \int_\omega f_{\xi_i \xi_j}(\nabla \overline{u})\, \overline{u}_{x_k x_j} \psi_{x_i}.$$

Setting

$$a_{ij}(x) = f_{\xi_i \xi_j}(\nabla \overline{u})$$

we have that $a_{ij} = a_{ji} \in L^\infty(\omega)$ and

$$\sum_{i,j=1}^n a_{ij}(x)\, \lambda_i \lambda_j \geq \gamma_5\, |\lambda|^2, \quad \text{a.e. in } \omega \text{ and } \forall \lambda \in \mathbb{R}^n.$$

Setting $v = \overline{u}_{x_k} \in W^{1,2}(\omega)$, we are in a position to apply Theorem 4.18 to deduce that $\overline{u} \in W^{2,2}(O) \cap C^{1,\alpha}(O)$ for some $0 < \alpha < 1$; as wished. ■

Exercise 4.7.2. It is clear, in view of Example 1.33 (iii), that $\overline{u} \in u_0 + W_0^{1,2}(\Omega; \mathbb{R}^n)$. It therefore remains to show that it satisfies the Euler-Lagrange equation. The proof is divided into three steps.

Step 1. We let

$$g^{ij}(u) = \frac{4}{n-2} \frac{u^i u^j}{1 + |u|^2}$$

and thus the function can be written as

$$f(u, \xi) = \sum_{i,j} \left(\xi_i^j \right)^2 + \left[\sum_{i,j} \left(\delta_{ij} + g^{ij}(u) \right) \xi_i^j \right]^2.$$

We therefore find

$$f_{\xi_i^j} = 2\xi_i^j + 2 \left[\sum_k \xi_k^k + \sum_{k,l} g^{kl}\xi_k^l \right] [\delta_{ij} + g^{ij}]$$

$$f_{u^j} = 2 \left[\sum_k \xi_k^k + \sum_{k,l} g^{kl}\xi_k^l \right] \left[\sum_{k,l} g^{kl}_{u^j}\xi_k^l \right].$$

We therefore have to prove that $\overline{u}$ satisfies the following system of equations

$$\sum_i \int_\Omega \left\{ f_{\xi_i^j} \left(\overline{u}, \nabla \overline{u} \right) \varphi^j_{x_i} + f_{u^j} \left(\overline{u}, \nabla \overline{u} \right) \varphi^j \right\} = 0, \quad \forall j = 1, \cdots, n$$

and for every $\varphi \in W_0^{1,2} \left(\Omega; \mathbb{R}^n \right)$.

We will show in Step 2 that $\overline{u}(x) = x/|x|$ satisfies the following two identities

$$\sum_{k,l} g^{kl}\left(\overline{u} \right) \frac{\partial \overline{u}^l}{\partial x_k} = 0 \quad \text{and} \quad \sum_{k,l} g^{kl}_{u^j}\left(\overline{u} \right) \frac{\partial \overline{u}^l}{\partial x_k} = 0 \tag{7.28}$$

for every $j = 1, \cdots, n$ and every $x \neq 0$. The Euler-Lagrange equation is then reduced to

$$\sum_i \int_\Omega \left\{ \frac{\partial \overline{u}^j}{\partial x_i} + \left(\sum_k \frac{\partial \overline{u}^k}{\partial x_k} \right) \left(\delta_{ij} + g^{ij}\left(\overline{u} \right) \right) \right\} \varphi^j_{x_i} = 0, \quad \forall j = 1, \cdots, n \tag{7.29}$$

and for every $\varphi \in W_0^{1,2} \left(\Omega; \mathbb{R}^n \right)$.

Note that $\overline{u} \in W^{1,2} \left(\Omega; \mathbb{R}^n \right) \cap C^\infty \left(\overline{\Omega} \setminus \{0\}; \mathbb{R}^n \right)$, g^{ij} are bounded and the elements in $\{.\}$ belong to $L^p(\Omega)$ with $p = 2 > n/(n-1)$ (recall that $n \geq 3$). Therefore, appealing to Exercise 1.4.16, (7.29) will be verified if we can establish (see Step 3) that, for every $x \neq 0$ and for every $j = 1, \cdots, n$,

$$\sum_i \frac{\partial}{\partial x_i} \left\{ \frac{\partial \overline{u}^j}{\partial x_i} + \left(\sum_k \frac{\partial \overline{u}^k}{\partial x_k} \right) \left(\delta_{ij} + g^{ij}\left(\overline{u} \right) \right) \right\} = 0. \tag{7.30}$$

The proof will therefore be complete once (7.28) and (7.30) are verified.

Step 2. Let us start by observing that, for every $x \neq 0$,

$$\frac{\partial \overline{u}^l}{\partial x_k} = \frac{\partial \overline{u}^k}{\partial x_l} = \frac{\delta_{kl}}{|x|} - \frac{x_k x_l}{|x|^3}.$$

This leads, for every $k = 1, \cdots, n$, to

$$\sum_l \overline{u}^l \frac{\partial \overline{u}^k}{\partial x_l} = \sum_l \overline{u}^l \frac{\partial \overline{u}^l}{\partial x_k} = \sum_l \frac{x_l}{|x|} \left[\frac{\delta_{kl}}{|x|} - \frac{x_k x_l}{|x|^3} \right] = \frac{x_k}{|x|^2} - \frac{x_k}{|x|^4} \sum_l (x_l)^2 = 0.$$

We therefore have

$$\sum_{k,l} g^{kl}(\overline{u}) \frac{\partial \overline{u}^l}{\partial x_k} = \frac{4}{(n-2)\left(1 + |\overline{u}|^2\right)} \sum_k \overline{u}^k \sum_l \overline{u}^l \frac{\partial \overline{u}^l}{\partial x_k} = 0$$

$$\begin{aligned}
\sum_{k,l} g^{kl}_{u^j}(\overline{u}) \frac{\partial \overline{u}^l}{\partial x_k} &= \frac{4}{n-2} \sum_k \left[\frac{\overline{u}^k}{1+|\overline{u}|^2} \right]_{u^j} \sum_l \overline{u}^l \frac{\partial \overline{u}^l}{\partial x_k} \\
&\quad + \frac{4}{n-2} \sum_l \left[\overline{u}^l \right]_{u^j} \sum_k \frac{\overline{u}^k}{1+|\overline{u}|^2} \frac{\partial \overline{u}^l}{\partial x_k} \\
&= 0
\end{aligned}$$

establishing (7.28).

Step 3. It remains to prove (7.30). Note first that

$$\sum_k \frac{\partial \overline{u}^k}{\partial x_k} = \sum_k \left[\frac{1}{|x|} - \frac{(x_k)^2}{|x|^3} \right] = \frac{n-1}{|x|}.$$

We therefore have

$$\begin{aligned}
\frac{\partial \overline{u}^j}{\partial x_i} + \left(\sum_k \frac{\partial \overline{u}^k}{\partial x_k} \right) \left(\delta_{ij} + g^{ij}(\overline{u}) \right) &= \frac{\delta_{ij}}{|x|} - \frac{x_i x_j}{|x|^3} + \frac{n-1}{|x|} \left(\delta_{ij} + \frac{2 x_i x_j}{(n-2)|x|^2} \right) \\
&= \frac{n\,\delta_{ij}}{|x|} + \frac{n\,x_i x_j}{(n-2)|x|^3}.
\end{aligned}$$

A straightforward computation leads, for every $x \neq 0$ and for every $j = 1, \cdots, n$, to

$$\sum_i \frac{\partial}{\partial x_i} \left\{ \frac{(n-2)\delta_{ij}}{|x|} + \frac{x_i x_j}{|x|^3} \right\} = \frac{\partial}{\partial x_j} \left[\frac{(n-2)}{|x|} + \frac{(x_j)^2}{|x|^3} \right] + x_j \sum_{i \neq j} \frac{\partial}{\partial x_i} \left[\frac{x_i}{|x|^3} \right]$$

$$= 0.$$

Combining the above two identities leads to (7.30), as wished. ∎

7.5   Chapter 5. Minimal surfaces

7.5.1   Generalities about surfaces

Exercise 5.2.1. (i) Elementary.

(ii) Apply (i) with $a = v_x$, $b = v_y$ and the definition of E, F and G.

(iii) Since $e_3 = (v_x \times v_y) / |v_x \times v_y|$, we have

$$\langle e_3; v_x \rangle = \langle e_3; v_y \rangle = 0.$$

Differentiating with respect to x and y, we deduce that

$$0 = \langle e_3; v_{xx} \rangle + \langle (e_3)_x ; v_x \rangle = \langle e_3; v_{xy} \rangle + \langle (e_3)_y ; v_x \rangle$$

$$= \langle e_3; v_{xy} \rangle + \langle (e_3)_x ; v_y \rangle = \langle e_3; v_{yy} \rangle + \langle (e_3)_y ; v_y \rangle$$

and the result follows from the definition of L, M and N.   ■

Exercise 5.2.2. (i) We have

$$v_x = (-y \sin x, y \cos x, a), \quad v_y = (\cos x, \sin x, 0)$$

$$e_3 = \frac{(-a \sin x, a \cos x, -y)}{\sqrt{a^2 + y^2}}$$

$$v_{xx} = (-y \cos x, -y \sin x, 0), \quad v_{xy} = (-\sin x, \cos x, 0), \quad v_{yy} = 0$$

and hence

$$E = a^2 + y^2, \quad F = 0, \quad G = 1, \quad L = N = 0, \quad M = \frac{a}{\sqrt{a^2 + y^2}}$$

which leads to $H = 0$, as wished.

(ii) A straightforward computation gives

$$v_x = \left(1 - x^2 + y^2, -2xy, 2x\right), \quad v_y = \left(2xy, -1 + y^2 - x^2, -2y\right)$$

$$e_3 = \frac{(2x, 2y, x^2 + y^2 - 1)}{(1 + x^2 + y^2)}$$

$$v_{xx} = (-2x, -2y, 2), \quad v_{xy} = (2y, -2x, 0), \quad v_{yy} = (2x, 2y, -2)$$

and hence

$$E = G = \left(1 + x^2 + y^2\right)^2, \quad F = 0, \quad L = -2, \quad N = 2, \; M = 0$$

which shows that, indeed, $H = 0$.   ■

Exercise 5.2.3. (i) Since $|v_x \times v_y|^2 = w^2 \left(1 + (w')^2\right)$, we obtain the result.

(ii) Observe that the function

$$f(w, \xi) = w\sqrt{1 + \xi^2}$$

is not convex over $(0, +\infty) \times \mathbb{R}$; although the function $\xi \to f(w, \xi)$ is strictly convex, whenever $w > 0$. We therefore only give necessary conditions for the existence of minimizers of (P_α) and hence we write the Euler-Lagrange equation associated with (P_α), namely

$$\frac{d}{dx}[f_\xi(w, w')] = f_w(w, w') \quad \Leftrightarrow \quad \frac{d}{dx}\left[\frac{ww'}{\sqrt{1 + (w')^2}}\right] = \sqrt{1 + (w')^2}. \quad (7.31)$$

Invoking Theorem 2.8, we find that any minimizer w of (P_α) satisfies

$$\frac{d}{dx}[f(w, w') - w'f_\xi(w, w')] = 0 \quad \Leftrightarrow \quad \frac{d}{dx}\left[\frac{w}{\sqrt{1 + (w')^2}}\right] = 0$$

which implies, if we let $a > 0$ be a constant,

$$(w')^2 = \frac{w^2}{a^2} - 1. \quad (7.32)$$

Before proceeding further, let us observe the following facts.

1) The function $w \equiv a$ is a solution of (7.32) but not of (7.31) and therefore it is irrelevant for our analysis.

2) To $a = 0$ corresponds $w \equiv 0$, which is also not a solution of (7.31) and moreover does not satisfy the boundary conditions $w(0) = w(1) = \alpha > 0$.

3) Any solution of (7.32) must verify $w^2 \geq a^2$ and, since $w(0) = w(1) = \alpha > 0$, thus verifies $w \geq a > 0$.

We can therefore search for solutions of (7.32) of the form

$$w(x) = a \cosh \frac{f(x)}{a}$$

where f satisfies, when inserted into the equation, $(f')^2 = 1$, which implies that either $f' \equiv 1$ or $f' \equiv -1$, since f is C^1. Thus, the solution of the differential equation is of the form

$$w(x) = a \cosh \frac{x + \mu}{a}.$$

Since $w(0) = w(1)$, we deduce that $\mu = -1/2$. Finally, since $w(0) = w(1) = \alpha$, every solution C^2 of (P_α) must be of the form

$$w(x) = a \cosh \frac{2x-1}{2a} \quad \text{and} \quad a \cosh \frac{1}{2a} = \alpha.$$

Summarizing, we see that, depending on the values of α, the Euler-Lagrange equation (7.31) may have 0, 1 or 2 solutions (in particular for α small, (7.31) has no C^2 solution satisfying $w(0) = w(1) = \alpha$ and hence (P_α) also has no C^2 minimizer). $\blacksquare$

Exercise 5.2.4. By hypothesis there exist a bounded connected open set with smooth boundary $\Omega \subset \mathbb{R}^2$ and a map $v \in C^2(\overline{\Omega}; \mathbb{R}^3)$ $(v = v(x, y)$, with $v_x \times v_y \neq 0$ in $\overline{\Omega})$ so that $\Sigma_0 = v(\overline{\Omega})$. Let

$$e_3 = \frac{v_x \times v_y}{|v_x \times v_y|}.$$

We then let, for $\epsilon \in \mathbb{R}$ and $\varphi \in C_0^\infty(\Omega)$,

$$v^\epsilon(x, y) = v(x, y) + \epsilon \varphi(x, y) e_3.$$

Finally, let $\Sigma^\epsilon = v^\epsilon(\overline{\Omega})$. Since Σ_0 is of minimal area and $\partial \Sigma^\epsilon = \partial \Sigma_0$, we should have

$$\iint_\Omega |v_x \times v_y| \leq \iint_\Omega |v_x^\epsilon \times v_y^\epsilon|. \tag{7.33}$$

Let

$$E = |v_x|^2, \quad F = \langle v_x; v_y \rangle \quad \text{and} \quad G = |v_y|^2$$

$$E^\epsilon = |v_x^\epsilon|^2, \quad F^\epsilon = \langle v_x^\epsilon; v_y^\epsilon \rangle \quad \text{and} \quad G^\epsilon = |v_y^\epsilon|^2.$$

We therefore get

$$E^\epsilon = |v_x + \epsilon \varphi (e_3)_x + \epsilon \varphi_x e_3|^2 = E + 2\epsilon [\varphi_x \langle v_x; e_3 \rangle + \varphi \langle v_x; (e_3)_x \rangle] + O(\epsilon^2)$$

$$F^\epsilon = F + \epsilon \left[\varphi_x \langle v_y; e_3 \rangle + \varphi_y \langle v_x; e_3 \rangle + \varphi \langle v_y; (e_3)_x \rangle + \varphi \langle v_x; (e_3)_y \rangle \right] + O(\epsilon^2)$$

$$G^\epsilon = G + 2\epsilon \left[\varphi_y \langle v_y; e_3 \rangle + \varphi \langle v_y; (e_3)_y \rangle \right] + O(\epsilon^2)$$

where $O(t)$ stands for a function f so that $|f(t)/t|$ is bounded in a neighborhood of $t = 0$. Appealing to the definition of L, M, N, Exercise 5.2.1 and to the fact that $\langle v_x; e_3 \rangle = \langle v_y; e_3 \rangle = 0$, we obtain

$$E^\epsilon G^\epsilon - (F^\epsilon)^2 = (E - 2\epsilon L \varphi)(G - 2\epsilon \varphi N) - (F - 2\epsilon \varphi M)^2 + O(\epsilon^2)$$

$$= EG - F^2 - 2\epsilon \varphi [EN - 2FM + GL] + O(\epsilon^2)$$

$$= (EG - F^2)[1 - 4\epsilon \varphi H] + O(\epsilon^2).$$

We therefore conclude that

$$\left| v_x^\epsilon \times v_y^\epsilon \right| = \left| v_x \times v_y \right| (1 - 2\,\epsilon\,\varphi\,H) + O\left(\epsilon^2\right)$$

and hence

$$\mathrm{Area}\left(\Sigma^\epsilon\right) = \mathrm{Area}\left(\Sigma_0\right) - 2\epsilon \iint_\Omega \varphi\,H\,\left| v_x \times v_y \right| + O\left(\epsilon^2\right). \qquad (7.34)$$

Using (7.33) and (7.34) (i.e. we perform the derivative with respect to ϵ) we get

$$\iint_\Omega \varphi\,H\,\left| v_x \times v_y \right| = 0, \quad \forall \varphi \in C_0^\infty\left(\Omega\right).$$

Since $\left| v_x \times v_y \right| > 0$ (due to the fact that Σ_0 is a regular surface), we deduce from the fundamental lemma of the calculus of variations (Theorem 1.24) that $H = 0$. ∎

7.5.2   The Douglas-Courant-Tonelli method

Exercise 5.3.1. We have

$$w_x = v_\lambda \lambda_x + v_\mu \mu_x\,, \quad w_y = v_\lambda \lambda_y + v_\mu \mu_y$$

and thus

$$|w_x|^2 = |v_\lambda|^2\,\lambda_x^2 + 2\lambda_x\mu_x\,\langle v_\lambda; v_\mu \rangle + \mu_x^2\,|v_\mu|^2$$
$$|w_y|^2 = |v_\lambda|^2\,\lambda_y^2 + 2\lambda_y\mu_y\,\langle v_\lambda; v_\mu \rangle + \mu_y^2\,|v_\mu|^2\,.$$

Since $\lambda_x = \mu_y$ and $\lambda_y = -\mu_x$, we deduce that

$$|w_x|^2 + |w_y|^2 = \left[|v_\lambda|^2 + |v_\mu|^2\right]\left[\lambda_x^2 + \lambda_y^2\right]$$

and thus

$$\iint_\Omega \left[|w_x|^2 + |w_y|^2\right] dx\,dy = \iint_\Omega \left[|v_\lambda|^2 + |v_\mu|^2\right]\left[\lambda_x^2 + \lambda_y^2\right] dx\,dy.$$

Changing variables in the second integral, bearing in mind that

$$\lambda_x\mu_y - \lambda_y\mu_x = \lambda_x^2 + \lambda_y^2\,,$$

we get the result, namely

$$\iint_\Omega \left[|w_x|^2 + |w_y|^2\right] dx\,dy = \iint_B \left[|v_\lambda|^2 + |v_\mu|^2\right] d\lambda\,d\mu. \quad ∎$$

7.5.3 Nonparametric minimal surfaces

Exercise 5.5.1. Set

$$f = \frac{1 + u_x^2}{\sqrt{1 + u_x^2 + u_y^2}}, \quad g = \frac{u_x u_y}{\sqrt{1 + u_x^2 + u_y^2}}, \quad h = \frac{1 + u_y^2}{\sqrt{1 + u_x^2 + u_y^2}}.$$

A direct computation shows that

$$f_y = g_x \quad \text{and} \quad g_y = h_x\,,$$

since

$$Mu = \left(1 + u_y^2\right) u_{xx} - 2 u_x u_y u_{xy} + \left(1 + u_x^2\right) u_{yy} = 0\,.$$

Setting

$$\varphi(x, y) = \int_0^x \int_0^y g(s, t)\, dt\, ds + \int_0^x \int_0^t f(s, 0)\, ds\, dt + \int_0^y \int_0^t h(0, s)\, ds\, dt$$

we find that

$$\varphi_{xx} = f, \quad \varphi_{xy} = g, \quad \varphi_{yy} = h$$

and hence that

$$\varphi_{xx}\varphi_{yy} - \varphi_{xy}^2 = 1.$$

The fact that φ is convex follows from the above identity, $\varphi_{xx} > 0$, $\varphi_{yy} > 0$ and Theorem 1.52. ∎

7.6 Chapter 6. Isoperimetric inequality

7.6.1 The case of dimension 2

Exercise 6.2.1. One can consult Hardy-Littlewood-Polya [65] page 185, for more details. Let $u \in X$ where

$$X = \left\{ u \in W^{1,2}(-1, 1) : u(-1) = u(1) \text{ with } \int_{-1}^1 u = 0 \right\}.$$

Define

$$z(x) = u(x + 1) - u(x)$$

and note that $z(-1) = -z(0)$, since $u(-1) = u(1)$. We deduce that we can find $\alpha \in (-1, 0]$ so that $z(\alpha) = 0$, which means that $u(\alpha + 1) = u(\alpha)$. We denote this common value by a (i.e. $u(\alpha + 1) = u(\alpha) = a$). Since $u \in W^{1,2}(-1, 1)$ it is easy to see that the function

$$v(x) = (u(x) - a)^2 \cot[\pi(x - \alpha)]$$

vanishes at $x = \alpha$ and $x = \alpha+1$ (this follows from Hölder inequality, see Exercise 1.4.3). We therefore have (recalling that $u(-1) = u(1)$ and $\cot[\pi(1-\alpha)] = \cot[\pi(-1-\alpha)]$)

$$\int_{-1}^{1} \left\{ (u')^2 - \pi^2 (u-a)^2 - (u' - \pi(u-a)\cot[\pi(x-\alpha)])^2 \right\} dx$$

$$= \pi \left[(u(x) - a)^2 \cot[\pi(x-\alpha)] \right]_{-1}^{1} = 0.$$

Since $\int_{-1}^{1} u = 0$, we get from the above identity that

$$\int_{-1}^{1} \left((u')^2 - \pi^2 u^2 \right) dx = 2\pi^2 a^2 + \int_{-1}^{1} (u' - \pi(u-a)\cot[\pi(x-\alpha)])^2 dx$$

and hence Wirtinger inequality follows. Moreover, we have equality in Wirtinger inequality if and only if $a = 0$ and, c denoting a constant,

$$u' = \pi u \cot[\pi(x-\alpha)] \quad \Leftrightarrow \quad u = c\sin[\pi(x-\alpha)]. \quad \blacksquare$$

Exercise 6.2.2. Since the minimum in (P) is attained by $u \in X$, we have, for any $v \in X \cap C_0^\infty(-1,1)$ and any $\epsilon \in \mathbb{R}$, that

$$I(u + \epsilon v) \geq I(u).$$

Therefore the Euler-Lagrange equation is satisfied, namely

$$\int_{-1}^{1} (u'v' - \pi^2 u\,v) = 0, \quad \forall v \in X \cap C_0^\infty(-1,1). \tag{7.35}$$

Let us transform it in a more classical way and choose a function $f \in C_0^\infty(-1,1)$ with $\int_{-1}^{1} f = 1$ and let $\varphi \in C_0^\infty(-1,1)$ be arbitrary. Set

$$v(x) = \varphi(x) - \left(\int_{-1}^{1} \varphi \right) f(x) \quad \text{and} \quad \lambda = -\frac{1}{\pi^2} \int_{-1}^{1} (u'f' - \pi^2 u f).$$

Observe that $v \in X \cap C_0^\infty(-1,1)$. Use (7.35), the fact that

$$\int_{-1}^{1} f = 1, \quad \int_{-1}^{1} v = 0$$

and the definition of λ to get, for every $\varphi \in C_0^\infty(-1,1)$,

$$\int_{-1}^{1} [u'\varphi' - \pi^2(u-\lambda)\varphi]$$

$$= \int \left[u'\left(v' + f'\int\varphi \right) - \pi^2 u \left(v + f\int\varphi \right) \right] + \pi^2\lambda\int\varphi$$

$$= \int (u'v' - \pi^2 u v) + \left[\int\varphi \right] \left[\pi^2\lambda + \int (u'f' - \pi^2 u f) \right] = 0.$$

The regularity of u (which is a minimizer of (P) in X) then follows (see Proposition 4.1) at once from the above equation. Since we know (from Theorem 6.1) that among smooth minimizers of (P) the only ones are of the form

$$u\,(x) = \alpha \cos(\pi x) + \beta \sin(\pi x)$$

we have the result. ∎

Exercise 6.2.3. We divide the proof into two steps.

Step 1. We start by introducing some notations. Since we work with fixed u, v, we drop the dependence on these variables in $L = L\,(u,v)$ and $M = M\,(u,v)$. However, we need to express the dependence of L and M on the intervals (α, β), where $a \le \alpha < \beta \le b$, and we therefore let

$$L\,(\alpha,\beta) = \int_\alpha^\beta \sqrt{(u')^2 + (v')^2} \quad \text{and} \quad M\,(\alpha,\beta) = \int_\alpha^\beta u\,v'$$

so that in these new notations

$$L\,(u,v) = L\,(a,b) \quad \text{and} \quad M\,(u,v) = M\,(a,b)\,.$$

We next let

$$O = \left\{ x \in (a,b) : \left(u'\,(x)\right)^2 + \left(v'\,(x)\right)^2 > 0 \right\}.$$

The case where $O = (a,b)$ has been considered in Step 1 of Theorem 6.4 (there we assumed that $(u')^2 + (v')^2 > 0$ in $[a, b]$, but the argument is still valid if we assume that it holds only on (a, b), since then $\varphi, \psi \in W^{1,\infty}_{per}\,(-1,1) \cap C^1\,(-1,1)$). If O is empty the result is trivial, so we assume from now on that this is not the case. Since the functions u' and v' are continuous, the set O is open. We can then find (see Theorem 9 of Chapter 1 in [38] or Theorem 6.59 in [67])

$$a \le a_i < b_i < a_{i+1} < b_{i+1} \le b, \quad \forall i \ge 1$$

$$O = \bigcup_{i=1}^\infty (a_i, b_i)\,.$$

In the complement of O, O^c, we have $(u')^2 + (v')^2 = 0$, and hence

$$L\,(b_i, a_{i+1}) = M\,(b_i, a_{i+1}) = 0. \tag{7.36}$$

Step 2. We then change the parametrization on every (a_i, b_i), exactly as in Step 1 of Theorem 6.4. We choose a multiple of the arc length, namely

$$\begin{cases} y = \eta\,(x) = -1 + 2\,\dfrac{L\,(a,x)}{L\,(a,b)} \\ \varphi\,(y) = u\,\left(\eta^{-1}\,(y)\right) \quad \text{and} \quad \psi\,(y) = v\,\left(\eta^{-1}\,(y)\right). \end{cases}$$

Note that this is well defined, since $(a_i, b_i) \subset O$. We then let

$$\alpha_i = -1 + 2\frac{L(a, a_i)}{L(a, b)} \quad \text{and} \quad \beta_i = -1 + 2\frac{L(a, b_i)}{L(a, b)}$$

so that

$$\beta_i - \alpha_i = 2\frac{L(a_i, b_i)}{L(a, b)}.$$

Furthermore, since $L(b_i, a_{i+1}) = 0$, we get

$$\beta_i = \alpha_{i+1} \quad \text{and} \quad \bigcup_{i=1}^{\infty} [\alpha_i, \beta_i] = [-1, 1].$$

We also easily find that, for $y \in (\alpha_i, \beta_i)$,

$$\sqrt{(\varphi'(y))^2 + (\psi'(y))^2} = \frac{L(a, b)}{2} = \frac{L(a_i, b_i)}{\beta_i - \alpha_i}$$

$$\varphi(\alpha_i) = u(a_i), \quad \psi(\alpha_i) = v(a_i), \quad \varphi(\beta_i) = u(b_i), \quad \psi(\beta_i) = v(b_i).$$

Note that $\varphi, \psi \in W^{1,\infty}(\alpha_i, \beta_i) \cap C^1(\alpha_i, \beta_i)$. From all the above observations, we infer that $\varphi, \psi \in W^{1,2}(-1, 1)$, with

$$\varphi(-1) = \varphi(1) \quad \text{and} \quad \psi(-1) = \psi(1)$$

and

$$L(a_i, b_i) = \frac{2}{L(a, b)} \int_{\alpha_i}^{\beta_i} \left((\varphi')^2 + (\psi')^2 \right) \tag{7.37}$$

$$M(a_i, b_i) = \int_{\alpha_i}^{\beta_i} \varphi \psi'. \tag{7.38}$$

We thus obtain, using (7.36), (7.37) and (7.38),

$$L(a, b) = \sum_{i=1}^{+\infty} L(a_i, b_i) = \frac{2}{L(a, b)} \int_{-1}^{1} \left((\varphi')^2 + (\psi')^2 \right)$$

$$M(a, b) = \sum_{i=1}^{+\infty} M(a_i, b_i) = \int_{-1}^{1} \varphi \psi'.$$

We therefore find, invoking Corollary 6.3, that

$$[L(u, v)]^2 - 4\pi M(u, v) = [L(a, b)]^2 - 4\pi M(a, b)$$

$$= 2 \int_{-1}^{1} \left((\varphi')^2 + (\psi')^2 \right) - 4\pi \int_{-1}^{1} \varphi \psi' \geq 0$$

as wished. The case of equality also follows from the corollary. ■

Exercise 6.2.4. Let $u \in C^1_{\text{per}}([-1,1])$ with $\int_{-1}^1 u = 0$. Define

$$v(x) = \pi \int_{-1}^x u(t) \, dt.$$

Note that $v \in C^1_{\text{per}}([-1,1])$ and, by Cauchy-Schwarz inequality (or Jensen inequality), that

$$[L(u,v)]^2 = \left[\int_{-1}^1 \sqrt{(u')^2 + (v')^2}\right]^2 \leq 2\int_{-1}^1 \left((u')^2 + (v')^2\right). \qquad (7.39)$$

It then follows, from the isoperimetric inequality (Theorem 6.4) and since $v' - \pi u = 0$, that

$$0 \leq [L(u,v)]^2 - 4\pi M(u,v) \leq 2\int_{-1}^1 \left((u')^2 + (v')^2\right) - 4\pi \int_{-1}^1 u v'$$

$$= 2\int_{-1}^1 (v' - \pi u)^2 + 2\int_{-1}^1 \left((u')^2 - \pi^2 u^2\right) = 2\int_{-1}^1 \left((u')^2 - \pi^2 u^2\right).$$

Thus Wirtinger inequality is proved. To establish the result when the inequality is an equality, we see from the above inequality and from Theorem 6.4 that

$$(u(x) - r_1)^2 + (v(x) - r_2)^2 = r_3^2, \quad \forall x \in [a,b]$$

where $r_1, r_2, r_3 \in \mathbb{R}$ are constants. Since equality must also hold in (7.39), we deduce (see Exercise 1.5.3) that

$$(u')^2 + (v')^2 = \text{constant}.$$

Combining these two observations and the fact that $v' - \pi u = 0$, we deduce that, indeed, equality holds in Wirtinger inequality if and only if $u(x) = \alpha \cos(\pi x) + \beta \sin(\pi x)$, for any $\alpha, \beta \in \mathbb{R}$. ■

7.6.2 The case of dimension n

Exercise 6.3.1. We clearly have

$$C = (\bar{a} + B) \cup (\bar{b} + A) \subset A + B.$$

It is also easy to see that

$$(\bar{a} + B) \cap (\bar{b} + A) = \{\bar{a} + \bar{b}\}.$$

Observe then that

$$M\left(C\right) = M\left(\overline{a} + B\right) + M\left(\overline{b} + A\right) - M\left[\left(\overline{a} + B\right) \cap \left(\overline{b} + A\right)\right] = M\left(A\right) + M\left(B\right)$$

and hence

$$M\left(A\right) + M\left(B\right) \leq M\left(A + B\right). \quad \blacksquare$$

Exercise 6.3.2. **(i)** We adopt the same notations as those of Theorem 5.12. By hypothesis there exist a bounded connected open set with smooth boundary $\Omega \subset \mathbb{R}^2$ and a map $v \in C^2\left(\overline{\Omega}; \mathbb{R}^3\right)$ $\left(v = v\left(x, y\right), \text{ with } v_x \times v_y \neq 0 \text{ in } \overline{\Omega}\right)$ so that $\partial A_0 = v\left(\overline{\Omega}\right).$

From the divergence theorem it follows that

$$M\left(A_0\right) = \frac{1}{3} \iint_\Omega \langle v; v_x \times v_y \rangle. \tag{7.40}$$

Let $\epsilon \in \mathbb{R}$, $\varphi \in C_0^\infty\left(\Omega\right)$ and

$$v^\epsilon\left(x, y\right) = v\left(x, y\right) + \epsilon\,\varphi\left(x, y\right) e_3$$

where $e_3 = \left(v_x \times v_y\right)/\left|v_x \times v_y\right|.$

We next consider

$$\partial A^\epsilon = \left\{v^\epsilon\left(x, y\right) : \left(x, y\right) \in \overline{\Omega}\right\} = v^\epsilon\left(\overline{\Omega}\right).$$

We have to evaluate $M\left(A^\epsilon\right)$ and we start by computing

$$
\begin{aligned}
v_x^\epsilon \times v_y^\epsilon &= \left(v_x + \epsilon\left(\varphi_x\, e_3 + \varphi\,(e_3)_x\right)\right) \times \left(v_y + \epsilon\left(\varphi_y\, e_3 + \varphi\,(e_3)_y\right)\right) \\
&= v_x \times v_y + \epsilon\left[\varphi\left((e_3)_x \times v_y + v_x \times (e_3)_y\right)\right] \\
&\quad + \epsilon\left[\varphi_x\, e_3 \times v_y + \varphi_y\, v_x \times e_3\right] + O\left(\epsilon^2\right)
\end{aligned}
$$

(where $O\left(t\right)$ stands for a function f so that $\left|f\left(t\right)/t\right|$ is bounded in a neighborhood of $t = 0$) which leads to

$$
\begin{aligned}
\langle v^\epsilon; v_x^\epsilon \times v_y^\epsilon \rangle &= \langle v + \epsilon\,\varphi\, e_3; v_x^\epsilon \times v_y^\epsilon \rangle \\
&= \langle v; v_x \times v_y \rangle + \epsilon\,\varphi \langle e_3; v_x \times v_y \rangle + \epsilon\,\varphi \left\langle v; (e_3)_x \times v_y + v_x \times (e_3)_y \right\rangle \\
&\quad + \epsilon \langle v; \varphi_x\, e_3 \times v_y + \varphi_y\, v_x \times e_3 \rangle + O\left(\epsilon^2\right).
\end{aligned}
$$

Observing that

$$\langle e_3; v_x \times v_y \rangle = \left|v_x \times v_y\right|$$

and returning to (7.40), we get, after integration by parts, (recalling that $\varphi = 0$ on $\partial \Omega$),

$$
\begin{aligned}
M\left(A^\epsilon\right) - M\left(A_0\right) &= \frac{\epsilon}{3} \iint_\Omega \varphi\{|v_x \times v_y| + \left\langle v; (e_3)_x \times v_y + v_x \times (e_3)_y \right\rangle \\
&\quad - (\langle v; e_3 \times v_y\rangle)_x - (\langle v; v_x \times e_3\rangle)_y\} + O\left(\epsilon^2\right) \\
&= \frac{\epsilon}{3} \iint_\Omega \varphi\left\{|v_x \times v_y| - \langle v_x; e_3 \times v_y\rangle - \langle v_y; v_x \times e_3\rangle\right\} + O\left(\epsilon^2\right).
\end{aligned}
$$

Since (see Exercise 5.2.1)

$$
\langle v_x; e_3 \times v_y\rangle = \langle v_y; v_x \times e_3\rangle = -|v_x \times v_y|,
$$

we obtain that

$$
M\left(A^\epsilon\right) - M\left(A_0\right) = \epsilon \iint_\Omega \varphi |v_x \times v_y| + O\left(\epsilon^2\right). \tag{7.41}
$$

(ii) We recall from (7.34) in Exercise 5.2.4 that we have

$$
L\left(\partial A^\epsilon\right) - L\left(\partial A_0\right) = -2\epsilon \iint_\Omega \varphi H |v_x \times v_y| + O\left(\epsilon^2\right). \tag{7.42}
$$

Combining (7.41), (7.42), the minimality of A_0 and a Lagrange multiplier α, we get

$$
\iint_\Omega (-2\varphi H + \alpha \varphi)|v_x \times v_y| = 0, \quad \forall \varphi \in C_0^\infty(\Omega).
$$

The fundamental lemma of the calculus of variations (Theorem 1.24) then implies that $H = $ constant (since ∂A_0 is a regular surface we have $|v_x \times v_y| > 0$). ∎

Exercise 6.3.3. The present exercise shows the relationship between the isoperimetric inequality and the Sobolev inequality for $W^{1,1}$ functions. We do not provide the best constant and do not show the fact that they are equivalent. These questions were first discussed in Federer-Fleming [50] and Fleming-Rishel [51] (for a more informal presentation see Osserman [92] and the bibliography quoted there). Let $\epsilon > 0$ be sufficiently small so that $A_\epsilon = A + \overline{B_\epsilon} \subset \Omega$. Define

$$
u_\epsilon(x) = \begin{cases} 1 & \text{if } x \in A \\ \operatorname{dist}(x, \partial A_\epsilon)/\epsilon & \text{if } x \in A_\epsilon \setminus A \\ 0 & \text{if } x \in \Omega \setminus A_\epsilon. \end{cases}
$$

Since the distance function is Lipschitz with Lipschitz constant 1, we find that $u_\epsilon \in C^{0,1}(\overline{\Omega})$, more precisely

$$
|u_\epsilon(x) - u_\epsilon(y)| \le \frac{1}{\epsilon}|x - y|, \quad \forall x, y \in \overline{\Omega}.
$$

Therefore, $u_\epsilon \in W_0^{1,\infty}(\Omega)$ (see Theorem 1.37 and Remark 1.38 (i)) and (see Exercise 1.4.14 (iii))

$$|\nabla u_\epsilon| = 0 \quad \text{a.e. in } A \cup (\Omega \setminus A_\epsilon) \quad \text{and} \quad |\nabla u_\epsilon| \le 1/\epsilon \quad \text{a.e. in } A_\epsilon \setminus A.$$

We hence have

$$\|u_\epsilon\|_{L^{\frac{n}{n-1}}} = \left(\int_\Omega |u_\epsilon|^{\frac{n}{n-1}} \right)^{\frac{n-1}{n}} \ge (M(A))^{\frac{n-1}{n}}$$

while

$$\|\nabla u_\epsilon\|_{L^1} = \int_\Omega |\nabla u_\epsilon| \le \frac{1}{\epsilon} [M(A_\epsilon) - M(A)].$$

Appealing to the fact that

$$\|u_\epsilon\|_{L^{\frac{n}{n-1}}} \le \gamma \|\nabla u_\epsilon\|_{L^1}$$

we get

$$(M(A))^{\frac{n-1}{n}} \le \frac{\gamma}{\epsilon} [M(A_\epsilon) - M(A)]$$

and thus, from Minkowski-Steiner formula, we have indeed obtained that

$$[L(\partial A)]^n - \frac{1}{\gamma^n} [M(A)]^{n-1} \ge 0. \quad \blacksquare$$

Bibliography

[1] Adams R.A., *Sobolev spaces*, Academic Press, New York, 1975. Second edition with Fournier J., Elsevier/Academic Press, Amsterdam, 2003.

[2] Akhiezer N.I., *The calculus of variations*, Blaisdell, New York, 1962.

[3] Alibert J.J. and Dacorogna B., An example of a quasiconvex function that is not polyconvex in two dimensions, *Arch. Rational Mech. Anal.* **117** (1992), 155-166.

[4] Almgren F.J., *Plateau's problem. An invitation to varifold geometry*, W.A. Benjamin, New York, 1966.

[5] Ambrosio L., Fusco N. and Pallara D., *Functions of bounded variation and free discontinuity problems*, Oxford University Press, Oxford, 2000.

[6] Antman S.S., *Nonlinear problems of elasticity*, Springer, Berlin, 1995.

[7] Ball J.M., Convexity conditions and existence theorems in non-linear elasticity, *Arch. Rational Mech. Anal.* **63** (1977), 337-403.

[8] Ball J.M. and Mizel V., One dimensional variational problems whose minimizers do not satisfy the Euler-Lagrange equations, *Arch. Rational Mech. Anal.* **90** (1985), 325-388.

[9] Bandle C., *Isoperimetric inequalities and applications*, Pitman, London, 1980.

[10] Berger M., *Geometry I and II*, Springer, Berlin, 1987.

[11] Blaschke W., *Kreis und Kugel*, Chelsea, New York, 1949.

[12] Bliss G., *Lectures on the calculus of variations*, University of Chicago Press, Chicago, 1951.

[13] Bolza O., *Lectures on the calculus of variations*, Chelsea, New York, 1946.

301

[14] Brézis H., *Analyse fonctionnelle, théorie et applications*, Masson, Paris, 1983. English translation: *Functional analysis, Sobolev spaces and partial differential equations*, Springer, New-York, 2010.

[15] Buttazzo G., *Semicontinuity, relaxation and integral represention in the calculus of variations*, Pitman, Longman, London, 1989.

[16] Buttazzo G., Ferone V. and Kawohl B., Minimum problems over sets of concave functions and related questions, *Math. Nachr.* **173** (1995), 71-89.

[17] Buttazzo G., Giaquinta M. and Hildebrandt S., *One dimensional variational problems*, Oxford University Press, Oxford, 1998.

[18] Buttazzo G. and Kawohl B., On Newton's problem of minimal resistance, *Math. Intell.* **15** (1992), 7-12.

[19] Carathéodory C., *Calculus of variations and partial differential equations of the first order*, Holden Day, San Francisco, 1965.

[20] Cesari L., *Optimization: theory and applications*, Springer, New York, 1983.

[21] Chern S.S., An elementary proof of the existence of isothermal parameters on a surface, *Proc. Amer. Math. Soc.* **6** (1955), 771-782.

[22] Ciarlet P., *Mathematical elasticity, Volume 1, Three dimensional elasticity*, North Holland, Amsterdam, 1988.

[23] Clarke F.H., *Optimization and nonsmooth analysis*, Wiley, New York, 1983.

[24] Courant R., *Dirichlet's principle, conformal mapping and minimal surfaces*, Interscience, New York, 1950.

[25] Courant R., *Calculus of variations*, Courant Institute Publications, New York, 1962.

[26] Courant R. and Hilbert D., *Methods of mathematical physics*, Wiley, New York, 1966.

[27] Crandall M.G., Ishii H. and Lions P.L., User's guide to viscosity solutions of second order partial differential equations, *Bull. Amer. Math. Soc.* **27** (1992), 1-67.

[28] Croce G. and Dacorogna B., On a generalized Wirtinger inequality, *Discrete Contin. Dyn. Syst. Ser. A* **9** (2003), 1329-1341.

[29] Csató G., Dacorogna B. and Kneuss O., *The pullback equation for differential forms,* Birkhäuser, Boston, 2012.

[30] Dacorogna B., Quasiconvexity and relaxation of nonconvex variational problems, *J. Funct. Anal.* **46** (1982), 102-118.

[31] Dacorogna B., *Weak continuity and weak lower semicontinuity of nonlinear functionals,* Springer, Berlin, 1982.

[32] Dacorogna B., *Direct methods in the calculus of variations,* Springer, Berlin, New York, 1989; 2nd edition, 2007.

[33] Dacorogna B. and Marcellini P., A counterexample in the vectorial calculus of variations, in *Material instabilities in continuum mechanics,* ed. Ball J.M., Oxford Science Publications, Oxford, 1988, 77-83.

[34] Dacorogna B. and Marcellini P., *Implicit partial differential equations,* Birkhäuser, Boston, 1999.

[35] Dacorogna B. and Murat F., On the optimality of certain Sobolev exponents for the weak continuity of determinants, *J. Funct. Anal.* **105** (1992), 42-62.

[36] Dacorogna B. and Pfister C.E., Wulff theorem and best constant in Sobolev inequality, *J. Math. Pures Appl.* **71** (1992), 97-118.

[37] Dal Maso G., *An introduction to Γ-convergence,* Progress in Nonlinear Differential Equations and their Applications, **8**, Birkhäuser, Boston, 1993.

[38] De Barra G., *Measure theory and integration,* Wiley, New York, 1981.

[39] De Giorgi E., *Teoremi di semicontinuità nel calcolo delle variazioni,* Istituto Nazionale di Alta Matematica, Roma, 1968-1969.

[40] De Giorgi E., *Selected papers,* Edited by L. Ambrosio, G. Dal Maso, M. Forti, M. Miranda and S. Spagnolo, Springer-Verlag, Berlin, 2006.

[41] Dierkes U., Hildebrandt S., Küster A. and Wohlrab O., *Minimal surfaces I and II,* Springer, Berlin, 1992.

[42] Dierkes U., Hildebrandt S. and Sauvigny F., *Minimal surfaces,* Springer, Berlin, 2010.

[43] Edmunds D.E. and Evans W.D., *Spectral theory and differential operators,* Oxford Science Publications, Oxford, 1987.

[44] Ekeland I., *Convexity methods in Hamiltonian mechanics*, Springer, Berlin, 1990.

[45] Ekeland I. and Temam R., *Analyse convexe et problèmes variationnels*, Dunod, Paris, 1974. English translation: *Convex analysis and variational problems*, North Holland, Amsterdam, 1976.

[46] Evans L.C., *Weak convergence methods for nonlinear partial differential equations*, American Mathematical Society, Providence, 1990.

[47] Evans L.C., *Partial differential equations*, American Mathematical Society, Providence, 1998.

[48] Evans L.C. and Gariepy R.F., *Measure theory and fine properties of functions*, Studies in Advanced Mathematics, CRC Press, Boca Raton, 1992.

[49] Federer H., *Geometric measure theory*, Springer, Berlin, 1969.

[50] Federer H. and Fleming W.H., Normal and integral currents, *Ann. Math.* **72** (1960), 458-520.

[51] Fleming W.H. and Rishel R., An integral formula for total gradient variation, *Arch. Math.* (Basel) **11** (1960), 218-222.

[52] Folland G.B., *Introduction to partial differential equations*, Princeton University Press, Princeton, 1976.

[53] Fusco N., Maggi F. and Pratelli A., The sharp quantitative isoperimetric inequality, *Ann. Math.* **168** (2008), 941-980.

[54] Gelfand I.M. and Fomin S.V., *Calculus of variations*, Prentice-Hall, Englewood Cliffs, 1963.

[55] Giaquinta M., *Multiple integrals in the calculus of variations and nonlinear elliptic systems*, Princeton University Press, Princeton, 1983.

[56] Giaquinta M. and Hildebrandt S., *Calculus of variations I and II*, Springer, Berlin, 1996.

[57] Gilbarg D. and Trudinger N.S., *Elliptic partial differential equations of second order*, Springer, Berlin, 1977.

[58] Giusti E., *Minimal surfaces and functions of bounded variations*, Birkhäuser, Boston, 1984.

[59] Giusti E., *Metodi diretti del calcolo delle variazioni*, Unione Matematica Italiana, Bologna, 1994. English translation: *Direct methods in the calculus of variations*, World Scientific, Singapore, 2003.

[60] Giusti E. and Miranda M., Un esempio di soluzioni discontinue per un problema di minimo relativo ad un integrale regolare del calcolo delle variazioni, *Boll. Un. Mat. Ital.* **1** (1968), 219-226.

[61] Goldstine H.H., *A history of the calculus of variations from the 17th to the 19th century*, Springer, Berlin, 1980.

[62] Gromov. M., Isoperimetric inequalities in Riemannian manifolds, in appendix to Milman V. and Schechtman G., *Asymptotic theory of finite-dimensional normed spaces*, Lecture Notes in Mathematics, **1200**, Springer-Verlag, Berlin, 1986.

[63] Hadamard J., Sur quelques questions du calcul des variations, *Bull. Soc. Math. France* **33** (1905), 73-80.

[64] Hadamard J., *Leçons sur le calcul des variations*, Hermann, Paris, 1910.

[65] Hardy G.H., Littlewood J.E. and Polya G., *Inequalities*, Cambridge University Press, Cambridge, 1961.

[66] Hestenes M.R., *Calculus of variations and optimal control theory*, Wiley, New York, 1966.

[67] Hewitt E. and Stromberg K., *Real and abstract analysis*, Springer, Berlin, 1965.

[68] Hildebrandt S. and Tromba A., *Mathematics and optimal form*, Scientific American Library, New York, 1984.

[69] Hildebrandt S. and Von der Mosel H., Plateau's problem for parametric double integrals: existence and regularity in the interior, *Comm. Pure Appl. Math.* **56** (2003), 926-955.

[70] Hörmander L., The boundary problems of physical geodesy, *Arch. Rational Mech. Anal.* **62** (1976), 1-52.

[71] Hörmander L., *Notions of convexity*, Birkhäuser, Boston, 1994.

[72] Hsiung C.C., *A first course in differential geometry*, Wiley, New York, 1981.

[73] Ioffe A.D. and Tihomirov V.M., *Theory of extremal problems*, North Holland, Amsterdam, 1979.

[74] John F., *Partial differential equations*, Springer, Berlin, 1982.

[75] Kawohl B., Recent results on Newton's problem of minimal resistance, *Nonlinear analysis and applications* (Warsaw, 1994), Gakuto Internat. Ser. Math. Sci. Appl. **7** (1996), 249-259.

[76] Kinderlherer D. and Stampacchia G., *Introduction to variational inequalities and their applications*, Academic Press, New York, 1980.

[77] Ladyzhenskaya O.A. and Uraltseva N.N., *Linear and quasilinear elliptic equations*, Academic Press, New York, 1968.

[78] Lebesgue H., *Leçons sur l'intégration et la recherche des fonctions primitives*, Gauthier-Villars, Paris, 1928.

[79] Lions J.L. and Magenes E., *Non-homogeneous boundary value problems and applications I, II, III*, Springer, Berlin, 1972.

[80] Lions P.L., *Generalized solutions of Hamilton-Jacobi equations*, Research Notes in Mathematics **69**, Pitman, London, 1982.

[81] Marcellini P., Non convex integrals of the calculus of variations, in: *Methods of nonconvex analysis*, ed. Cellina A., Lecture Notes in Mathematics **1446**, Springer, Berlin, 1990, 16-57.

[82] Marcellini P. and Sbordone C., Semicontinuity problems in the calculus of variations, *Nonlinear Anal.*, **4** (1980), 241-257.

[83] Mawhin J. and Willem M., *Critical point theory and Hamiltonian systems*, Springer, Berlin, 1989.

[84] Monna A.F., *Dirichlet's principle: a mathematical comedy of errors and its influence on the development of analysis*, Oosthoeck, Utrecht, 1975.

[85] Morrey C.B., Quasiconvexity and the lower semicontinuity of multiple integrals, *Pacific J. Math.* **2** (1952), 25-53.

[86] Morrey C.B., *Multiple integrals in the calculus of variations*, Springer, Berlin, 1966.

[87] Morse M., *The calculus of variations in the large*, American Mathematical Society, New York, 1934.

[88] Necas J., *Les méthodes directes en théorie des équations elliptiques*, Masson, Paris, 1967.

[89] Nitsche J.C., *Lecture on minimal surfaces*, Cambridge University Press, Cambridge, 1989.

[90] Ornstein D., A non-inequality for differential operators in the L^1 norm, *Arch. Rational Mech. Anal.* **11** (1962), 40-49.

[91] Osserman R., *A survey on minimal surfaces*, Van Nostrand, New York, 1969.

[92] Osserman R., The isoperimetric inequality, *Bull. Amer. Math. Soc.* **84** (1978), 1182-1238.

[93] Pars L., *An introduction to the calculus of variations*, Heinemann, London, 1962.

[94] Payne L., Isoperimetric inequalities and their applications, *SIAM Rev.* **9** (1967), 453-488.

[95] Pisier G., *The volume of convex bodies and Banach space geometry*, Cambridge University Press, Cambridge, 1989.

[96] Polya G. and Szegö G., *Isoperimetric inequalities in mathematical physics*, Princeton University Press, Princeton, 1951.

[97] Porter T.I., A history of the classical isoperimetric problem, in *Contributions to the calculus of variations (1931-1932)*, ed. Bliss G.A. and Graves L.M., University of Chicago Press, Chicago, 1933.

[98] Rockafellar R.T., *Convex analysis*, Princeton University Press, Princeton, 1970.

[99] Rudin W., *Real and complex analysis*, McGraw-Hill, New York, 1966.

[100] Rudin W., *Functional analysis*, McGraw-Hill, New York, 1973.

[101] Rund H., *The Hamilton-Jacobi theory in the calculus of variations*, Van Nostrand, Princeton, 1966.

[102] Stein E.M. and Shakarchi R., *Fourier analysis*, Princeton University Press, Princeton, 2003.

[103] Struwe M., *Plateau's problem and the calculus of variations*, Princeton University Press, Princeton, 1988.

[104] Struwe M., *Variational methods: applications to nonlinear partial differential equations and Hamiltonian systems*, Springer, Berlin, 1990.

[105] Sverak V., Rank one convexity does not imply quasiconvexity, *Proc. Royal Soc. Edinburgh* **120A** (1992), 185-189.

[106] Tonelli L., *Fondamenti di calcolo delle variazioni, I, II*, Zanichelli, Bologna, 1921.

[107] Troutman J.L., *Variational calculus with elementary convexity*, Springer, New York, 1983

[108] Webster R., *Convexity*, Oxford University Press, Oxford, 1994.

[109] Weinstock R., *Calculus of variations with applications to physics and engineering*, McGraw-Hill, New York, 1952.

[110] Willem M., *Minimax theorems*, Progress in Nonlinear Differential Equations and their Applications, **24**, Birkhäuser , Boston, 1996.

[111] Willem M., *Principes d'analyse fonctionnelle*, Cassini, Paris, 2007.

[112] Young L.C., *Lectures on the calculus of variations and optimal control theory*, W.B. Saunders, Philadelphia, 1969.

[113] Zeidler E., *Nonlinear functional analysis and its applications, I, II, III, IV*, Springer, New York, 1985-1988.

[114] Ziemer W.P., *Weakly differentiable functions*, Springer, New York, 1989.

Index